INSIGHT INTO FUZZY MODELING

INSIGHT INTO FUZZY MODELING

VILÉM NOVÁK, IRINA PERFILIEVA, AND ANTONÍN DVOŘÁK
University of Ostrava, Institute for Research and Applications of Fuzzy Modeling, the Czech Republic

Published by John Wiley & Sons, Inc., Hoboken, New Jersey. Published simultaneously in Canada

For general information on our other products and services or for technical support, please contact our Customer Care Department within the United States at (800) 762-2974, outside the United States at (317) 572-3993 or fax (317) 572-4002.

Wiley also publishes its books in a variety of electronic formats. Some content that appears in print may not be available in electronic formats. For more information about Wiley products, visit our web site at www.wiley.com.

Library of Congress Cataloging-in-Publication Data:

Names: Novák, Vilém, 1951- author. Perfilieva, Irina, 1953- author. Dvořák, Antonín, 1970 - author.
Title: Insight into fuzzy modeling / Vilém Novák, Irina Perfilieva, and Antonín Dvořák, University of Ostrava, Czech Republic.
Description: Hoboken, New Jersey : John Wiley & Sons, Inc., 2016. | Includes bibliographical references and index.
Identifiers: LCCN 2015040313 (print) | LCCN 2015047241 (ebook) | ISBN 9781119193180 (hardback) | ISBN 9781119193197 (pdf) | ISBN 9781119193203 (epub)
Subjects: LCSH: Simulation methods. | Fuzzy mathematics. | Fuzzy systems–Mathematical models. | BISAC: TECHNOLOGY & ENGINEERING / Electronics / General.
Classification: LCC T57 .N68 2016 (print) | LCC T57 (ebook) | DDC 511.3/13–dc23
LC record available at http://lccn.loc.gov/2015040313

Typeset in 10/12pt TimesLTStd by SPi Global, Chennai, India

Printed in the United States of America

10 9 8 7 6 5 4 3 2

To our children David, Vitalik, Martin, Anna, and Jaroslav

CONTENTS IN BRIEF

CONTENTS

PREFACE

Fuzzy modeling is a special branch of mathematical modeling that has two goals: (i) to construct models based on information that can be given not only in numbers but also, imprecisely, usually in a form of expressions of natural language; (ii) to construct models with less computational demands, which are more robust, that is, little sensitive to changes in the input data.

In comparison with classical models, the fuzzy ones are closer to human way of thinking. For example, when processing images, classical methods work with single pixels. People, however, do not see pixels but larger and usually imprecisely delineated parts of the image. This is the main reason why it is so difficult to develop methods that are as powerful as the human eye. It happens quite often that some objective measure says that a given image is good but the human eye sees it differently and says “no”.

The fuzzy modeling methods are developed with the idea to capture the way how people grasp and manipulate available information. Therefore, fuzzy models make it possible to solve highly nontrivial problems. On one side, these problems come from areas where mathematics has not yet (or very little) contributed, for example, psychology, geography, or geology. The fuzzy modeling methods, however, are successful also in solving classical problems, such as control, time series forecasting, image processing, or classical mathematical problems such as approximation of functions, solution of differential equations, or signal processing. In all cases, these methods manifest the above-mentioned properties — less computational demands and robustness.

Our book is specific from several points of view. First of all, the reader will find in it a consistent and well-established notation and terminology. Furthermore, we made maximum efforts to explain the basic ideas of the presented methods and to avoid misleading terminology occurring in some older books (e.g., we avoid erroneous terms such as “Mamdani implication”).

One of the main contributions of our book is that it contains original results obtained both in the theory as well as in the applications. These results cover especially newly developed theories and also original and, in our opinion, more precise explanation of older known results (this concerns especially relational interpretation of the widely used fuzzy IF-THEN rules).

A special focus is placed on two original theories: the fuzzy natural logic, which, besides others, includes mathematical model of special parts of linguistic semantics, and the theory of fuzzy transform. None of these theories has been yet published in the book form. Our aim is to demonstrate the power of both theories and their increasing potential in applications. We therefore split the book into two parts.

The first part contains extensive presentation of the theory of fuzzy modeling. Let us remark that the development of this theory is in large extent motivated by necessity to solve concrete problems.

The second part presents selected applications in three important areas: control and decision-making, image processing, and time series analysis and forecasting. It is interesting that the latter application combines both above-mentioned theories.

Recall that the book is focused on fuzzy modeling — the essential part of soft computing. Therefore, methods based on fuzzy set theory are preferred. We deliberately omitted other well-known and important theories that are usually mentioned in connection with soft computing, namely, the theory of artificial neural nets is missing because it is deeply elaborated in numerous specialized books and papers that we suggest to read. We think that peripheral presentation of this theory in one chapter of the book focused on fuzzy modeling would be more confusing than beneficial. For the same reasons, we omitted optimization techniques such as evolutionary algorithms or particle swarm optimization that can also be found in numerous specialized books and papers.

We want to thank our colleagues for their help when preparing this book: Martin Dyba, Michal Holčapek, Petr Hurtík, Viktor Pavliska, Marek Vajgl, Radek Valášek and Pavel Vlašánek. Last but not least, we also thank the workers of John Wiley & Sons publishing house.

VILÉM NOVÁK, IRINA PERFILIEVA, AND ANTONÍN DVOŘÁK

Ostrava, Czech Republic
August 2015

ACKNOWLEDGMENTS

We want to thank to the European Regional Development Fund in the IT4Innovations Centre of Excellence project (CZ.1.05/1.1.00/02.0070) and to the NPU II project LQ1602 "IT4Innovations excellence in science" provided by the MŠMT for their support.

ABOUT THE COMPANION WEBSITE

This book is accompanied by a companion website:

www.wiley.com/go/novak/fuzzy/modeling

The website includes:

- executable files with programs realizing methods described in the book
- manuals to these programs
- selected demonstration problems
- updates

PART I

FUNDAMENTALS OF FUZZY MODELING

This part consists of six chapters in which we first explain the role of indeterminacy in human life, which led to the development of special mathematical theories such as probability theory, fuzzy set theory, and fuzzy logic. In Chapter 2, we give an overview of the basic notions of the latter two theories. The main contribution of this part is contained in the three subsequent chapters in which we explain the theory of fuzzy IF-THEN rules, show that they have two possible interpretations enabling different kinds of applications, and explain the theory of fuzzy transform. A strong accent is put to the description of the model of semantics of a special class of linguistic expressions that are used on many places in this book. The theoretical part is finished by brief description of the main principles of fuzzy cluster analysis.

Insight into Fuzzy Modeling, First Edition. Vilém Novák, Irina Perfilieva, and Antonín Dvořák.

Companion Website: www.wiley.com/go/novak/fuzzy/modeling

1

WHAT IS FUZZY MODELING

1.1 INDETERMINACY IN HUMAN LIFE

Fuzzy modeling is a group of special mathematical methods that make it possible to include in the model imprecise or vaguely formulated expert information that is often characterized using natural language. The developed models (we call them fuzzy models) are very successful because they provide solution in situations when traditional mathematical models fail—either due to their non-adequacy, or due to their inability to utilize the full available information.

Note that the idea to include imprecise information in our models contradicts to what has always been required: as high precision as possible. There is, however, a good reason for doing it, namely, we face a discrepancy between relevance and precision. The so-called *principle of incompatibility* formulated by L. A. Zadeh in [149] says the following:

> As a complexity of system increases, our ability to make absolute, precise, and significant statements about the system's behavior diminishes. At some moment, there will be trade-off between precision and relevance. Increase in precision can be gained only through decrease in relevance; increase in relevance can be gained only through the decrease in precision.

For example, from the description of an enterprise in several sentences, we may learn about its main activity, size, total number of its employees, its business successes, and problems. But we will know nothing about individual people, specific machines, and their parts. To describe everything in detail, we would need much

Insight into Fuzzy Modeling, First Edition. Vilém Novák, Irina Perfilieva, and Antonín Dvořák.

Companion Website: www.wiley.com/go/novak/fuzzy/modeling

more sentences, numbers, tables, and so on. But then the amount of information exponentially increases. We would thus learn more, but any detail would concern only a small part of the enterprise. The requirement to describe the whole enterprise in full detail would lead to a big pile of thick books that, however, nobody would be able to read. And if yes, to understand the content, he/she would need natural language, which means that he/she would have to return to imprecise characterization. Otherwise, he/she would be lost in the abundance of irrelevant details.

We can see that to express relevant information, we need natural language. This is the only and very accomplished tool that enable us to work effectively with vague concepts.

Is full precision achievable? We argue that full precision is only our illusion and is not achievable, even in principle. Otherwise, we could obtain the same result independently on the chosen precision. But this is, in general, impossible. For example, let us compare two containers according to their volume. If their volume is absolutely the same, then we obtain the same number independently if we measure in m^3, mm^3, or in arbitrary fractions such as billionths, quadrillionths, 10^{-120} of m^3, and so on. But this is impossible because at the level of atoms or even elementary particles, we would not be able to distinguish which of the latter belongs to the body of the container and which does not. We conclude that the struggle for limit precision brings us to contradiction.

Let us emphasize that vagueness is inseparable feature of the semantics of natural language. We argue that it is not its weakness but its strength. Natural language is used in almost any human activity. For example, if we want to learn driving a car, we need a teacher who explains us—in natural language—what should we do, for example, "slow down a little", "now accelerate but not too much", and so on. Though such commands are vague, they are sufficient for us to be able to learn driving.

The main theories applied in fuzzy modeling are (mathematical) fuzzy logic and the fuzzy set theory. When facing vagueness, we may ask why we speak about fuzzy sets and fuzzy logic and do not consider techniques of probability and statistics?

The probability theory provides a mathematical model of uncertainty that is met when considering an event that has not yet occurred and we do not know whether it will indeed occur or not. Such an event can be, for example, a result of an experiment we are going to realize. Uncertainty is thus a lack of information about occurrence of some event.[1]

The basic concept in probability theory is a *probability distribution*. This gives us information about occurrence of events from more to less likely ones. Further important concept is *independence of events*. If they are independent, then the probability of their simultaneous occurrence is equal to *product* of their respective probabilities.

On the other hand, let us consider, for example, a cupboard full of red dresses. Then to answer whether the given dress is "red" requires to characterize *truth* of the statement "the color of the given dress is red". This cannot be probability because to be red color is a property, not an occurring event. Moreover, the class of all wave lengths representing red color cannot be a set because we are not able to specify precisely the borderline between "redness" and "non-redness".

[1]This holds also in the case when we know that some event has occurred but we do not know which one.

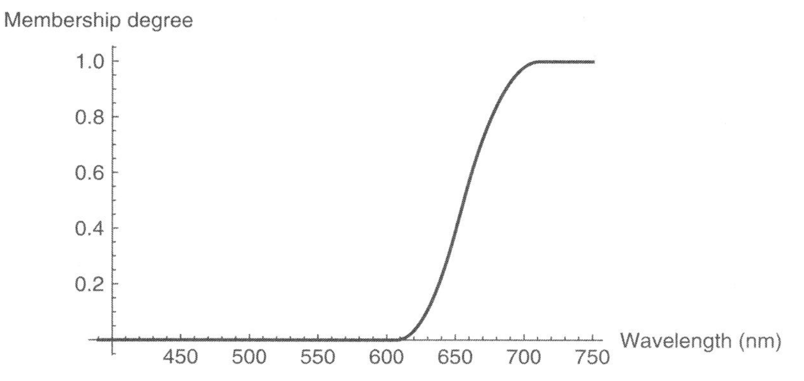

Figure 1.1 Fuzzy set modeling the meaning of "red color".

We can model the meaning of "red" using the concept of a fuzzy set. A fuzzy set A is a function

$$A : U \longrightarrow [0, 1],$$

where U is a set called *universe*. Each element $x \in U$ is assigned a membership degree $A(x) \in [0, 1]$ which is a *truth value*[2] of the proposition saying that $x \in A$. The value $A(x) = 1$ means that x *belongs* to A ($x \in A$). The value $A(x) = 0$ means that x *does not belong* to A ($x \notin A$). All other values mean only partial belonging to the fuzzy set A. To stress that A is a fuzzy set on U, we often write $A \underset{\sim}{\subset} U$.

If now we want to model what does it mean "red", we first define the universe of wave lengths that cover visible spectrum of light. People are able to see wave lengths from the interval $[390, 750]$ nm. Then "red" can be modeled by a fuzzy set $R \underset{\sim}{\subset} [390, 750]$ depicted in Figure 1.1. This means that light of wave length shorter than 600 nm is not red at all. Then the degree of "redness" increases with the increase in wavelengths up to full redness.

Of course, one may ask what is the probability of taking a red dress out of the cupboard. In this case, we face a combination of uncertainty and vagueness because the considered event is vaguely specified. We can thus summarize that there is a general concept called *indeterminacy*.[3] It has at least two distinguished facets: *vagueness* and *uncertainty*. Vagueness can be mathematically modeled using the fuzzy set theory, while uncertainty is mathematically modeled using the probability theory.[4] Of course, in reality, we often face both these facets together. For example, we can ask: "What is the probability that a tall man will come to our party?"[5]

Let us emphasize that indeterminacy cannot be removed. On one hand, it turns out that laws of nature inherently include uncertainty and it is not possible even in

[2]The interval $[0, 1]$ of truth values can be replaced by any other ordered set having the greatest and the smallest elements.

[3]One can also find this concept under the name "uncertainty in the wider sense".

[4]Note that there are also other theories of uncertainty, for example, possibility or belief theory.

[5]We can paraphrase this discussion as follows: the probability theory gives us an answer to the question "what event will occur?" and the fuzzy set theory gives us an answer to the question "what is the occurring event?"

principle to know all aspects causing occurrence of some event. On the other hand, vagueness is related to our way of regarding the world around us and its properties.

We argue that the presence of vagueness is the only way to familiarize with a new situation, or to communicate. Imagine, for example, that when parking a car, we would have at disposal instructions such as "turn the steering wheel by $19°25'32''$ to the left and move by 368.1256 mm back". Following such instructions would require great effort to make sufficiently precise measurements and to move accordingly. However, this would, in fact, be wasting of time because in practice, we do not need so precise parking position. It is sufficient to follow only vague instructions such as "turn the steering wheel a little to the left and move slightly back". Finally, note that we always face imprecision even when high precision is required, for example, when programming precise manipulating robots; the difference is only in the considered scale, that is, "small" could mean, for example, values around 1.3 mm or less.

1.2 FUZZY MODELING: WITH AND WITHOUT WORDS

The attempt to utilize the imprecise information in mathematical models led to the development of fuzzy modeling techniques. Recall that mathematical models manipulate with variables. In traditional models, values of the considered variable are taken from some set of numbers called a universe. Traditional mathematical models manipulate directly with its elements. In a fuzzy model, however, variables may represent fuzzy subsets of the universe. Hence, fuzzy models require partitioning of the universe into parts, for which it is specific that they need not be precisely formed and can overlap.

One of the very important modeling methods is *cluster analysis*. Its idea is the following: for a given set V of some elements, find its partition into c sets of subsets $R_j \subset V, j = 1,\ldots,c$, called *clusters*, in such a way that if two objects x, y belong to the same cluster R_j, then they are similar, while if they belong to different clusters, then they are not similar. For example, sizes of shoes represent subsets of lengths of human feet; the length of feet of people having, for example, size 6 is between 241 and 250 mm, for size 7 it is between 251 and 259 mm, and so on.

The classical cluster analysis provides partitioning into disjoint clusters, that is, we require that

$$U = R_1 \cup \cdots \cup R_c \quad \text{and} \quad R_i \cap R_j = \emptyset \quad \text{if } i \neq j.$$

This is often not realistic, because, as everybody knows, people often fit to more than one size of shoes. To cope with problems like this, we need generalization of the classical cluster analysis to the fuzzy one where the crisp clusters are replaced by fuzzy (possibly overlapping) ones. The fuzzy cluster analysis is described in Chapter 6.

The most important tool in fuzzy modeling are *fuzzy IF-THEN rules*. These are special expressions, which characterize relations among parts of two or more universes. For example, let us consider an electric boiler and two universes: values of electric current (A) and temperature (°C). Then the following is a typical fuzzy IF-THEN rule:

$$\mathscr{R}: \ \mathsf{IF} \text{ electric current is very strong } \mathsf{THEN} \text{ temperature is high} \tag{1.1}$$

By this rule, part of the universe of values of electric current characterized by the expression "very strong" is related to the part of the universe of degrees characterized as "high temperature". In practice, we usually have more such rules at disposal. A set $\mathscr{R}_1, \ldots, \mathscr{R}_m$ of rules (1.1) is called a *linguistic description.*

The reader has certainly met rules of this form already, namely, in programming languages. These are, however, *crisp* rules that do not allow any imprecision. On the other hand, IF-THEN rules used by people are almost always vague. The reason is that they contain vague natural language expressions that are central for human thinking. In this book, we will describe the way how semantics of certain class of natural language expressions can be modeled mathematically. The possibility to grasp the meaning of IF-THEN rules in the form close to human thinking should be considered as a great scientific accomplishment.

We manipulate with fuzzy IF-THEN rules by means of a scheme that reminds classical modus ponens rule from logic. Therefore, it is called *generalized modus ponens*:

Condition:	IF error is *big* AND change of error is *small* THEN control action is *very big*
Observation:	error is *roughly big* AND change of error is *small*
Conclusion:	control action is *big*

This means that if we know both the condition and the observation, we can deduce what we should do. For example, we deduce whether we should brake or turn a regulator cock or do something else. The tools of fuzzy modeling described in this book enable us to transform a linguistic description into an algorithm whose result is an action.

It is surprising how strong is the application potential of fuzzy IF-THEN rules (which are, in fact, quite simple expressions). The applications started in process control, but it is possible to describe by means of them very wide class of decision-making problems carried out by people. It starts from common events (crossing the street, dressing) and leads to important decisions requiring expert knowledge, for example, in medicine, management, and technology.

The rules of the form (1.1) resemble sentences of natural language. In this book, we will describe two ways how such rules can be interpreted:

(i) relational interpretation,
(ii) linguistic interpretation.

In case (i), we take an IF-THEN rule $\mathscr{R}$ in (1.1) as a rough characterization of some dependence. The linguistic expressions occurring in the rule are, in fact, taken only as names (codes) of some fuzzy sets. The rule as well as the whole linguistic description are interpreted by a fuzzy relation, which is a fuzzy set in a Cartesian product of several universes. The role of natural language is here auxiliary and no model of its semantics is considered. On the other hand, the methods developed on the basis of this interpretation of fuzzy IF-THEN rules provide well-substantiated tools for approximation of continuous functions. The relational interpretation and elaboration of fuzzy IF-THEN rules are in detail described in Chapter 3.

In case (ii), we take an IF-THEN rule $\mathscr{R}$ in (1.1) as a conditional clause formulated in natural language and the linguistic description is construed as a special text characterizing the decision situation. This interpretation is related to a paradigm of *computing with words and perceptions*, which was proposed by Zadeh in [152] (cf. also [153]). This is a methodology, in which objects of computation are words and propositions drawn from a natural language. Several publications appeared since then (e.g., [144, 154], and other ones).

To stress that fuzzy IF-THEN rules are taken as conditional linguistic clauses, we will call them *fuzzy/linguistic* IF-THEN rules. The methods for their elaboration belong among mathematical tools developed in the so-called *fuzzy natural logic* (FNL). This is a class of mathematical theories with the goal to model natural human thinking that is characterized by the use of natural language. The central position in FNL is played by the theory of *evaluative linguistic expressions*, that is, the theory of the semantics of linguistic expressions such as "nice", "very deep", "more or less strong", and "extremely quick". Such expressions occur in the rules $\mathscr{R}$ in (1.1), which are then used by people in various situations when it is necessary to make a decision, to evaluate some product (e.g., very good, strong, not safe), and in various other occasions.

As mentioned, the *linguistic description* is understood as a special text characterizing, for example, a sophisticated control strategy, decision-making, or behavior of a complex system. Applying a special reasoning method (called *perception-based logical deduction*), we can form models that effectively utilize expert knowledge and mimic the way how people behave when facing complicated decision situations. Interpretation and elaboration of fuzzy/linguistic IF-THEN rules, linguistic descriptions, and reasoning on the basis of them are described in detail in Chapter 5.

A special and very effective method of fuzzy modeling is *fuzzy transform* (*F-transform*). This method gains still more attention for its fascinating applications in diverse areas.

The basic concept of the F-transform is that of a *fuzzy partition* of an interval of real numbers $[a, b] \subset \mathbb{R}$. This is a finite set of fuzzy sets $A_k \underset{\sim}{\subset} [a, b]$, $k = 0, \ldots, n$ that fulfill special conditions. Using a fuzzy partition, a real continuous function f : $[a, b] \longrightarrow [c, d]$ is transformed into a finite vector $[F_0[f], \ldots, [F_n[f]]$ of *components*. This procedure is called a *direct phase*. Then (possibly after some computations) we can transform the vector of components back to a space of continuous functions. The result is a function $\hat{f} : [a, b] \longrightarrow [c, d]$, which *approximates* the original function f. This procedure is called an *inverse phase*. Parameters of the F-transform can be set in such a way that $\hat{f}$ has desired interesting properties. This opens the door to various kinds of applications. In this book, we will describe applications of the F-transform in image processing and in analysis and forecasting of time series. It can also be applied in numerical solution of differential equations, signal processing, data mining, and elsewhere. The fuzzy transform is described in detail in Chapter 4.

There are many applications of fuzzy modeling. The most distinguished ones are in control—we speak about *fuzzy control*. There are several reasons for applying fuzzy control in practice. One of them is the relative ease of its design. For example, when Hubble telescope had to be repaired, the automatic arm that took the telescope in and out of the spaceship was controlled using fuzzy controller. Its development took about 2 weeks. The same control developed in parallel using classical proportional-integral-derivative (PID) controller was in 2 weeks far from being finished. In fact, the complexity of design of fuzzy controller depends very little on the complexity of the controlled process.

Surprisingly, there are even processes whose satisfactory automatic control can be realized using fuzzy controller only. A typical example is the control of purification process in the sewage treatment plant. One can hardly find a mathematical model on the basis of which classical control can be designed. Therefore, this facility is usually controlled by people who apply their practical experience. We can express the latter using rules such as (1.1) and, consequently, utilize it using fuzzy modeling techniques.

A very important property of fuzzy models is their *robustness*. This means that they are little sensitive to external disturbances. For example, one of the very successful applications of fuzzy control based on fuzzy/linguistic IF-THEN rules is control of five aluminum smelting furnaces in Al Invest company in a small village Břidličná in the Czech Republic. The control is subject to large disturbances caused by repeated opening, recharging, charging, and closing of the smelting furnace. However, the fuzzy control works without disruption already for many years (more details about this application are presented in Chapter 7). Thanks to all these properties, fuzzy control and other fuzzy modeling techniques have become standard tools used by companies in their production now. We can find fuzzy controllers in automatic washing machines, dishwashers, cars, and other products.

It should be emphasized that robustness is a typical feature not only of fuzzy control but also of applications of fuzzy modeling in general. For example, when applying fuzzy modeling methods to character recognition, we need only few patterns (see [90]), while in classical solutions, we need hundreds of them. In applications of the F-transform, robustness manifests itself on little dependence of the result on the choice of the initial conditions. For example, when solving differential equations, the result is similar to the classical regularization method. This means that the result is almost independent on the choice of initial values (cf. [57, 108, 117]).

The range of applications of fuzzy modeling methods is very wide because they enable us to work with imprecise information that is often available in the form of sentences of natural language only. Of course, this does not mean that classical methods should be abandoned. On the contrary, they should be combined with methods of fuzzy modeling in such a way that each of the methods is used in a situation when it can lead to the best result. Namely, classical methods are convenient if precise mathematical model of our problem can be found. On the other hand, if the available information is imprecise and we need robustness, then methods based on fuzzy modeling should be preferred.

2

OVERVIEW OF BASIC NOTIONS

In this chapter, we briefly overview some basic mathematical notions and basic concepts of the fuzzy set theory and fuzzy logic. We refer the readers interested in more detailed information to the special literature (see References).

2.1 RELATIONS, FUNCTIONS, ORDERED SETS

We suppose that the reader knows the concept of set and basic operations with sets. In this section, we recall only few basic facts.

Sets will be denoted by capital letters $A, B, \ldots, U, V, \ldots$ If an element u *belongs* to a set U, then we write $u \in U$. By $\mathbb{N}$, we denote the set of all natural numbers and by $\mathbb{R}$ the set of all real ones.

2.1.1 Relations

Let U and V be arbitrary sets. Their *Cartesian product* is a set of ordered pairs

$$U \times V = \{\langle u, v\rangle \mid u \in U, v \in V\}.$$

A *binary relation* R between elements of U and V is a subset of the Cartesian product

$$R \subseteq U \times V.$$

If elements u, v are in relation R, that is, $\langle u, v\rangle \in R$, then we often write uRv. If $U = V$, then we say that R is a binary relation on U.

Insight into Fuzzy Modeling, First Edition. Vilém Novák, Irina Perfilieva, and Antonín Dvořák.

Companion Website: www.wiley.com/go/novak/fuzzy/modeling

If $R \subseteq U \times V$ and $S \subseteq V \times W$ are binary relations, then their *composition* $R \circ S$ is a binary relation in $U \times W$ defined by

$$R \circ S = \{\langle u, w\rangle \mid (\exists v \in V)((\langle u, v\rangle \in R \text{ and } \langle v, w\rangle \in S)\}.$$

Functions. A function F is a special binary relation $F \subseteq U \times V$ (we say that F is a *function from* U to V) which has the following properties:

(i) To each $u \in U$, there is $v \in V$ such that $\langle u, v\rangle \in F$.
(ii) If $\langle u, v\rangle \in F$ and $\langle u, w\rangle \in F$, then $v = w$. (uniqueness)

It is usual to write $v = F(u)$ instead of $\langle u, v\rangle \in F$ and say that $v \in V$ is assigned to $u \in U$. If F is a function from U to V, then we write $F : U \longrightarrow V$.

The set U is called the *domain* of F and often denoted by *dom*(F). The set of all $v \in V$ which are assigned to some $u \in U$ is called the *range* of F and denoted by rng(F). Note that, in general, rng$(F) \subseteq V$.

A function is

(i) *injective* (injection), if $v = F(u)$ and $v = F(u')$ imply $u = u'$.
(ii) *surjective* (surjection), if to each $v \in V$, there is $u \in U$ such that $v = F(u)$, that is, if rng$(F) = V$.
(iii) *bijective* (bijection) if it is both injective as well as surjective.
(iv) *identical* (identity) if $U = V$ and $f(u) = u$ for all $u \in U$. This function is usually denoted by 1_U.

If $f : U \longrightarrow V$ and $g : V \longrightarrow U$ are functions such that $g(f(u)) = u$, that is, the composition $f \circ g = 1_U$, then g is the *inverse function* to f and is denoted by f^{-1}.

Important is a class of real continuous functions $F : \mathbb{R} \longrightarrow \mathbb{R}$. A useful characterization of them is the so-called *modulus of continuity*. Let $F : [a, b] \longrightarrow \mathbb{R}$ be a continuous function and $h > 0$. Then the modulus of continuity of F is defined by

$$\omega(h, F) = \max_{\substack{|x-y|<h \\ x,y\in[a,b]}} |F(x) - F(y)|. \tag{2.1}$$

Note that the modulus of continuity measures, in a certain sense, smoothness of F with respect to the parameter h. If we fix h, then smaller changes in the course of F lead to smaller modulus of continuity (2.1).

Ordered sets. Let L be a set. A partial ordering $\leq$ on L is a binary relation on L that has the following properties:

(i) $a \leq a$ for all $a \in L$. (reflexivity)
(ii) If $a \leq b$ and $b \leq a$, then $a = b$ for all $a, b \in L$. (antisymmetry)
(iii) If $a \leq b$ and $b \leq c$, then $a \leq c$ for all $a, b, c \in L$. (transitivity)

A pair $(L, \leq)$ is called a *partially ordered set*.Note that there may be incomparable elements in $(L, \leq)$, that is, elements $a, b \in L$ for which neither $a \leq b$ nor $b \leq a$. An ordered set $(L, \leq)$ is *linearly ordered* (also called a *chain*) if for all $a, b \in L$ either $a \leq b$ or $b \leq a$.

A sharp inequality $<$ is transitive, irreflexive, and asymmetric, that is, (i) does not hold for any $a \in L$ and $a < b$ implies that $b < a$ does not hold.

Let $(L, \leq)$ be a partially ordered set. An element $d \in L$ is *maximal* if for any $a \in L$, $d \leq a$ implies $d = a$. It is *minimal* if $a \leq d$ implies $a = d$. Note that there can be more maximal as well as minimal elements. The element d is the *greatest* if $a \leq d$ and the *smallest* if $d \leq a$ holds for all $a \in L$. Every greatest (smallest) element is maximal (minimal) but not vice versa. If a set has the greatest (smallest) element, then it is unique.

An ordered set can have maximal (minimal) elements but no greatest (smallest) one. On the other hand, none of these elements may exist. For example, there is no maximal (and so, no greatest) and no minimal (and so, no smallest) element in the interval of real numbers $(0.3, 0.7)$.

Let $a, b \in L$. Then the *upper bound* of the set $\{a, b\}$ is any element $d \in L$ such that $a \leq d$ as well as $b \leq d$. Similarly, the *lower bound* of the set $\{a, b\}$ is any element $d \in L$ such that $d \leq a$ as well as $d \leq b$. A minimal element of the set of all upper bounds of the set $\{a, b\}$ is *supremum* of a and b and is denoted by $a \vee b$. Notice that in a linearly ordered set (e.g., in the set of real numbers), supremum of two elements is equal to the greater of them.

Similarly, $a \wedge b$ denotes *infimum* of the elements a and b, that is, maximal element of the set of all lower bounds e of the set $\{a, b\}$, and in a linearly ordered set, it is equal to the smaller of them.

We can extend the operations of supremum and infimum to sets (possibly infinite). For example, let $(0.3, 0.7)$ be an interval of real numbers. Then,

$$\bigvee \{a \mid a \in (0.3, 0.7)\} = 0.7, \qquad \bigwedge \{a \mid a \in (0.3, 0.7)\} = 0.3.$$

A partially ordered set $(L, \leq)$ for which it holds that to any elements $a, b \in L$ their supremum as well as infimum belong to L is called a *lattice*. Operations $\vee$ and $\wedge$ in a lattice are called *join* and *meet*, respectively. An alternative definition is that a lattice is an algebra $\langle L, \vee, \wedge \rangle$ such that the following holds for all $a, b, c \in L$:

(i) $a \vee b = b \vee a$ as well as $a \wedge b = b \wedge a$, (commutativity)

(ii) $a \vee (b \vee c) = (a \vee b) \vee c$ as well as $a \wedge (b \wedge c) = (a \wedge b) \wedge c$, (associativity)

(iii) $a \vee (b \wedge a) = a$ as well as $a \wedge (b \vee a) = a$. (absorption)

EXAMPLE 2.1 In Figure 2.1, three partially ordered sets are depicted. In the diagrams, if there is an arrow from x to y, then $x \leq y$.

The set (a) has two maximal and two minimal elements but it is not a lattice. (b) is a five-element chain which is also a lattice and has the greatest element e and the smallest element a. (c) is a lattice with the greatest element e and smallest a but it is not chain. In the diagram, one can also see supremum and infimum. For example, in

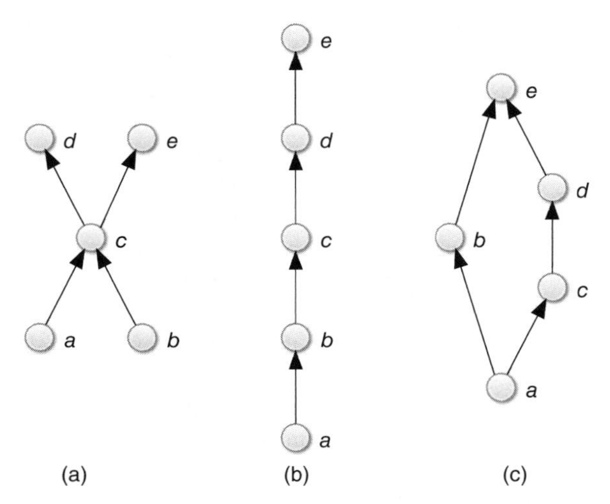

Figure 2.1 Three partially ordered sets. (a) has two maximal and two minimal elements. (b) is a five-element chain. (c) is a lattice with the greatest element e and smallest a.

diagram (a), we have $d \wedge e = c = a \vee b$, but $d \vee e$ and $a \wedge b$ do not exist. Similarly, in diagram (c), $b \vee d = e$ and $b \wedge d = a$. □

A special kind of ordering of couples of elements is the *linear lexicographic ordering* defined as follows: let $(A, \leq), (B, \leq)$ be two linearly ordered sets. Then a linear lexicographic ordering of couples (a, b) where $a \in A$ and $b \in B$ is defined by

$$(a, b) \leq (c, d) \quad \text{iff} \quad a < b, \text{ or } a = b \text{ and } c \leq d. \tag{2.2}$$

For example, $(2, 4) \leq (2, 5) \leq (3, 1)$.

2.2 FUZZY SETS AND FUZZY RELATIONS

In Chapter 1, we already mentioned the concept of fuzzy set. In this chapter, we describe the theory of fuzzy sets in more detail.

2.2.1 The Concept of a Fuzzy Set

Fuzzy sets, in a certain sense, generalize the classical concept of a set. Their motivation stems from the following idea: Imagine that somebody wants us to specify a set of heights of all *tall* people. First, we can say that the height of any tall man/woman lays between 160 and 240 cm. Therefore, we start with the set $U = [160, 240]$ (cm). In further reasoning, however, we will be confronted with insurmountable difficulties, namely, we find out that we are not able to specify *tall people* precisely. For example, if we decide that the height of a tall man/woman is at least 175 cm, then we can immediately ask: "What about the height 174.6 cm?" We are not able to distinguish man/woman 175 and 174.6 cm tall by naked eye only. But, according to our decision, the first man/woman is tall, and the second one is not. We arrive at clearly counterintuitive conclusion.[1] Therefore, in a common practice, we cannot understand numbers to be fully precise but rather to be imprecise, for example, "approximately 150". Alternatively, we can use vague linguistic expressions such as "small" and "very tall".

The general definition of a fuzzy set is the following.

Definition 2.1 *Let U be a set called* universe. *A* fuzzy set *A is a function*

$$A : U \longrightarrow [0, 1]. \tag{2.3}$$

The function (2.3) is also called a membership function *of the fuzzy set A. If $u \in U$, then the number $A(u) \in [0, 1]$ is called a* membership degree *of u in the fuzzy set A.*

It follows from this definition that a fuzzy set is *identified* with its membership function. The membership degree $A(u)$ for $u \in U$ is the *degree of truth* of the statement "u belongs to A". Hence, membership degrees are, in fact, degrees of truth.

EXAMPLE 2.2 In Chapter 1, we already gave an example of fuzzy set of wavelengths representing red color. Another typical example is the fuzzy set of "heights of small people" which can be characterized, for example, by the following membership function defined in the universe $U = [55, 270]$ (possible heights of adult people in centimeters):

$$A_{small}(u) = \begin{cases} 1, & \text{if } u \leq 165, \\ 0, & \text{if } u > 185, \\ 1 - \frac{1}{2}\left(\frac{u - 165}{10}\right)^2, & \text{if } 165 < u < 175, \\ \frac{1}{2}\left(\frac{185 - u}{10}\right)^2, & \text{if } 175 \leq u \leq 185. \end{cases}$$

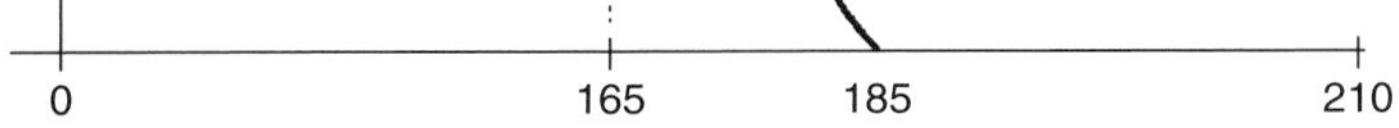

Figure 2.2 Schematic depiction of a possible fuzzy set of small people.

This fuzzy set is schematically depicted in Figure 2.2. According to this example, we can say that, for example, the proposition "165 cm is a height of a small man/woman"

[1] Such a crisp decision procedure can have even fatal consequence. For example, the threshold for enlisting men in the army was always at the height 150 cm. However, if we try to reach absolute precision, we would enlist a boy with the height 150.05 cm but not the boy with the height 149.95 cm, although we are not able to distinguish these two heights by naked eye or standard measurement.

is true in the degree 1, while "190 cm is a height of a small man/woman" is true in the degree 0, that is, it is false (such a man/woman is not small at all). Of course, any person smaller than 165 cm is also small in the degree 1, while any person taller than 185 cm is not small at all. □

Note that the universe can be a set of any kind. For example, a set of people, plants, or, quite often, a set of real numbers $\mathbb{R}$. The latter case is important especially in modeling because numbers represent results of some measurements or actions.

Remark 2.1 *In the literature, the membership degree of an element u to a fuzzy set A is often denoted by $\mu_A(u)$. But, this is inaccurate, confusing, and, moreover, quite inconvenient, because, for example, for more complexly defined fuzzy sets, a nontransparent expression may occur in the subscript (e.g., $\mu_{(A\cup B)\cap(B\cup\overline{D})}$). Therefore, we will not use the symbol μ in this book.*

A fuzzy set generalizes a classical set in the following sense: In the classical set theory, we can define the so-called *characteristic function* $\chi_A : U \longrightarrow \{0, 1\}$ of a set A with respect to U by

$$\chi_A(u) = \begin{cases} 1, & \text{if } u \in A, \\ 0, & \text{if } u \notin A. \end{cases}$$

Thus, $\chi_A(u) = 1$ if an element u *belongs* to the set A and $\chi_A(u) = 0$ otherwise. We can immediately see that the membership function is a generalization of the characteristic function of a classical set.

Recall that $A \underset{\sim}{\subseteq} U$ denotes a fuzzy set A from Definition 2.1 defined in the universe U. If we want to write down a fuzzy set on a finite universe U explicitly, we use the following notation:

$$A = \left\{{}^{a_1}\!/u_1, \ldots, {}^{a_n}\!/u_n\right\}, \tag{2.4}$$

where $u_1, \ldots, u_n \in U$ are elements, which are assigned membership degrees $a_1, \ldots, a_n \in (0, 1]$. Note that elements with membership degree 0 are not included in (2.4).

EXAMPLE 2.3 Let a universe U be $\{0, 1, \ldots, 10\}$. Then

$$A = \left\{{}^{0.4}\!/1, {}^{0.7}\!/2, {}^{0.5}\!/4, {}^{1}\!/6, {}^{1}\!/7, {}^{0.1}\!/9\right\} \tag{2.5}$$

is a fuzzy set on U to which the number 1 belongs in membership degree 0.4, 2 in membership degree 0.7, etc. Numbers from U with membership degree equal to 0 are not written in (2.5). This is clear because they do not belong to the fuzzy set A. □

If a universe is not finite, then elements of a fuzzy set cannot be written down in the list (2.4). In this case, we write fuzzy sets as

$$A = \left\{{}^{A(u)}\!/u \;\middle|\; u \in U\right\}, \tag{2.6}$$

where $A(u)$ is a specific function, for example, a partially linear (triangle) or quadratic function (provided that $U \subseteq \mathbb{R}$, where $\mathbb{R}$ is the set of real numbers).

Remark 2.2 *In older literature, we can also meet the denotation*

$$A = \int_{u \in \mathbb{R}} \mu_A(u)/u.$$

The integral sign, however, is inaptly used here in the sense of union and not in its original sense. We will not use this denotation in this book.

Below are few important concepts of fuzzy set theory.

Support.

$$\text{Supp}(A) = \{u \mid u \in U,\ A(u) > 0\},$$

that is, the support of a fuzzy set A is the *set* of all elements of the universe U, whose membership degree to A is greater than zero.

EXAMPLE 2.4 The support of the fuzzy set A in (2.5) is

$$\text{Supp}(A) = \{1, 2, 4, 6, 7, 9\}.$$

□

a-cut. Let $a \in [0, 1]$. Then

$$A_a = \{u \mid u \in U,\ A(u) \geq a\} \tag{2.7}$$

that is, the a-cut[2] is the *set* of elements of the universe U with membership degrees greater than or equal to the given degree a. We obtain this set from the fuzzy set A by means of "cutting off" all elements with membership degrees smaller than a.

EXAMPLE 2.5 Let A be the fuzzy set from Example 2.3. Then

$$A_{0.5} = \{2, 4, 6, 7\},$$
$$A_{0.7} = \{2, 6, 7\}.$$

□

The following simple property holds for a-cuts of fuzzy sets:

$$\text{If } a \leq b, \text{ then } A_b \subseteq A_a.$$

Kernel.

$$\text{Ker}(A) = \{u \mid u \in U,\ A(u) = 1\}$$

that is, a kernel[3] is a *set* of elements from U that definitely belong to fuzzy set A. They represent *typical elements* (prototypes) for the given fuzzy set. For example, typically tall people can be, say, all people taller than 190 cm.

[2]One can meet the term α-cut in the literature.

[3]Sometimes it is also called "core".

EXAMPLE 2.6 The kernel of the fuzzy set A from (2.5) is

$$\mathrm{Ker}(A) = \{6, 7\} . \qquad \square$$

Obviously, the kernel of a fuzzy set A is its 1-cut.

We say that a fuzzy set is *normal* if $\mathrm{Ker}(A) \neq \emptyset$. Fuzzy sets, which are not normal, are called *subnormal*. Obviously, all elements of a subnormal fuzzy set have their membership degrees smaller than 1.

Empty fuzzy set. This is a fuzzy set that contains no elements:

$$\emptyset = \left\{ {}^{0}\!/u \,\middle|\, u \in U \right\} .$$

We see that the empty fuzzy set coincides with the classical empty set.

Fuzzy singleton. An important role is played by the *one-element fuzzy set* (commonly called *fuzzy singleton*)

$$\left\{ {}^{a}\!/u \right\} , \tag{2.8}$$

which is a fuzzy analogy of the classical one-element set. Expression (2.8) means that only the element $u \in U$ belongs to the given fuzzy singleton with a membership degree $a > 0$. If $a = 1$, then the fuzzy singleton coincides with the classical one-element set.

To simplify our language, we will call the fuzzy set (2.8) with the membership degree $1 > a > 0$ a *general fuzzy singleton*, and if $a = 1$, then we will call it a *fuzzy singleton*. The above concept is important in fuzzy control where the result of an individual measurement can be understood just as a fuzzy singleton.

EXAMPLE 2.7 Let a temperature measured in a furnace be 950 °C. Then we can understand it as a fuzzy singleton

$$\left\{ {}^{1}\!/950\,^{\circ}\mathrm{C} \right\} . \qquad \square$$

Convex fuzzy set. This concept is meaningful if the universe is a subset of the set of real numbers. We will meet convex fuzzy sets in the definition of fuzzy numbers.

A fuzzy set $A \underset{\sim}{\subset} U \subseteq \mathbb{R}$ is *convex*, if for all elements $u, v \in U$ and any $0 \leq \lambda \leq 1$, it holds that

$$A(\lambda u + (1 - \lambda)v) \geq A(u) \wedge A(v).$$

It can be shown that a fuzzy set is convex if and only if all its a-cuts are connected intervals. A comparison of a convex and nonconvex fuzzy set is depicted in Figure 2.3.

Set of all fuzzy sets in the universe U. This is the set

$$\mathscr{F}(U) = \left\{ A \mid A \underset{\sim}{\subset} U \right\} . \tag{2.9}$$

From the mathematical point of view, $\mathscr{F}(U)$ is the set of all mappings $U \longrightarrow [0, 1]$, that is,

$$\mathscr{F}(U) = \{A \mid A : U \longrightarrow [0, 1]\} = [0, 1]^{U}.$$

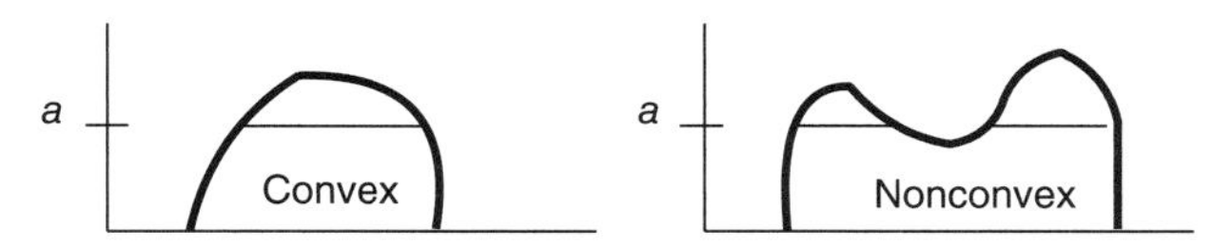

Figure 2.3 Convex and nonconvex fuzzy set.

Distance between fuzzy sets. Let $A, B \subsetapprox U$ be fuzzy sets, U be a finite universe, and $p \geq 1$ be a number (usually a natural one). Then, *p-distance of fuzzy sets* A, B is the number

$$d^p(A,B) = \left(\sum_{u \in U} |A(u) - B(u)|^p \right)^{\frac{1}{p}}. \tag{2.10}$$

Usually, we put $p = 1$ or $p = 2$. Higher numbers p can hardly be meaningful in practice. This definition is sufficient for practical purposes. If we want to extend it to infinite universes, we must replace the sum by integral. However, this requires additional conditions on the integrability of membership functions of the fuzzy sets A and B. We will not deal with this extended concept in this book.

2.2.2 Operations with Fuzzy Sets

Analogously, as in the case of classical sets, we can define basic operations of intersection, union, and complement also for fuzzy sets. However, unlike classical sets, we can define many other ones that in classical set theory are either senseless or equivalent with some of the basic operations. This fact substantially extends the expressive power of fuzzy set theory.

Intersection. An *intersection* of fuzzy sets is defined using one of the three possible operations for $a, b \in [0, 1]$:

(i) *Minimum* $\min\{a, b\} = a \wedge b$.

(ii) *Algebraic product* $a \cdot b$.

(iii) *Łukasiewicz conjunction*

$$a \otimes b = \max\{0, a + b - 1\}. \tag{2.11}$$

Remark 2.3 *The last operation is named after famous Polish logician Jan Łukasiewicz (1878–1956), who wrote fundamental works about many-valued logic in the 1920s of the 20th century. This operation plays a substantial role in many logical considerations and is even more important than the usual operation of minimum.*

Using the operations considered above, *intersection* of fuzzy sets A and B is a fuzzy set C with the membership function defined in one of the following three ways:

(i) *Minimum intersection:* $C = A \cap B$, where

$$C(u) = A(u) \wedge B(u), \quad u \in U. \tag{2.12}$$

(ii) *Product intersection:* $C = A \cap_P B$, where

$$C(u) = A(u) \cdot B(u), \qquad u \in U. \tag{2.13}$$

(iii) *Łukasiewicz intersection:* $C = A \otimes B$, where

$$C(u) = A(u) \otimes B(u) = \max\{0, A(u) + B(u) - 1\}, \quad u \in U. \tag{2.14}$$

If we replace the interval $[0, 1]$ by $\{0, 1\}$ and the membership functions A and B by the corresponding characteristic functions χ_A, χ_B, then all three operations reduce to the ordinary intersection of sets.

EXAMPLE 2.8 Let the universe U be the same as in Example 2.3, and let

$$A = \left\{{}^{0.4}/1, {}^{0.7}/2, {}^{0.5}/4, {}^{1}/6, {}^{1}/7, {}^{0.1}/9\right\}, \tag{2.15}$$

$$B = \left\{{}^{0.2}/1, {}^{0.7}/2, {}^{0.9}/3, {}^{0.6}/4, {}^{1}/6, {}^{0.8}/7, {}^{0.3}/9\right\}. \tag{2.16}$$

Then

$$A \cap B = \left\{{}^{0.2}/1, {}^{0.7}/2, {}^{0.5}/4, {}^{1}/6, {}^{0.8}/7, {}^{0.1}/9\right\},$$

where, for example, the membership degree 0.2 in the fuzzy singleton ${}^{0.2}/1$ has been obtained as $0.4 \wedge 0.2 = 0.2$.

The product intersection is

$$A \cap_P B = \left\{{}^{0.08}/1, {}^{0.49}/2, {}^{0.3}/4, {}^{1}/6, {}^{0.8}/7, {}^{0.03}/9\right\},$$

and the Łukasiewicz intersection is

$$A \otimes B = \left\{{}^{0.4}/2, {}^{0.1}/4, {}^{1}/6, {}^{0.8}/7\right\},$$

where the membership degree 0.1 in the fuzzy singleton ${}^{0.1}/4$ has been obtained as $0.5 \otimes 0.6 = \max\{0, 0.5 + 0.6 - 1\} = 0.1$. □

All three operations used in (2.12)–(2.14) naturally correspond to the logical connective "AND" (logical conjunction). Therefore, we can use the intersection if we want to characterize, for example, the fuzzy set of all people, who are "clever and young". We define fuzzy sets of young people and clever people and then construct their intersection.

Note that the Łukasiewicz intersection is more strict than the other two ones. We can use it in situations where fuzzy sets A and B are in some sense in mutually negative relationship, or if we are not sure what is the relationship between them. For example, "very big and very thin trees"—to be a "big tree" contradicts in a certain extent to the fact that the tree is also "thin". The Łukasiewicz intersection is "safer" in the following sense. If for some $u \in U$ should hold that $(A \cap B)(u) = 0$, then the risk that it would not be equal to 0 is minimal for the Łukasiewicz intersection.[4]

[4]We can have $(A \otimes B)(u) = 0$ even if both $A(u) > 0$ as well as $B(u) > 0$.

Union. A *union* of fuzzy sets is defined using one of the three possible operations for $a, b \in [0, 1]$:

(i) *Maximum* $\max\{a, b\} = a \vee b$.
(ii) *Product disjunction* $a \odot b = a + b - a \cdot b$.
(iii) *Łukasiewicz disjunction*

$$a \oplus b = \min\{1, a + b\} . \tag{2.17}$$

Then *union* of fuzzy sets A and B is a fuzzy set C with membership function defined in one of three ways:

(i) *Maximum union:* $C = A \cup B$, where

$$C(u) = A(u) \vee B(u), \quad u \in U. \tag{2.18}$$

(ii) *Product union:* $C = A \cup_P B$, where

$$C(u) = A(u) \odot B(u) = A(u) + B(u) - A(u) \cdot B(u), \quad u \in U. \tag{2.19}$$

(iii) *Łukasiewicz union:* $C = A \uplus B$, where

$$C(u) = A(u) \oplus B(u) = \min\{1, A(u) + B(u)\} , \quad u \in U. \tag{2.20}$$

As in the case of intersection, if we replace the interval $[0, 1]$ by $\{0, 1\}$ and the membership functions A and B by the corresponding characteristic functions χ_A, χ_B, then all three operations reduce to the ordinary union of sets.

EXAMPLE 2.9 For fuzzy sets (2.15) and (2.16) from Example 2.8, we get

$$A \cup B = \{{}^{0.4}/1, {}^{0.7}/2, {}^{0.9}/3, {}^{0.6}/4, {}^{1}/6, {}^{1}/7, {}^{0.3}/9\} ,$$

where, for example, the membership degree 0.4 in the fuzzy singleton ${}^{0.4}/1$ has been obtained as $0.4 \vee 0.2 = 0.4$.

The product union is

$$A \cup_P B = \{{}^{0.52}/1, {}^{0.91}/2, {}^{0.8}/4, {}^{1}/6, {}^{1}/7, {}^{0.37}/9\} ,$$

and the Łukasiewicz union is

$$A \uplus B = \{{}^{0.6}/1, {}^{1}/2, {}^{0.9}/3, {}^{1}/4, {}^{1}/6, {}^{1}/7, {}^{0.4}/9\} ,$$

where, for example, the membership degree 1 in the fuzzy singleton ${}^{1}/2$ is obtained as $0.7 \oplus 0.7 = \min\{1, 0.7 + 0.7\} = 1$. □

All three operations used in (2.18)–(2.20) correspond to logical disjunction (connective "OR"). For example, we use union if we want to characterize all people who are "young or middle-aged". We define a fuzzy set of young people and a fuzzy set of middle-aged people and carry out their union.

Now it should be clear why a fuzzy set can be understood as a union of fuzzy singletons. The fuzzy set A from Example 2.8 can obviously be written as

$$A = \{{}^{0.4}/1\} \cup \{{}^{0.7}/2\} \cup \{{}^{0.5}/4\} \cup \{{}^{1}/6\} \cup \{{}^{1}/7\} \cup \{{}^{0.1}/9\}\,.$$

Similar principle should also be used if it happens during computations that we obtain more than one fuzzy singleton with identical support. Then, the result is the fuzzy singleton with maximal membership degree.

EXAMPLE 2.10 Let us suppose that our intermediate computations produce the following class:[5]

$$C = \{{}^{0.4}/1, {}^{0.7}/1, {}^{0.9}/1, {}^{0.3}/2, {}^{0.4}/2, {}^{0.6}/2, {}^{1}/3, {}^{0.5}/4, {}^{1}/4, {}^{0.1}/5\}\,. \tag{2.21}$$

On the basis of the principle described above, the fuzzy set (2.21) is equal to

$$C = \{{}^{0.9}/1, {}^{0.6}/2, {}^{1}/3, {}^{1}/4, {}^{0.1}/5\}\,. \tag{2.22}$$

□

Complement. The *complement* $\overline{A}$ of a fuzzy set A is the fuzzy set with the following membership function:

$$\overline{A}(u) = 1 - A(u), \quad u \in U. \tag{2.23}$$

Note that this is the usual and most often used definition of a complement in fuzzy set theory introduced originally by Zadeh in [146].

EXAMPLE 2.11 For the fuzzy set A in (2.15) from Example 2.8, we get

$$\overline{A} = \{{}^{0.6}/1, {}^{0.3}/2, {}^{1}/3, {}^{0.5}/4, {}^{1}/5, {}^{1}/8, {}^{0.9}/9, {}^{1}/10\}\,,$$

where, for example, the membership degree 0.6 in fuzzy singleton ${}^{0.6}/1$ is obtained as

$$1 - 0.4 = 0.6\,.$$

□

The complement $\overline{A}$ of a fuzzy set A is the fuzzy set of all elements that do not possess the property characterizing A. For example, "not young people" is a fuzzy set of people, which have the property of being "not young". As we cannot, for a given age, always positively decide, whether it corresponds to "young", there are people both young as well as not young in some degree. On that account, we generally obtain

$$A \cap \overline{A} \neq \emptyset$$

and

$$A \cup \overline{A} \neq U.$$

[5]Strictly speaking, (2.21) is not fuzzy set in the sense of Definition 2.1, because a membership function can assign to one element of the universe one membership degree only.

It means that the so-called *law of excluded middle* is violated. On the other hand, for Łukasiewicz intersection, this law *does hold*, because

$$A \otimes \overline{A} = \emptyset,$$
$$A \uplus \overline{A} = U.$$

As we work with all the operations of fuzzy logic, we can generally conclude that the law of excluded middle *holds* in fuzzy set theory (and in fuzzy logic as well). Note that the range of available operations in fuzzy set theory is much wider than that in classical set theory.

Fuzzy sets $A, B \subsetneqq U$ are *disjoint* if

$$A \cap B = \emptyset.$$

Furthermore, A and B are *weakly disjoint* if

$$A \otimes B = \emptyset.$$

It follows from the definition of both operations of intersection that if fuzzy sets $A, B \subsetneqq U$ are disjoint, then for any element $u \in U$, it holds that $A(u) = 0$ or $B(u) = 0$. If they are weakly disjoint, then it holds only that $A(u) + B(u) \leq 1$.

Fuzzy implications and residuum of fuzzy sets. Important class of operations on truth values are *fuzzy implications*, which generalize classical implications. There are many of them, but the following are the most distinguished:[6]

(i) *Łukasiewicz implication*

$$a \rightarrow b = \min\{1, 1 - a + b\}, \quad a, b \in [0, 1]. \tag{2.24}$$

(ii) *Gödel implication*

$$a \xrightarrow{G} b = \begin{cases} 1, & \text{if } a \leq b, \\ b & \text{otherwise,} \end{cases} \quad a, b \in [0, 1]. \tag{2.25}$$

(iii) *Product implication*

$$a \xrightarrow{P} b = \begin{cases} 1, & \text{if } a \leq b, \\ \dfrac{b}{a} & \text{otherwise.} \end{cases} \quad a, b \in [0, 1]. \tag{2.26}$$

The most important of the implications introduced above is the Łukasiewicz one whose behavior corresponds well to intuition and it provides well-working applications. The main reason for this claim is the fact that Łukasiewicz implication is continuous in both arguments. Therefore, it always behaves smoothly without abrupt changes or jumps. Moreover, it can be proved that any other continuous implication

[6]It is possible to introduce fuzzy implications axiomatically, see, for example, [5].

on $[0, 1]$ is *isomorphic* with the Łukasiewicz one. Therefore, the latter is the only plausible continuous implication operation on $[0, 1]$.

Using the implication operation, we define the following special operation on fuzzy sets. The *residuum* of fuzzy sets A and B is a fuzzy set C with the membership function defined in one of the following three ways:

(i) *Łukasiewicz residuum* $C = A \ominus B$, where

$$C(u) = A(u) \rightarrow B(u) = \min\{1, 1 - A(u) + B(u)\}, \qquad u \in U. \tag{2.27}$$

(ii) *Gödel residuum* $C = A \ominus_G B$, where

$$C(u) = A(u) \overset{G}{\rightarrow} B(u), \quad u \in U. \tag{2.28}$$

(iii) *Product residuum* $C = A \ominus_P B$, where

$$C(u) = A(u) \overset{P}{\rightarrow} B(u), \quad u \in U. \tag{2.29}$$

EXAMPLE 2.12 For fuzzy sets (2.15) and (2.16) from Example 2.8, we get ($\ominus$ is the Łukasiewicz residuum)

$$A \ominus B = \{^{0.8}/1, ^{1}/2, ^{1}/3, ^{1}/4, ^{1}/5, ^{1}/6, ^{0.8}/7, ^{1}/8, ^{1}/9, ^{1}/10\},$$

where, for example, the membership degree 0.8 in the fuzzy singleton $^{0.8}/1$ is obtained as

$$0.4 \rightarrow 0.2 = 1 \wedge (1 - 0.4 + 0.2) = 0.8.$$

□

Notice that we can define the complement of a fuzzy set A using the Łukasiewicz residuum by

$$\overline{A} = A \ominus \emptyset.$$

Fuzzy equivalence and similarity of fuzzy sets. Fuzzy implication can be used for the definition of fuzzy equivalence that generalizes classical logical equivalence as follows:

$$a \leftrightarrow b = (a \rightarrow b) \wedge (b \rightarrow a). \tag{2.30}$$

Using (2.30), we obtain the following formulas for specific fuzzy equivalences:

(i) *Łukasiewicz equivalence*

$$a \leftrightarrow b = 1 - |a - b|, \quad a, b \in [0, 1]. \tag{2.31}$$

(ii) *Gödel equivalence*

$$a \overset{G}{\leftrightarrow} b = \begin{cases} 1, & \text{if } a = b, \\ a \wedge b & \text{otherwise.} \end{cases} \qquad a, b \in [0, 1]. \tag{2.32}$$

(iii) *Product equivalence*

$$a \overset{P}{\leftrightarrow} b = \begin{cases} 1, & \text{if } a = b, \\ \dfrac{a \wedge b}{a \vee b} & \text{otherwise,} \end{cases} \qquad a, b \in [0, 1]. \qquad (2.33)$$

The most important is the Łukasiewicz equivalence, the other two are mentioned here only for the sake of completeness. Note that they can be taken as graded equalities between truth values—the farther they are, the smaller is their equality. As a special case, $1 \leftrightarrow 0 = 0$, which means that the fuzzy equality between 1 and 0 is equal to 0.

Fuzzy equivalence can be used for a definition of *similarity of fuzzy sets* as follows (we may also speak about *fuzzy equality of fuzzy sets*).

Let $A, B \underset{\sim}{\subseteq} U$. Then we define the degree of similarity between A and B by

$$(A \Leftrightarrow B) = \bigwedge_{u \in U} (A(u) \leftrightarrow B(u)). \qquad (2.34)$$

Note that if $A = B$, then $(A \Leftrightarrow B) = 1$.

EXAMPLE 2.13 Let the fuzzy sets A, B be defined as in (2.15) and (2.16) in Example 2.8. Let us compute their similarity using (2.34). Note that if the corresponding membership degrees are the same, then the result of $a \leftrightarrow a$ is equal to 1. Therefore, we should consider only the cases when the membership degrees are different. Then the degree of similarity between the fuzzy sets A and B is

$$(A \Leftrightarrow B) = (0.2 \leftrightarrow 0.4) \wedge (0 \leftrightarrow 0.9) \wedge (0.5 \leftrightarrow 0.6) \wedge (1 \leftrightarrow 0.8) \wedge$$
$$(0.1 \leftrightarrow 0.3) = 0.1.$$

If we use (2.32) or (2.33), we obtain $(A \Leftrightarrow B) = 0$. The reason is the occurrence of the couple $0 \leftrightarrow 0.9$ because then $0 \wedge 0.9 = 0$. One can see that this result is somewhat counterintuitive.

Note that if there is an element u with the membership degrees $A(u) = 1$ and $B(u) = 0$ (or vice versa), then $0 \leftrightarrow 1 = 0$ and, consequently, it always holds that the degree of similarity $(A \Leftrightarrow B)$ is equal to 0. □

Triangular norms and generalization of operations with fuzzy sets. The operations with fuzzy sets described above can be taken as fundamental ones. It is clear that they generalize operations with classical sets.

Important direction in the theoretical development of fuzzy set theory is related to the possibility to define new additional operations. They are realized by means of the so-called t-norms.[7] Let us examine this notion in more detail.

A binary function $\mathbf{T} : [0, 1] \times [0, 1] \longrightarrow [0, 1]$ is called a t-norm if it fulfills the following properties for all $a, b, c \in [0, 1]$:

[7] The letter "t" means *triangular*. Triangular norms (t-norms) were originally proposed in order to generalize metrics, and triangular inequality is one of the important properties of metrics.

(a) $\mathbf{T}(a, b) = \mathbf{T}(b, a)$, (*commutativity*)
(b) $\mathbf{T}(a, \mathbf{T}(b, c)) = \mathbf{T}(\mathbf{T}(a, b), c)$, (*associativity*)
(c) If $a \le b$, then $\mathbf{T}(a, c) \le \mathbf{T}(b, c)$, (*monotonicity*)
(d) $\mathbf{T}(0, a) = 0$ and $\mathbf{T}(1, a) = a$. (*boundary conditions*)

t-norms possess a lot of interesting properties. They can be, for example, continuous or noncontinuous, but also *Archimedean*, which means that each sequence $\{a_n \mid n \in \mathbb{N}\}$, such that $a_1 < 1$ and $a_{n+1} = \mathbf{T}(a_n, a_n)$, converges to 0. A lot of other interesting properties have been studied (see [61]).

In the following text, we will write

$$a \, \mathbf{T} \, b$$

instead of $\mathbf{T}(a, b)$.

The basic t-norms are

$$a \, M \, b = a \wedge b = \min(a, b), \tag{2.35}$$

$$a \, P \, b = a \cdot b, \tag{2.36}$$

$$a \, \mathbf{T}_\infty \, b = a \otimes b = 0 \vee (a + b - 1), \tag{2.37}$$

$$a \, W \, b = \begin{cases} a \wedge b, & \text{if } a \vee b = 1, \\ 0 & \text{otherwise.} \end{cases} \tag{2.38}$$

The t-norm in (2.35) is the operation of minimum used in the definition of the minimum intersection of fuzzy sets. Furthermore, (2.36) is the (normal) product of real numbers and (2.37) is the *Łukasiewicz t-norm*. Recall that it was used for the definition of Łukasiewicz intersection. Finally, (2.38) is the so-called *drastic product*.

The following holds for any t-norm $\mathbf{T}$:

$$a \, W \, b \le a \, \mathbf{T} \, b \le a \, M \, b, \tag{2.39}$$

where $a, b \in [0, 1]$. Hence, drastic product is the smallest and minimum is the largest among all t-norms.

The Łukasiewicz t-norm $\mathbf{T}_\infty$ has an important property called *nilpotency*: for any $a < 1$, there is $n > 0$ such that

$$\underbrace{a \otimes \ldots \otimes a}_{n\text{-times}} = 0. \tag{2.40}$$

This property is used in Section 7.4.1.

If a t-norm $\mathbf{T}$ is given, we can define the corresponding *t-conorm* $\mathbf{S} : [0, 1] \times [0, 1] \longrightarrow [0, 1]$ as

$$a \, \mathbf{S} \, b = 1 - (1 - a)\mathbf{T}(1 - b). \tag{2.41}$$

Obviously, we can obtain a t-norm from a t-conorm using the formula

$$a \, \mathbf{T} \, b = 1 - (1 - a) \, \mathbf{S} \, (1 - b).$$

Basic t-conorms corresponding to t-norms (2.35)–(2.38) above, respectively, are

$$a\,N\,b = a \vee b = \max(a,b), \tag{2.42}$$

$$a\,Q\,b = a + b - a \cdot b, \tag{2.43}$$

$$a\,\mathbf{S}_\infty\,b = 1 \wedge (a+b), \tag{2.44}$$

$$a\,V\,b = \begin{cases} a \vee b, & \text{if } a \wedge b = 0, \\ 1 & \text{otherwise.} \end{cases} \tag{2.45}$$

Similarly as above, (2.42) is the maximum operation and $\mathbf{S}_\infty(a,b)$ is the Łukasiewicz disjunction. They were used in the definition of the union of fuzzy sets.

The following holds for any t-conorm $\mathbf{S}$:

$$a\,N\,b \le a\,\mathbf{S}\,b \le a\,V\,b, \tag{2.46}$$

where $a, b \in [0,1]$. Hence, drastic sum V is the largest and maximum is the smallest among all t-conorms.

t-norms can be defined in various ways. There is a very important family of t-norms, called *Frank family of t-norms*. It also includes the above-defined t-norms and is defined by the formula

$$a\,\mathbf{T}_s\,b = \log_s\left[1 + \frac{(s^a-1)\cdot\left(s^b-1\right)}{s-1}\right],$$

where s is a parameter that fulfills $s \in (0,\infty) - \{1\}$. We get $\mathbf{T}_0 = \min$ and $\mathbf{T}_1 = P$ as special limit cases.

The corresponding Frank t-conorms with the same parameter s are defined by

$$a\,\mathbf{S}_s\,b = 1 - \log_s\left[1 + \frac{\left(s^{1-a}-1\right)\cdot\left(s^{1-b}-1\right)}{s-1}\right].$$

Hamacher t-norms form another interesting family defined by

$$a\,\mathbf{T}^h_s\,b = \frac{a \cdot b}{s + (1-s)\cdot(a+b-a\cdot b)}.$$

Again, s is a parameter, where $s \in (0,\infty)$.

In fuzzy logic, t-norms serve as interpretations of the *logical conjunction*. If, for example, φ is a proposition whose truth degree is $a \in [0,1]$ and ψ is a proposition whose truth degree is $b \in [0,1]$, then the truth degree of the proposition "φ and at the same time ψ" is computed as $a\,\mathbf{T}\,b$, where $\mathbf{T}$ is some t-norm. Its selection is given by various considerations (see special literature, e.g., [44, 99]). Similarly, t-conorms are possible interpretations of the *logical disjunction*. Analogously as above, the truth degree of the proposition "φ or ψ" is computed by means of $\mathbf{S}(a,b)$, where $\mathbf{S}$ is some t-conorm.

Using t-norms, we can define the class of *generalized intersections* of fuzzy sets A and B by

$$C = A \cap_T B \;\text{ iff }\; C(u) = A(u)\,\mathbf{T}\,B(u), \quad u \in U. \tag{2.47}$$

Similarly, t-conorms are used for definition of *generalized unions* of fuzzy sets:

$$C = A \cup_S B \text{ iff } C(u) = A(u)\, \mathbf{S}\, B(u), \quad u \in U. \tag{2.48}$$

In fuzzy logic, it is possible to define more implication operations. These operations are related to t-norms. In our case, (2.24) is related to the Łukasiewicz t-norm (2.11), (2.25) to minimum, and (2.26) to product.[8]

2.2.3 Fuzzy Numbers

Fuzzy numbers are special expressions of natural language whose meaning is identified with special fuzzy sets defined on the set of real numbers $\mathbb{R}$. When speaking about fuzzy numbers in the sequel, we will usually have in mind these fuzzy sets.

Fuzzy numbers are assumed to have a special shape similar to that depicted in Figure 2.4. Intuitively, a fuzzy number represents an inexact value, that is, a value that can be described by words such as "about z_o" and "roughly z_0". Typical examples are "about 5", "roughly 1205", "approximately 1 m", etc. Such numbers are by no means rare. On the contrary, in real life, we work almost exclusively with numbers that are fuzzy. Only exceptionally we have exact numbers in mind.

For example, when measuring width of a table using a usual tape measure, then the result can be, say, 98 cm. However, it actually means "about 98 cm", because our measurement is rough. We cannot be sure that the width is not, for example, 98.015 cm. Notice that results of our measurements are always inexact, even if we use the best measuring device to be found. If we measure in nanometers, then automatically there exists some imprecision measured in picometers. Hence, we can always treat results of measurements as fuzzy numbers.

In some sciences, for example, in geology, all measurements are very fuzzy and all acquired numbers are understood as such. For example, if a geologist says that some mineral was found "500 meters" deep, then he/she will be perfectly happy, if the actual measured depth will be, say, 529 m. He/she will consider this measured value as a perfect fit with "500 meters".

From the mathematical point of view, fuzzy numbers are convex fuzzy sets (see Section 2.2.1) with continuous membership function, one-element kernel and a support being a bounded interval. This means that a fuzzy number $Z \underset{\sim}{\subset} \mathbb{R}$ has the

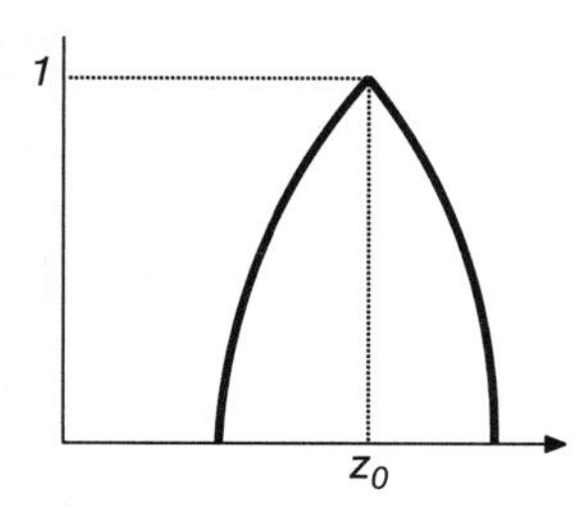

Figure 2.4 Membership function of the fuzzy number "about z_0".

[8]This implication is sometimes called Goguen implication.

following properties:

$$\mathrm{Ker}(Z) = \{z_0\} \quad \text{and} \quad \mathrm{Supp}(Z) = [z_1, z_2],$$

where $z_0, z_1, z_2 \in \mathbb{R}$ and $z_1 < z_0 < z_2$.[9] As already stated in the discussion on convex fuzzy sets, all a-cuts Z_a for $a \in (0, 1]$, and also the support are intervals in $\mathbb{R}$. Note that some authors also accept nonbounded support (e.g., when it is modeled by a Gaussian function as below).

The most often used type of fuzzy numbers are the so-called *triangular fuzzy numbers*. They are nothing else than fuzzy numbers whose membership function has a triangular shape. The general formula is

$$Z(x, a, b, c) = \begin{cases} 0, & x < a \text{ or } x > c, \\ \dfrac{x-a}{b-a}, & a \leq x < b, \\ \dfrac{c-x}{c-b}, & b < x \leq c, \\ 1, & x = b, \end{cases} \tag{2.49}$$

where a, b, c, $a < b < c$, are parameters depicted in Figure 2.5(a).

Usually, parameters a and c are located symmetrically around the value b. This means that such a membership function forms an isosceles triangle.

One can also meet the so-called *Gaussian fuzzy number*. Its membership function is

$$Z(x, c, \sigma) = \exp\left(-\frac{(c-x)^2}{2\sigma^2}\right), \tag{2.50}$$

where c is the kernel and σ is the standard deviation. This fuzzy number is depicted in Figure 2.5(b).

Extension principle. We can also introduce basic arithmetics with fuzzy numbers, that is, operations of addition, subtraction, multiplication, division and several more

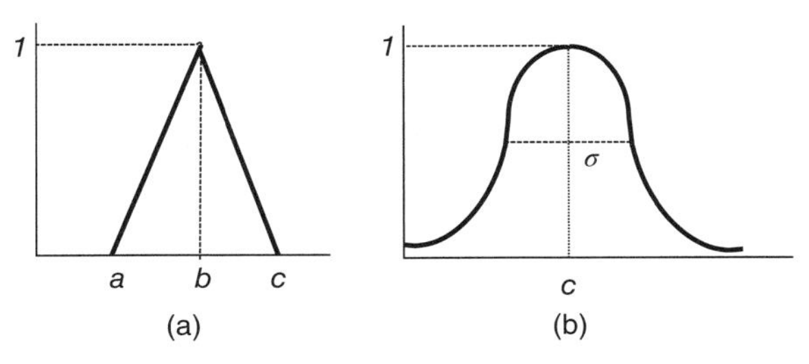

Figure 2.5 Membership function of a fuzzy number: (a) with triangular membership function and (b) with Gaussian membership function.

[9]In some applications, for example, in the geology example above, we can consider fuzzy numbers with kernels as intervals, too. Hence, kernel of the fuzzy number "500 meters" can be, for example, [480, 520] (in meters).

complex operations. The arithmetics of fuzzy numbers is based on the so-called *extension principle*.

Let $\bullet$ be some arithmetical operation (e.g., addition or multiplication) and $Z_1, Z_2 \subsetneqq \mathbb{R}$ be fuzzy numbers. The extension principle permits to extend the operation $\bullet$ to operation $\odot$ on fuzzy numbers as follows:

$$(Z_1 \odot Z_2)(z) = \bigvee_{\substack{z=x \bullet y \\ x,y \in \mathbb{R}}} (Z_1(x) \wedge Z_2(y)). \tag{2.51}$$

According to this formula, the result of the operation $\odot$ is the fuzzy number $Z_1 \odot Z_2$ consisting of elements $z \in \mathbb{R}$ obtained as follows: for each z, we find $x, y \in \mathbb{R}$ such that $z = x \bullet y$, then find minimum of $Z_1(x)$ and $Z_2(y)$, and finally take the membership degree $(Z_1 \odot Z_2)(z)$ as supremum (maximum) of all such membership degrees.

Formula (2.51) is not suitable for direct computation. Therefore, it is usually transformed to a more suitable form using special properties. Because a-cuts (2.7) of fuzzy numbers are intervals, it is possible to transform operations on fuzzy numbers to operations on their a-cuts. For example, the addition $Z_1 \oplus Z_2$ of two fuzzy numbers Z_1 and Z_2 can be computed using the formula

$$(Z_1 \oplus Z_2)_a = (Z_1)_a + (Z_2)_a, \quad a \in [0, 1]. \tag{2.52}$$

This means that we consider all a-cuts of both summed fuzzy numbers. The result are a-cuts of the fuzzy number, which is the sum of Z_1 and Z_2. Recall that the following holds for intervals of real numbers:

$$[a, b] + [c, d] = [a + c, b + d].$$

We can also use explicit formulas derived from the extension principle (2.51) for the computation of operations on fuzzy numbers. For example, for addition and multiplication, we get

$$(Z_1 \oplus Z_2)(z) = \bigvee_{y \in \mathbb{R}} (Z_1(y) \wedge Z_2(z - y)), \tag{2.53}$$

$$(Z_1 \odot Z_2)(z) = \bigvee_{y \in \mathbb{R}-\{0\}} \left(Z_1(y) \wedge Z_2\left(\frac{z}{y}\right)\right). \tag{2.54}$$

The structure of fuzzy numbers is not so rich as the structure of real numbers. Therefore, many properties that hold for real numbers do not hold for fuzzy numbers. Computations with fuzzy numbers and their properties are described in detail, for example, in books [32, 48, 68, 75]. For a recent approach based on the so-called MI-groups, see [50].

EXAMPLE 2.14 Let two fuzzy numbers be given:

$$(\text{“about 2”})(x) = \begin{cases} x - 1, & x \in [1, 2], \\ 3 - x, & x \in [2, 3], \\ 0 & \text{otherwise,} \end{cases}$$

$$(\text{“about 5”})(y) = \begin{cases} y-4, & y \in [4,5], \\ 6-y, & y \in [5,6], \\ 0 & \text{otherwise.} \end{cases}$$

Then, using formula (2.53), we get

$$(\text{“about 2”+“about 5”})(z) = \begin{cases} \frac{z}{2} - 2.5, & z \in [5,7], \\ 4.5 - \frac{z}{2}, & z \in [7,9], \\ 0, & \text{otherwise.} \end{cases}$$

This result can be interpreted as “roughly 7”. Notice that the support of the resulting fuzzy number has been expanded in comparison with supports of its summands. □

2.2.4 Fuzzy Partition and Fuzzy Covering

Important notions are those of fuzzy partition and fuzzy covering. Let U be a set and let

$$A_1, \ldots, A_n \underset{\approx}{\subseteq} U \tag{2.55}$$

be fuzzy sets.

Fuzzy covering. We say that fuzzy sets (2.55) form a *fuzzy covering* of U if

$$U \subseteq \text{Supp}\left(\bigcup_{i=1}^{n} A_i\right),$$

that is, if any element of the universe belongs to at least one of the fuzzy sets $A_1, \ldots, A_n$ with a nonzero membership degree.

Fuzzy partition. We say that fuzzy sets (2.55) form a *fuzzy partition* if they are normal, convex,

$$U = \text{Supp}\left(\bigcup_{i=1}^{n} A_i\right), \tag{2.56}$$

and it holds that

$$A_i \cap A_j = \emptyset \tag{2.57}$$

for all $i \neq j$. Condition (2.57) does not permit overlapping of fuzzy sets. This is not in an accordance with our intuition, which tells us that there are always elements with common properties in at least some degree. Therefore, (2.57) is often weakened to

$$A_i \otimes A_j = \emptyset. \tag{2.58}$$

This leads to the condition

$$A_i(u) + A_j(u) \leq 1 \tag{2.59}$$

for all $u \in U$. A typical fuzzy partition is schematically depicted in Figure 2.6. Fuzzy partition is a very important concept in fuzzy modeling. It is used in various

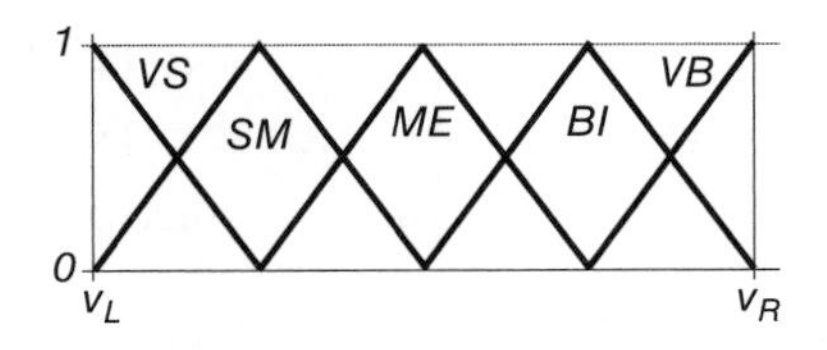

Figure 2.6 Example of a fuzzy partition of the interval $[v_L, v_R]$.

approximation techniques described in the following chapters. In Chapter 4, we will return to this concept and study it in more detail.

2.2.5 Cartesian Product and Fuzzy Relations

Cartesian product. Let $A \underset{\sim}{\subset} U, B \underset{\sim}{\subset} V$ be fuzzy sets. Their *Cartesian product* is a fuzzy set $A \times B \underset{\sim}{\subset} U \times V$ with the membership function

$$(A \times B)(\langle u, v \rangle) = A(u) \wedge B(v) \tag{2.60}$$

for all $u \in U$ and $v \in V$, where $\langle u, v \rangle$ is the ordered pair of the elements u and v.

EXAMPLE 2.15 The Cartesian product of fuzzy sets (2.15) and (2.16) from Example 2.8 is the fuzzy set

$$A \times B = \{{}^{0.2}/\langle 1,1\rangle, {}^{0.4}/\langle 1,2\rangle, {}^{0.4}/\langle 1,3\rangle, \ldots, {}^{0.2}/\langle 2,1\rangle, {}^{0.7}/\langle 2,2\rangle,$$
$$\ldots, {}^{0.1}/\langle 9,1\rangle, {}^{0.1}/\langle 9,2\rangle, \ldots, {}^{0.1}/\langle 9,9\rangle\},$$ □

In formula (2.60), the operation of minimum $\wedge$ is used. In general, any suitable t-norm can be used instead. Hence, we define a *generalized Cartesian product* by

$$\left(A \overset{\mathbf{T}}{\times} B\right)(\langle u, v \rangle) = A(u)\, \mathbf{T}\, B(v), \quad u \in U, v \in V, \tag{2.61}$$

where $\mathbf{T}$ is, for example, some of the t-norms considered above. Quite often, the ordinary product is used, that is,

$$\left(A \dot{\times} B\right)(\langle u, v \rangle) = A(u) \cdot B(v), \quad u \in U, v \in V. \tag{2.62}$$

EXAMPLE 2.16 The generalized Cartesian product (2.62) of fuzzy sets (2.15) and (2.16) from Example 2.8 is the fuzzy set

$$A \dot{\times} B = \{{}^{0.08}/\langle 1,1\rangle, {}^{0.28}/\langle 1,2\rangle, {}^{0.36}/\langle 1,3\rangle, \ldots, {}^{0.14}/\langle 2,1\rangle,$$
$${}^{0.49}/\langle 2,2\rangle, \ldots, {}^{0.02}/\langle 9,1\rangle, {}^{0.07}/\langle 9,2\rangle, \ldots, {}^{0.03}/\langle 9,9\rangle\}.$$

It can be seen from this example that the Cartesian product defined in this way captures some kind of interaction between the fuzzy sets A and B. □

Fuzzy relations. A general fuzzy relation (possibly n-ary) is a fuzzy set on (classical) Cartesian product of universes $U_1, \ldots, U_n$:

$$R \underset{\sim}{\subset} U_1 \times \cdots \times U_n.$$

The membership degree $R(\langle u_1, \dots, u_n \rangle)$ expresses a degree in which the elements $u_i \in U_i$, $i = 1, \dots, n$, are in the relation R. We will write $R(u_1, \dots, u_n)$ instead of $R(\langle u_1, \dots, u_n \rangle)$. A fuzzy relation imprecisely characterizes some kind of relation among elements from the universes $U_1, \dots, U_n$.

EXAMPLE 2.17 In mathematics, the relationship "approximately equal" is quite often used (we denote it by the symbol $\approx$). From our point of view, it is a fuzzy relation that can be modeled, for example, by the membership function

$$\|u \approx v\| = e^{-c|u-v|},$$

where $u, v \in \mathbb{R}$ and $c > 0$ is an appropriate constant. One can see that the more u differs from v, the smaller is the degree $\|u \approx v\|$. Obviously, $u = v$ iff $\|u \approx v\| = 1$. □

EXAMPLE 2.18 Let three sets $U = \{1, \dots, 10\}$, $V = \{a, \dots, j\}$, and $W = \{\alpha, \dots, \epsilon\}$ be given. We can specify a *ternary* fuzzy relation

$$R \underset{\sim}{\subset} U \times V \times W = \{1, \dots, 10\} \times \{a, \dots, j\} \times \{\alpha, \dots, \epsilon\}$$

in the following way:

$$R = \{{}^{0.2}/\langle 1, a, \alpha \rangle, {}^{0.4}/\langle 1, a, \beta \rangle, {}^{0.7}/\langle 1, b, \epsilon \rangle, {}^{0.2}/\langle 2, h, \alpha \rangle,$$
$$\dots, {}^{0.9}/\langle 2, i, \beta \rangle, {}^{1}/\langle 9, d, \alpha \rangle, {}^{0.1}/\langle 9, j, \epsilon \rangle, {}^{1}/\langle 10, j, \delta \rangle\},$$

This fuzzy relation says that, for example, the elements 1, b, and ϵ are in mutual relationship (relation) R in the degree 0.7. In other words, it is 70% true that these elements are in that relationship. For elements 9, d, and α, it is 100% true that they are in the relation R. □

Clearly, the Cartesian product of two fuzzy sets is a special binary fuzzy relation. If the number of elements in the support of R is finite then it is possible to write down binary fuzzy relations in the matrix form.

EXAMPLE 2.19 The fuzzy relation given by the Cartesian product in Example 2.15 can be written as

$R =$

	1	2	3	4	5	6	7	8	9	10
1	0.2	0.4	0.4	0.4	0	0.4	0.4	0	0.3	0
2	0.2	0.7	0.7	0.6	0	0.7	0.7	0	0.3	0
3	0	0	0	0	0	0	0	0	0	0
4	0.2	0.5	0.5	0.5	0	0.5	0.5	0	0.3	0
5	0	0	0	0	0	0	0	0	0	0
6	0.2	0.7	0.9	0.6	0	1	0.8	0	0.3	0
7	0.2	0.7	0.9	0.6	0	1	0.8	0	0.3	0
8	0	0	0	0	0	0	0	0	0	0
9	0.1	0.1	0.1	0.1	0	0.1	0.1	0	0.1	0
10	0	0	0	0	0	0	0	0	0	0

The outer rows or columns filled by zeros can be omitted. □

If necessary, we will freely combine both types of notation of fuzzy relations in the sequel.

Let $R \lesssim U_1 \times \cdots \times U_n$ be a fuzzy relation. Let us divide the set of indices $I = \{1, \ldots, n\}$ into two disjoint subsets $I = I_1 \cup I_2$ and let us denote $I_1 = \{i_1, \ldots, i_p\}$ and $I_2 = \{j_1, \ldots, j_q\}$ (obviously, $p + q = n$). Then, the *projection* of R on $U_{i_1} \times \cdots \times U_{i_p}$ is a fuzzy relation with the membership function

$$\left(\underset{U_{i_1}\times\cdots\times U_{i_p}}{\text{proj}} R\right)\left(c_{j_1}, \ldots, c_{j_q}\right) = \bigvee_{u_{i_1}\in U_{i_1},\ldots,u_{i_p}\in U_{i_p}} R. \tag{2.63}$$

EXAMPLE 2.20 Let us consider the ternary fuzzy relation from Example 2.18. We construct its projection on the set W. If we denote $U = U_1$, $V = U_2$, and $W = U_3$, then $I_1 = \{3\}$ and $I_2 = \{1, 2\}$. According to formula (2.63), we get

$$\underset{W}{\text{proj}}(R) = \left\{{}^{1}/\alpha, {}^{0.9}/\beta, {}^{1}/\delta, {}^{0.7}/\epsilon\right\},$$

where, for example, the membership degree in the fuzzy singleton ${}^{0.9}/\beta$ is determined by all fuzzy singletons containing the element β. In our example, these elements are ${}^{0.4}/\langle 1, a, \beta\rangle$ and ${}^{0.9}/\langle 2, i, \beta\rangle$, and we obtain $0.9 = 0.4 \vee 0.9$. □

Fuzzy relations are very important because they allow us to work with inexactly known relations among various phenomena—temperature, color, quantity of energy, and so on. Usually, we use binary fuzzy relations that are easier to work with.

Image of a fuzzy set in a fuzzy relation. Let $R \lesssim U \times V$ be a binary fuzzy relation and $A \lesssim U$ be a fuzzy set. *Image* $R''A$ of the fuzzy set A *in the fuzzy relation* R is the fuzzy set on V with the following membership function:

$$R''A(v) = \bigvee_{u\in U} (A(u) \wedge R(u, v)), \quad v \in V. \tag{2.64}$$

Intuitively, the membership degree $R''A(v)$ in (2.64) is determined as the greatest membership degree from the connection of degrees $A(u)$ with degrees in which there are elements $u \in U$ in fuzzy relation R with elements $v \in V$. This operation forms a basis of the so-called *Max-Min Inference Rule* proposed by E. H. Mamdani and S. Assilian (see Section 3.2.3). Note that it is a generalization of the classical definition of an image of a set in a relation.[10]

EXAMPLE 2.21 Let $R \lesssim \{1, \ldots, 10\} \times \{a, \ldots, f\}$ be fuzzy relation given by

$$\begin{aligned} R = \{&{}^{0.2}/\langle 1, a\rangle, {}^{0.4}/\langle 1, d\rangle, {}^{0.7}/\langle 1, f\rangle, {}^{0.2}/\langle 2, b\rangle, \\ &{}^{0.9}/\langle 2, d\rangle, {}^{1}/\langle 9, a\rangle, {}^{0.1}/\langle 9, d\rangle, {}^{1}/\langle 10, f\rangle\}, \end{aligned} \tag{2.65}$$

[10]If $A \subseteq U$ is a set and $R \subseteq U \times V$ is a binary relation, then the image of the set A in relation R is $R''A = \{y \mid (\exists x)\,(x \in A \;\&\; \langle x, y\rangle \in R\}$.

and $A \subsetneq \{1, \ldots, 10\}$ be the fuzzy set (2.15). Using (2.64) we obtain:

$$R''A = \left\{{}^{0.2}/_{a}, {}^{0.2}/_{b}, {}^{0.7}/_{d}, {}^{0.4}/_{f}\right\},$$

where, for example, the membership degree 0.2 in the fuzzy singleton ${}^{0.2}/_{a}$ is computed as

$$(0.4 \wedge 0.2) \vee (0.1 \wedge 1) = 0.2 \vee 0.1 = 0.2,$$

because the element a is associated with two elements having nonzero membership degrees, namely, to 1 with the degree 0.4 in ${}^{0.4}/_{1}$ and to 9 with the degree 0.1 in ${}^{0.1}/_{9}$. □

If supports of a fuzzy set and a fuzzy relation are finite, then we can compute the image of the fuzzy set in the fuzzy relation by a kind of matrix multiplication, where the addition and multiplication are replaced by the maximum and minimum operations, respectively.

EXAMPLE 2.22 If we use matrix "multiplication", then the solution of Example 2.21 looks as follows:

$$A = \begin{bmatrix} 0.4 & 0.7 & 0 & 0.5 & 0 & 1 & 1 & 0 & 0.1 & 0 \end{bmatrix}$$

$$R = \begin{bmatrix} 0.2 & 0 & 0 & 0.4 & 0 & 0.7 \\ 0 & 0.2 & 0 & 0.9 & 0 & 0 \\ 0 & 0 & 0 & 0 & 0 & 0 \\ 0 & 0 & 0 & 0 & 0 & 0 \\ 0 & 0 & 0 & 0 & 0 & 0 \\ 0 & 0 & 0 & 0 & 0 & 0 \\ 0 & 0 & 0 & 0 & 0 & 0 \\ 0 & 0 & 0 & 0 & 0 & 0 \\ 1 & 0 & 0 & 0.1 & 0 & 0 \\ 0 & 0 & 0 & 0 & 0 & 1 \end{bmatrix}$$

$$R''A = \begin{bmatrix} 0.2 & 0.2 & 0 & 0.7 & 0 & 0.4 \end{bmatrix},$$

where, for example, the first element in $R''A$ is obtained as

$$(0.4 \wedge 0.2) \vee (0.7 \wedge 0) \vee (0 \wedge 0) \vee (0.5 \wedge 0) \vee (0 \wedge 0) \vee (1 \wedge 0)$$
$$\vee (1 \wedge 0) \vee (0 \wedge 0) \vee (0.1 \wedge 1) \vee (0 \wedge 0) = 0.2\,.$$

□

As a special case of (2.64), let A be a fuzzy singleton $A = \left\{{}^{1}/_{u_0}\right\}$ for some $u_0 \in U$. Then, the computation (2.64) of the fuzzy set $B = R''\left\{{}^{1}/_{u_0}\right\}$ reduces to

$$B(v) = R(u_0, v), \quad v \in V. \tag{2.66}$$

We obtain the same result if we define B as a fuzzy set B_{u_0} of all $v \in V$ which are in relation R with u_0, that is,

$$B_{u_0} = \{{}^{r}/v \mid r = R(u_0, v), v \in V\}. \tag{2.67}$$

It is clear that (2.66) and (2.67) are identical. We can view (2.67) (and also (2.66)) as a "cut" of the fuzzy relation R at the point u_0.

Composition of fuzzy relations. Let $R \lesssim U \times V$ and $S \lesssim V \times W$ be two binary fuzzy relations. Then a *composition* of the fuzzy relations R and S is the fuzzy relation $R \circ S \lesssim U \times W$ defined by the membership function

$$(R \circ S)(u, w) = \bigvee_{v \in V} (R(u, v) \wedge S(v, w)), \quad u \in U, w \in W. \tag{2.68}$$

This definition is a generalization of the classical definition of composition of relations (see Section 2.1.1).

EXAMPLE 2.23 Let R be the fuzzy relation (2.65) from Example 2.21 and $S \lesssim \{a, \dots, f\} \times \{\alpha, \dots, \epsilon\}$ be the fuzzy relation

$$S = \{{}^{0.5}/\langle a, \alpha\rangle, {}^{1}/\langle b, \gamma\rangle, {}^{0.2}/\langle b, \epsilon\rangle, {}^{0.9}/\langle c, \beta\rangle, {}^{0.9}/\langle c, \gamma\rangle,$$
$${}^{1}/\langle d, \alpha\rangle, {}^{0.1}/\langle f, \beta\rangle, {}^{1}/\langle f, \epsilon\rangle\}.$$

Using (2.68), we obtain a new fuzzy relation $T \lesssim \{1, \dots, 10\} \times \{\alpha, \dots, \epsilon\}$

$$T = R \circ S = \{{}^{0.4}/\langle 1, \alpha\rangle, {}^{0.1}/\langle 1, \beta\rangle, {}^{0.7}/\langle 1, \epsilon\rangle, {}^{0.9}/\langle 2, \alpha\rangle, {}^{0.2}/\langle 2, \gamma\rangle,$$
$${}^{0.2}/\langle 2, \epsilon\rangle, {}^{0.5}/\langle 9, \alpha\rangle, {}^{0.1}/\langle 10, \beta\rangle, {}^{1}/\langle 10, \epsilon\rangle\}.$$

□

Similarly as above, if supports of binary fuzzy relations are finite, then we can compute their composition by matrix "multiplication", where the addition and multiplication are replaced by the maximum and minimum, respectively.

EXAMPLE 2.24 Let R and S be fuzzy relations from Example 2.23. Then their composition in matrix notation looks as follows:

$$R = \begin{bmatrix} 0.2 & 0 & 0 & 0.4 & 0 & 0.7 \\ 0 & 0.2 & 0 & 0.9 & 0 & 0 \\ 0 & 0 & 0 & 0 & 0 & 0 \\ 0 & 0 & 0 & 0 & 0 & 0 \\ 0 & 0 & 0 & 0 & 0 & 0 \\ 0 & 0 & 0 & 0 & 0 & 0 \\ 0 & 0 & 0 & 0 & 0 & 0 \\ 0 & 0 & 0 & 0 & 0 & 0 \\ 1 & 0 & 0 & 0.1 & 0 & 0 \\ 0 & 0 & 0 & 0 & 0 & 1 \end{bmatrix} \qquad S = \begin{bmatrix} 0.5 & 0 & 0 & 0 & 0 \\ 0 & 0 & 1 & 0 & 0.2 \\ 0 & 0.9 & 0.9 & 0 & 0 \\ 1 & 0 & 0 & 0 & 0 \\ 0 & 0 & 0 & 0 & 0 \\ 0 & 0.1 & 0 & 0 & 1 \end{bmatrix}$$

$$R \circ S = \begin{bmatrix} 0.4 & 0.1 & 0 & 0 & 0.7 \\ 0.9 & 0 & 0.2 & 0 & 0.2 \\ 0 & 0 & 0 & 0 & 0 \\ 0 & 0 & 0 & 0 & 0 \\ 0 & 0 & 0 & 0 & 0 \\ 0 & 0 & 0 & 0 & 0 \\ 0 & 0 & 0 & 0 & 0 \\ 0 & 0 & 0 & 0 & 0 \\ 0.5 & 0 & 0 & 0 & 0 \\ 0 & 0.1 & 0 & 0 & 1 \end{bmatrix}.$$

□

Remark 2.4

(i) The image $R''A$ (2.64) of a fuzzy set $A \subsetneqq U$ in a fuzzy relation $R \subsetneqq U \times V$ can also be understood as a special case of the composition of fuzzy relations as follows:

$$B = A \circ R, \tag{2.69}$$

where the fuzzy set A is understood as a unary fuzzy relation. The membership function of B in (2.69) is computed again using the formula (2.64). This is a more commonly used way of how the computation of an image of a fuzzy set in a fuzzy relation is written.

(ii) In the same way, if we consider $A = \{1/u_0\}$ for some $u_0 \in U$, then the composition (2.69) gives precisely the formula (2.66) (and also (2.67)).

2.2.6 Fuzzy Equality and Extensional Fuzzy Sets

Fuzzy equality. In Example 2.17, we mentioned the fuzzy relation "approximately equal". This fuzzy relation is an example of a more general notion called *fuzzy equality*. It is a mathematization of inexactly characterized equality or similarity between two objects. For example, two people can be similar in various extent; it is perfectly natural to say about two siblings that one is more similar to his/her mother than the other one. If we express the extent of similarity as the degree of truth of the proposition that given pair of objects is similar to each other, we obtain a binary fuzzy relation with some special properties.

First of all, we say that two objects are hundred percent similar if there is no characteristics on the basis of which we can say that they are different (this does not mean that they are identical!). Furthermore, if x is similar to y in some degree a, then, obviously, y is similar to x in the same degree. This means that the relation of fuzzy equality is symmetric. The last property is transitivity, which is in fuzzy set theory characterized by interpretation of the connective "and" in the phrase

if x is similar to y and y is similar to z then x is similar to z

by some t-norm. Hence, we can formally define fuzzy equality as follows.

The *fuzzy equality* $\approx$ on U is a fuzzy relation $\approx \subsetapprox U \times U$ with the following properties:

(a) $\| u \approx u \| = 1,$ (reflexivity)

(b) $\| u \approx v \| = \| v \approx u \|,$ (symmetry)

(c) $(\| u \approx v \| \ \mathbf{T} \ \| v \approx z \|) \leq \| u \approx z \|$ (transitivity)

for all $u, v, z \in U$, where $\| u \approx v \|$ denotes a membership (truth) degree of elements $u, v \in U$ in the fuzzy equality, that is, $\|u \approx v\| = a$ means that u and v are similar (equal) in the degree $a \in [0, 1]$.

EXAMPLE 2.25 In Example 2.17, we already mentioned one often used fuzzy equality that is transitive with respect to the product t-norm. Another example is the following:

$$\| u \approx v \| = 1 - (1 \wedge |u - v|), \quad u, v \in \mathbb{R}. \tag{2.70}$$

This is a fuzzy equality on $\mathbb{R}$ that is transitive with respect to the Łukasiewicz t-norm $\mathbf{T}_\infty$. Reflexivity and symmetry are obvious. Let us check the transitivity. We require that

$$\| u \approx v \| \ \mathbf{T}_\infty \ \| v \approx z \| = 0 \vee (1 - (1 \wedge |u - v|) + 1 - (1 \wedge |v - z|) - 1) \leq$$
$$\leq 1 - (1 \wedge |u - z|) = \| u \approx z \|$$

holds. This inequality easily follows from the well-known triangle inequality

$$|u - z| \leq |u - v| + |v - z|.$$

□

Evidently, classical equality is a special case of fuzzy equality if we define

$$\| u = v \| = \begin{cases} 1 & \text{for} \quad u = v, \\ 0 & \text{for} \quad u \neq v. \end{cases}$$

Extensional fuzzy sets. This notion is connected with the notion of fuzzy equality. A fuzzy set is extensional if together with its arbitrary element it contains all the elements approximately equal to the former. Again, it is a very natural notion, which is trivially fulfilled in classical set theory. Namely, if we know that $u \in A$, then from $v = u$, it immediately follows that $v \in A$. Thence, we obtain the following formal definition.

A fuzzy set $A \subsetapprox U$ is *extensional* with respect to a fuzzy relation $\approx$ and a t-norm $\mathbf{T}$ if for any two elements $u, v \in U$, it holds that

$$A(u) \mathbf{T} \ \| u \approx v \| \leq A(v).$$

If we realize that the t-norm is an interpretation of the logical conjunction in fuzzy logic, then this formula is indeed a generalization of the property mentioned above. It says that the truth degree of "v belongs to the fuzzy set A" is greater than or equal to the truth degree of "v is approximately equal to u and, at the same time, u belongs to A".

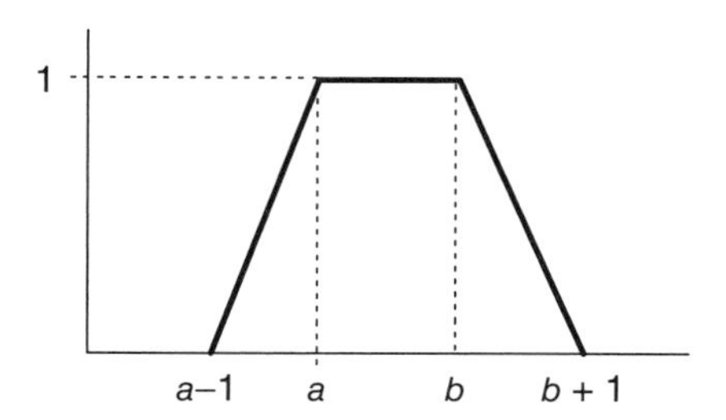

Figure 2.7 Trapezoidal fuzzy set which is extensional with respect to fuzzy relation $\|u \approx v\| = 1 - (1 \wedge |u - v|)$ and the Łukasiewicz t-norm $\mathbf{T}_\infty$.

EXAMPLE 2.26 Let $\approx$ be the fuzzy equality from Example 2.25, and let us consider the Łukasiewicz t-norm $\mathbf{T}_\infty$. Then the fuzzy set $A \subsetapprox \mathbb{R}$ with the membership function

$$A(u) = \begin{cases} 1, & a \le u \le b, \\ 0 \vee (1 - a + u), & u \le a, \\ 0 \vee (1 - u + b), & b \le u \end{cases}$$

is extensional with respect to $\approx$. This fuzzy set has a simple trapezoidal shape of the membership function depicted in Figure 2.7. □

Fuzzy points. Special extensional fuzzy sets are fuzzy points defined as follows. Let U be a universe and $\approx$ a fuzzy equality on U. Let $u_0 \in U$ be a fixed point. Then the fuzzy set

$$A_{u_0}(v) = \|v \approx u_0\| \tag{2.71}$$

is called a *fuzzy point*.

If a fuzzy set $A \subsetapprox U$ is normal, then it can always be seen as a fuzzy point of any element $u_0 \in \mathrm{Ker}(A)$ under the fuzzy equality $\approx$ on U given by the membership function

$$\|u \approx v\| = A(u) \leftrightarrow A(v), \qquad u, v \in U. \tag{2.72}$$

We can extend this property to families of fuzzy sets. Indeed, let $\mathbf{T}$ be either of minimum, product, or Łukasiewicz t-norm (2.35)–(2.37) (see also Section 2.2.2), and let $\leftrightarrow_T$ be the corresponding fuzzy equivalence (2.31)–(2.33). Furthermore, let $A_1, \ldots, A_n \subsetapprox U$ be a family of normal fuzzy sets and $u_1, \ldots, u_n$ be points such that $u_i \in \mathrm{Ker}(A_i)$, $i = 1, \ldots, n$. Finally, we suppose that the fuzzy sets $A_1, \ldots, A_n$ fulfill the following condition:

$$\bigvee_{u \in U} (A_i(u) \,\mathbf{T}\, B_j(u)) \le \bigwedge_{u \in U} (A_i(u) \leftrightarrow_T B_j(u)) \tag{2.73}$$

for all $i, j \in \{1, \ldots, n\}$. Then each A_i is a fuzzy point determined by the membership function

$$A_i(v) = \|u_i \approx v\|, \qquad v \in U, \tag{2.74}$$

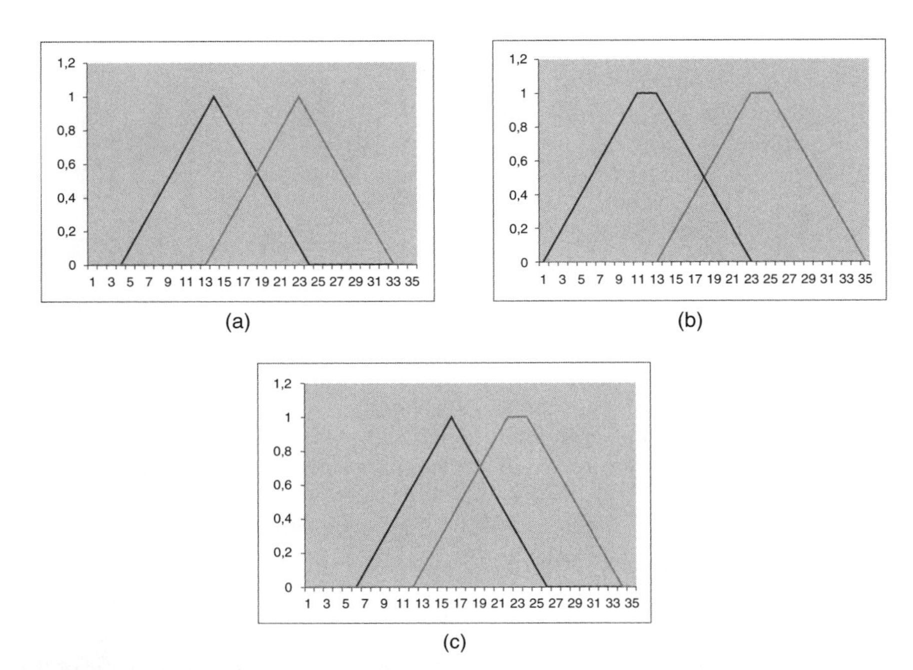

Figure 2.8 Schematic depiction of fuzzy sets with respect to condition (2.73): the couples of fuzzy sets (a) and (b) fulfill it while (c) does not.

where $\approx$ is a fuzzy equality on U defined by the membership function

$$\| u \approx v \| = \bigwedge_{i=1}^{n} (A_i(u) \leftrightarrow_T A_i(v)), \qquad u, v \in U. \tag{2.75}$$

For proofs and details, see [58, 109, 119].

Condition (2.73) requires that the family of fuzzy sets forms a kind of fuzzy partition in correspondence with proper shapes of the fuzzy sets. This situation is depicted in Figure 2.8. The couples of fuzzy sets in (a) and (b) fulfill (2.73), while those in (c) do not. The reason is that those in (c) overlap too much and, at the same time, the kernel of one of the fuzzy sets is too wide. This is not the case in (b) where the fuzzy sets, though they have also wide kernels, are sufficiently distant.

If all the fuzzy sets have triangular shape, then condition (2.73) is always fulfilled. One can see that condition (2.73) is not too strong. Hence, we can often assume that the given family of fuzzy sets is a system of fuzzy points.

Remark 2.5 *(a) The fuzzy set in Example 2.26 is extensional with respect to (2.70) but it is not a fuzzy point with respect to it. It is a fuzzy point, though, with respect to the fuzzy equality (2.72).*

(b) If we consider $U = \mathbb{R}$ and interpret $\|u \approx v\|$ by (2.70), then the fuzzy set

$$A_{u_0}(v) = \left\{ \|u \approx v\| / v \,\middle|\, v \in U \right\}$$

is a symmetric triangular fuzzy set with the kernel $u_0 \in U$.

2.3 ELEMENTS OF MATHEMATICAL FUZZY LOGIC

Mathematical fuzzy logic (MFL) is a generalization of classical mathematical logic. It has a similar role for fuzzy mathematics and its applications as has classical logic for classical mathematics. Therefore, its results are also essentially relevant for fuzzy modeling. In this book, however, we do not need to study MFL in detail. Therefore, we only briefly overview some of its main concepts. The interested reader may consult special books [23, 24, 44, 93, 99].

2.3.1 Structure of Truth Degrees in Mathematical Fuzzy Logic

In this section, we briefly overview fundamental structures of truth values, which determine the development of MFL. In the previous section, we saw a rich repertoire of possibilities how operations with fuzzy sets can be defined. We have noted that these operations are related to logical operations on truth degrees. In fact, they are determined by the considered algebraic structures.

Recall that in classical logic, the truth values form the Boolean algebra

$$\mathscr{L}_B = \langle \{\mathbf{0}, \mathbf{1}\}, \wedge, \vee, \neg, \mathbf{0}, \mathbf{1} \rangle, \tag{2.76}$$

where the operations $\wedge$ and $\vee$ are classical interpretations of conjunction and disjunction, respectively, and $\neg$ is the classical interpretation of negation. They are defined by the well-known truth tables:

$\vee$	**0**	**1**
0	**0**	**1**
1	**1**	**1**

$\wedge$	**0**	**1**
0	**0**	**0**
1	**0**	**1**

$\neg$	
0	**1**
1	**0**

In MFL, it is assumed that the structure of truth degrees forms a commutative *residuated lattice*, which is an algebra

$$\langle L, \wedge, \vee, \otimes, \rightarrow, \mathbf{0}, \mathbf{1} \rangle, \tag{2.77}$$

where L is a *support* (a set of truth values), $\vee$ and $\wedge$ are operations of supremum and infimum (see Section 2.2), $\mathbf{0}$ is the smallest, and $\mathbf{1}$ the greatest element of L. Furthermore, $\otimes$ is a *multiplication*, that is, a binary operation on L that is commutative, associative, and fulfills the law $a \otimes \mathbf{1} = a$, $a \in L$. $\rightarrow$ is a binary operation of *residuation* that is tied with $\otimes$ via the so-called *adjunction* condition

$$a \otimes b \leq c \text{ iff } a \leq b \rightarrow c \tag{2.78}$$

for all $a, b, c \in L$. The operations of meet $\wedge$ and multiplication $\otimes$ serve as interpretations of conjunction, while the residuum operation serves as an interpretation of implication. Therefore, the operation $\rightarrow$ is often called simply *implication*.

Residuated lattices generalize Boolean algebras in the sense that every Boolean algebra is a residuated lattice but not vice versa. Though the support L can be an arbitrary set, the most important for practical applications is the interval $L = [0, 1]$. Then, the multiplication $\otimes$ is a left continuous t-norm.

Classical examples of the residuation operation defined on $L = [0, 1]$ are the Łukasiewicz implication (2.24), Gödel implication (2.25), and product implication (2.26). It can easily be checked that they are adjoint (i.e., they fulfill the equivalence (2.78)) with the respective t-norms, namely, Gödel implication with minimum, product implication with ordinary product, and Łukasiewicz implication with Łukasiewicz t-norm (2.37).

Remark 2.6 *Residuation operations defined on* $[0, 1]$ *are special cases of the, so-called,* fuzzy implications *(cf. [5]). We can define more kinds of implication functions, for example, S-implications that generalize the classical definition* $a \to b = \neg a \vee b$. *Their importance for MFL is marginal, however.*

The implication (residuation) operation has many important properties. For example, let $a \leq b$. Then

$$a \to c \geq b \to c, \quad \mathbf{0} \to a = \mathbf{1},$$
$$c \to a \leq c \to b, \quad \mathbf{1} \to a = a,$$

where $a, b, c \in L$. The first inequality says that implication reverses the ordering in the first argument. The second inequality says that in the second argument, the ordering is preserved. This is in accordance with the behavior of the classical implication, for which it holds that $\mathbf{1} \to \mathbf{0} = \mathbf{0}$. It can be easily checked that the fuzzy implication function gives the same results for arguments $\mathbf{0}$ and $\mathbf{1}$ as the classical one.

Other important property is that the implication represents ordering, namely,

$$a \leq b \quad \Leftrightarrow \quad a \wedge b = a \quad \Leftrightarrow \quad a \vee b = b \quad \Leftrightarrow \quad a \to b = \mathbf{1}.$$

For any $a \in L$, it holds that $\mathbf{0} \leq a$ and $a \leq \mathbf{1}$.

The most distinguished algebra of truth values, which is essentially important also for fuzzy modeling, is the *standard Łukasiewicz MV-algebra*

$$\mathscr{L} = \langle [0, 1], \otimes, \oplus, \neg, 0, 1 \rangle. \tag{2.79}$$

The operations $\otimes, \oplus$ were already introduced in Section 2.2.2, namely, Łukasiewicz conjunction (2.11) and disjunction (2.17), respectively, and $\neg$ is the negation operation

$$\neg a = 1 - a$$

for all $a \in [0, 1]$. Furthermore, 0 and 1 are zero and unit in $\mathscr{L}$, that is, the elements that fulfill

$$a \otimes 1 = a, \quad a \otimes 0 = 0,$$
$$a \oplus 1 = 1, \quad a \oplus 0 = a,$$

$a, b \in [0, 1]$.

In the standard Łukasiewicz MV-algebra, we can define additional operations:

$$a \vee b = (a \otimes \neg b) \oplus b \qquad \text{(maximum)}, \tag{2.80}$$

$$a \wedge b = \neg(\neg a \vee \neg b) \qquad \text{(minimum)}, \tag{2.81}$$

$$a \rightarrow b = \neg a \oplus b \qquad \text{(Łukasiewicz implication)}, \tag{2.82}$$

$a, b \in [0, 1]$. If we expand these equations by means of (2.11) and (2.17), we obtain the commonly known operations of maximum and minimum and, moreover, equation (2.24) for the Łukasiewicz implication. Note that at the same time, it holds that $\neg a = a \rightarrow 0$.

It is easy to see that the standard Łukasiewicz MV-algebra can be seen also as a residuated lattice

$$\mathscr{L} = \langle [0, 1], \vee, \wedge, \otimes, \rightarrow, 0, 1 \rangle, \tag{2.83}$$

which is equivalent to (2.79). This algebra possesses many interesting properties, which rank it among the most important structures studied in many-valued logic. It was rightly chosen as the basic structure for the development of mathematical fuzzy logic. The reader can get acquainted with this field in books [44, 99].

2.3.2 Logical Inference

Logical inference can be understood as a process during which we derive some facts from other known facts using special rules, which imitate rules of human thinking (reasoning). They are called *deduction rules*. The *facts* are some statements about reality formulated in a certain formal language. These statements are called *formulas* or *propositions*. Deduction, that is, derivation of a conclusion, is realized by means of formulas.

The most important deduction rule is the rule of *modus ponens*. Formally, we can write it as follows:

$$\frac{A, A \Rightarrow B}{B}, \tag{2.84}$$

where A and B are formulas. The rule (2.84) says the following: if we know that a fact denoted by a formula A holds true and we also know that it is true that a fact B follows from the fact A, we conclude that the fact B also holds true. This rule pertains to basic rules of human thinking.[11]

EXAMPLE 2.27 Suppose that a broker learned the following: if value of some stock drops under 100 points, then it is necessary to sell all shares. Later, he/she learned that value of one share is 95 points. Then he/she immediately concludes that he/she must sell all the shares. Schematically, using the modus ponens rule (2.84), we can describe this situation as follows.

[11] This rule can be traced already to the works of Aristotle (more than 2000 years ago), though it is not explicitly formulated in them.

Let $A(x)$ be a formula "$x < 100$", where x denotes the value of one share. We write $A(x = n)$ if the value of one share is n. Furthermore, we define a formula B as $B := sell$. Then we can rewrite scheme (2.84) as

$$\frac{A(x), A(x) \Rightarrow B}{B}.$$

Because $95 < 100$, the formula $A(x = 95)$ is true. Hence, we conclude that the formula B is also true, that is, we should sell our shares. □

This example demonstrates not only the rule of the modus ponens but also its often inappropriate sharpness. For example, what shall we do if the value of the stock were $x = 99.99$ or $x = 100.01$? The crisp reasoning tells us to sell in the first case, but is it indeed a good decision?

Deduction proceeds with formulas interpreted in certain *models*, which are mathematical structures representing various aspects of reality.

EXAMPLE 2.28 Let us imagine that we have an industrial furnace at our command and suppose that we are interested in its inside temperature. The fact that this temperature was measured can be expressed by a formula $T(x)$. We read this formula as "temperature in the furnace is x". However, such a formula can be related to any furnace. Our concrete furnace is a *model* and we interpret the formula $T(x)$ in it. Having in mind our concrete furnace, we can measure the temperature inside, which is, say, 800 °C. Using the formula T, we can write $T(x = 800)$ and ask if the formula $T(x = 800)$ is true in our furnace (model). If the measured temperature is indeed 800 °C, we conclude that this formula is true. Clearly, setting $x = 700$, we see that the formula $T(x = 700)$ is not true in our furnace. But, it can be true in another furnace, that is, in another model. □

We can see that one formula can be interpreted in many models (many furnaces), and ask whether it is actually true in any model or not. Because mathematics requires certainty and generality, we have to consider all possible models. In order to work with them, we must understand this concept in a more abstract way. This means that we do not talk about concrete furnaces or other objects but only about numbers (and mathematical structures), that is, we consider temperatures only, without dealing with the question whether they refer to furnaces or to other objects.

The *truth* of formulas is the main interest in logic. In classical logic, there are only two truth values—*truth* (**1**) and *falsity* (**0**). In our example, we can unequivocally say that the formula $T(x = 800)$ is true in our furnace. But, how to solve a situation when we learn a formula $T(x \text{ is } big)$? Apparently, two truth values will not suffice any more. As already explained, there are infinitely many truth values in fuzzy logic filling the interval [0, 1]. In classical logic, we are interested in formulas that are true or false. In fuzzy logic, formulas are assigned some truth value from [0, 1] and, in many situations, we want it to be as high as possible. The most interesting for us, of course, is the truth degree 1.

2.3.3 Formal Systems of MFL

In this section, we very briefly and informally outline what is MFL[12] to give the interested reader the hint for study of the special literature.

In the same way as in classical logic, we distinguish syntax and semantics of MFL.

(i) *Syntax:* We start with a language J which is a set of some symbols. Then we form sequences of symbols that we call *formulas*. Furthermore, we specify *inference rules* using which we derive new formulas from the already given ones (example of such a rule is the modus ponens (2.84)). Finally, we introduce *axioms*, which are special designated formulas. Using axioms and inference rules, we construct *proofs*, which are finite sequences of formulas each of which is either an axiom or is derived from the previous formulas in the proof using an inference rule. The last formula in the proof is a *theorem* proved by the proof in concern. A set of all formulas provable from a given set of axioms is called a *theory* and usually denoted by T. We may thus formally write $T \vdash A$, which means that a formula A is provable in the theory T, that is, there exists its proof in the sense above.

(ii) *Semantics:* In logic, this means to assign interpretation to formulas, especially to assign truth values to them. Therefore, the first step is choosing a proper algebra of truth values. In MFL, this is a special kind of a residuated lattice (2.77). Further step is to assign operations from this algebra to proper connectives introduced in the syntax. For example, the connective $\boldsymbol{\vee}$ (disjunction of formulas) is assigned the operation $\vee$, $\Rightarrow$ (implication) is assigned the residuation $\rightarrow$, etc. It is specific for MFL that it has two conjunctions: the "ordinary" conjunction $\boldsymbol{\wedge}$ is interpreted by $\wedge$ and the strong conjunction **&** is interpreted by the multiplication $\otimes$. Further steps in the definition of semantics depend on the order of fuzzy logic introduced below.
After having defined basic truth assignment, we define a model $\mathcal{M}$ of a theory T, which is a truth valuation of all formulas such that $\mathcal{M}(A) = 1$ whenever A is an axiom of T. Then we say that a formula A is true in a degree $a \in L$ in a theory T if a is the infimum of the truth values $\mathcal{M}(A)$ for all models $\mathcal{M}$ of T.

Further classification of fuzzy logics follows classical logic. Namely, we distinguish the following:

(i) *Propositional fuzzy logic:* The language of this logic contains only propositional variables and connectives. Formulas in this logic represent propositions that are assigned truth values.

(ii) *Predicate first-order fuzzy logic:* The language of this logic contains symbols for objects (we call them terms) and also quantifiers $\forall$ and $\exists$. Then, for

[12]Let us remark that MFL is also called *fuzzy logic in narrow sense* (FLn) to make it distinct from L. A. Zadeh's concept of "fuzzy logic in wide sense".

example, a formula $(\forall x)A(x)$ (or $(\exists x)A(x)$) says that $A(x)$ holds for all objects x (there exists an object x for which $A(x)$ holds).

(iii) *Higher-order fuzzy logic:* This is the most complicated system of fuzzy logic whose language contains formulas of certain *types*.[13] Then a formula A_α where α is a type may denote an arbitrary kind of object including truth values, properties, properties of properties, etc.

According to the chosen algebra of truth values, we distinguish MTL (monoidal t-norm-based logic), BL (basic logic), product, Gödel, Łukasiewicz and many other kinds of fuzzy logics, both propositional as well as first-order ones. More details can be found in books [23, 24, 44] and many papers. The higher-order fuzzy logic was elaborated in detail in papers [79, 85].

All the mentioned logics have traditional style of syntax similar to the syntax of classical logic. A further generalization leads to the so-called *fuzzy logic with evaluated syntax* in which the basic concept is that of *evaluated formula* $^a/A$, where $a \in L$ is an initial evaluation saying that a formula is true at least in the degree a. In this logic, we consider a fuzzy set of axioms and introduce the concept of a *provability degree*, that is, $T \vdash_a A$ says that a formula A is provable in the degree a. This logic is explained in detail in book [99].

One of the possible extensions of MFL is *fuzzy natural logic* (FNL).[14] The general goal of FNL is to develop a model of human thinking whose typical feature is the use of natural language (see [81, 82, 83, 84, 86, 89]). It includes, in particular, the theory of evaluative linguistic expressions (see [83]), the theory of fuzzy/linguistic IF-THEN rules and approximate reasoning (see [92]), the theory of generalized intermediate quantifiers and their syllogisms (see [71, 72]), and some other special formal theories. We will present selected methods of FNL, which have direct applications in fuzzy modeling, in Chapter 5.

2.3.4 The Concept of Fuzzy IF-THEN Rule

In Chapter 1, we mentioned the important concept of fuzzy IF-THEN rule.

Definition 2.2 *A* fuzzy IF-THEN rule *is an expression of the form*

$$\mathscr{R} := \mathsf{IF}\ X_1\ \mathit{is}\ \mathscr{A}_1\ \mathsf{AND}\ \cdots\ \mathsf{AND}\ X_n\ \mathit{is}\ \mathscr{A}_n\ \mathsf{THEN}\ Y\ \mathit{is}\ \mathscr{B}, \tag{2.85}$$

where $X_1, \ldots, X_n, Y$ *are* variables *and* $\mathscr{A}_1, \ldots, \mathscr{A}_n, \mathscr{B}$ *represent properties of their values. A part of IF-THEN rule which follows IF, that is, in (2.85), the expression*

$$X_1\ \mathit{is}\ \mathscr{A}_1\ \mathsf{AND}\ \cdots\ \mathsf{AND}\ X_n\ \mathit{is}\ \mathscr{A}_n, \tag{2.86}$$

is called antecedent. *A part which follows THEN, that is, in (2.85), the expression*

$$Y\ \mathit{is}\ \mathscr{B},$$

is called consequent.

[13]We also speak about *fuzzy type theory* (FTT).

[14]Fuzzy natural logic was originally called *fuzzy logic in broader sense* (FLb-logic). This term, however, is not very fortunate because it is often mistaken with Zadeh's "fuzzy logic in wide sense".

The variables $X_1, \dots, X_n, Y$ can attain values being elements selected from certain universal sets $U_1, \dots, U_n, V$, respectively. In fuzzy modeling, we usually assume that $U_1 = \cdots = U_n = V = \mathbb{R}$.

EXAMPLE 2.29 In process control, we can consider variables

$$\begin{aligned} X_1 &:= \textit{error}, \\ X_2 &:= \textit{change of error}, \\ Y &:= \textit{change of control action}. \end{aligned}$$

Then, an example of fuzzy IF-THEN rule $\mathscr{R}$ is

IF *error* is *small* AND *change of error* is *very small*
THEN *change of control action* is *rather small*.

There are two independent variables in this rule, namely, X_1 is *error*, and X_2 is *change of error*, and one dependent variable Y is *change of control action*. The properties $\mathscr{A}_1, \mathscr{A}_2, \mathscr{B}$ are characterized using linguistic expressions $\mathscr{A}_1 :=$ "small", $\mathscr{A}_2 :=$ "very small" and $\mathscr{B} :=$ "rather small", respectively. □

Fuzzy IF-THEN rules express our knowledge about existence of some relationship among phenomena. A set (or system) of them represents a certain type of general knowledge about these phenomena. We can also speak about *granular knowledge* and individual rules are understood as knowledge granules.

As already suggested, one fuzzy IF-THEN rule is not sufficient for characterization of more complex relationships.

Definition 2.3 *A* linguistic description *is a finite set of fuzzy IF-THEN rules* $LD = \{\mathscr{R}_1, \dots, \mathscr{R}_m\}$ *with common variables:*

$$\begin{aligned} \mathscr{R}_1 &:= \text{IF } X_1 \textit{ is } \mathscr{A}_{11} \text{ AND } \cdots \text{ AND } X_n \textit{ is } \mathscr{A}_{1n} \text{ THEN } Y \textit{ is } \mathscr{B}_1, \\ \mathscr{R}_2 &:= \text{IF } X_1 \textit{ is } \mathscr{A}_{21} \text{ AND } \cdots \text{ AND } X_n \textit{ is } \mathscr{A}_{2n} \text{ THEN } Y \textit{ is } \mathscr{B}_2, \\ &\dots \dots\dots\dots\dots\dots\dots\dots\dots\dots\dots\dots \\ \mathscr{R}_m &:= \text{IF } X_1 \textit{ is } \mathscr{A}_{m1} \text{ AND } \cdots \text{ AND } X_n \textit{ is } \mathscr{A}_{mn} \text{ THEN } Y \textit{ is } \mathscr{B}_m. \end{aligned} \tag{2.87}$$

For the sake of simplicity, we will in the sequel suppose that $n = 1$, that is, that fuzzy IF-THEN rules have the form

$$\mathscr{R} := \text{IF } X \text{ is } \mathscr{A} \text{ THEN } Y \text{ is } \mathscr{B}. \tag{2.88}$$

When analyzing fuzzy IF-THEN rules, we can distinguish three levels of details:

(i) *Surface level*. This is the outer form of the rule that may look as a conditional expression of natural language.

(ii) The *level of formal syntax*. Any rule, and consecutively the whole linguistic description, is assigned a certain formula D of a chosen formal system.
(iii) *Semantic level*. On this level, the formula D is properly interpreted.

Relational interpretation of fuzzy IF-THEN rules. This interpretation is based on the assumption that each fuzzy IF-THEN rule represents a fuzzy relation that characterizes existing relation among objects.

(i) On the surface level, the IF-THEN rule $\mathscr{R}$ is a code written in a form that enables us to understand its deeper structure.
(ii) On the level of formal syntax, the expressions "X is $\mathscr{A}$","Y is $\mathscr{B}$" are assigned certain predicates $A(X), B(Y)$, respectively, and the formula D assigned to the rule $\mathscr{R}$ (as well as to the whole linguistic description) is a formula of predicate fuzzy logic.
(iii) On the semantic level, the formula D is assigned a special fuzzy relation characterizing dependence between objects from the corresponding universes.

Note that this interpretation is useful especially in cases when we want to describe some existing function but we have only rough information about it. Our task is to approximate its course as best as possible. Therefore, we also speak about *fuzzy approximation*. The relational interpretation is explained in detail in Chapter 3.

Logical/linguistic interpretation of fuzzy IF-THEN rules. This interpretation is based on the methods of *fuzzy natural logic*. Its main idea lies in the assumption that fuzzy IF-THEN rules are, in fact, special conditional sentences of natural language.

(i) On the surface level, the IF-THEN rule $\mathscr{R}$ is a conditional sentence of natural language.
(ii) On the level of formal syntax, the expressions "X is $\mathscr{A}$","Y is $\mathscr{B}$" are assigned formulas of higher-order fuzzy logic and so is also the formula D assigned to the IF-THEN rule $\mathscr{R}$.
(iii) On the semantic level, the formula D is assigned a special set of functions over fuzzy relations.

This interpretation should be applied when (possibly expert) information formulated in natural language is available. The computer provides results as if "it understands" natural language. The logical–linguistic interpretation is explained in detail in Chapter 5.

Let us emphasize that both aforementioned interpretations are not equivalent and should be used in different situations. Therefore, it is important to understand them well.

3

FUZZY IF-THEN RULES IN APPROXIMATION OF FUNCTIONS

Fuzzy approximation is a class of methods that applies fuzzy modeling methods to approximation of functions. These techniques assume that an approximated function is not known precisely, either because of insufficient available information or because its course or its description is too complicated. Therefore, only approximate description of it is at disposal. The fuzzy approximation methods provide fairly simple algorithms and they are at the same time very *robust*, that is, little sensitive to changes in the input conditions.

The basic fuzzy approximation techniques are fuzzy IF-THEN rules under relational interpretation, Takagi–Sugeno rules, and the F-transform. In this chapter, we describe the first two techniques. The technique of F-transform is more general, and we will deal with it in Chapter 4.

3.1 RELATIONAL INTERPRETATION OF FUZZY IF-THEN RULES

Throughout this section, we assume that we are given a continuous function $f : \mathbb{R} \longrightarrow \mathbb{R}$. Moreover, we assume that our information about its course is imprecise and limited. To cope with this situation, we will use fuzzy IF-THEN rules, each of which characterizes the behavior of f in some inexactly specified region. A set of IF-THEN rules then provides us with a more or less rough idea about the course of f.

The subsequent step is to construct (using fuzzy IF-THEN rules) a function f^A that approximates the function f (in a certain precisely given sense). This idea is depicted in Figure 3.1. The given function f is covered by a finite set of "patches", each of which is determined by some IF-THEN rule. The authors of this idea are E. H. Mamdani and S. Assilian [67], who followed the works of L. A. Zadeh.

Insight into Fuzzy Modeling, First Edition. Vilém Novák, Irina Perfilieva, and Antonín Dvořák.

Companion Website: www.wiley.com/go/novak/fuzzy/modeling

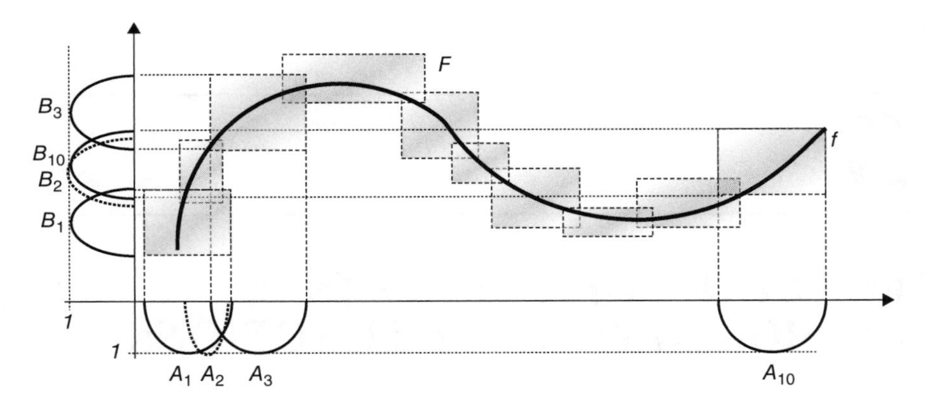

Figure 3.1 Schematic depiction of a fuzzy function F represented by a fuzzy graph that approximates a crisp function f. Below x-axis and to the left of y-axis, membership functions of the fuzzy sets A_j and B_j, respectively, $j = 1,\ldots, 10$, are depicted.

3.1.1 Finite Functions and Their Description

Before we describe this construction in more detail, we will explain how finite functions can be described using formal logic.

Let U, V be sets, and let a finite function $f : U \to V$ be given by the table

$$f : \begin{array}{c||c|c|c|c} x & u_1 & u_2 & \ldots & u_m \\ \hline y & v_1 & v_2 & \ldots & v_m \end{array} \tag{3.1}$$

where $u_i \in U, v_i \in V, i = 1,\ldots, m$. This table fully characterizes the function f. It can be described in words in two ways:

(a) The variable x has the value u_1 *and* y has the value v_1, *or*, x has u_2 *and* y has v_2, etc.

(b) *If* the variable x has the value u_1, *then* the variable y has the value v_1, *and, if* x has u_2, *then* y has v_2, etc.

In formal logic, we can write the above descriptions either as a *disjunction of conjunctions*

$$\begin{aligned} &x = u_1 \text{ AND } y = v_1 \text{ OR} \\ &\ldots\ldots\ldots\ldots\ldots\ldots \text{ OR} \\ &x = u_m \text{ AND } y = v_m, \end{aligned} \tag{3.2}$$

or as a *conjunction of implications*

$$\begin{aligned} &\text{IF } x = u_1 \text{ THEN } y = v_1 \text{ AND} \\ &\ldots\ldots\ldots\ldots\ldots\ldots \text{ AND} \\ &\text{IF } x = u_m \text{ THEN } y = v_m. \end{aligned} \tag{3.3}$$

In a concise form, we can write (3.2) and (3.3) as

$$\mathrm{DNF}(x,y) = \bigvee_{i=1}^{m} ((x = u_i) \wedge (y = v_i)), \tag{3.4}$$

$$\mathrm{CNF}(x,y) = \bigwedge_{i=1}^{m} ((x = u_i) \Rightarrow (y = v_i)). \tag{3.5}$$

The symbols $\bigvee$, $\bigwedge$, and $\Rightarrow$ denote the logical disjunction, logical conjunction, and logical implication, respectively.

The formula DNF(x, y) is called *disjunctive normal form* and CNF(x, y) is called *conjunctive normal form*. These two forms of description of the function f are in classical logic *equivalent* in the following sense.

Proposition 3.1 *Let $f : U \longrightarrow V$ be a finite function given by table (3.1). Then, for all $u_i \in U$, $i = 1,\dots,m$, it holds that*

$$\mathrm{DNF}(u_i,f(u_i)) \text{ iff } \mathrm{CNF}(u_i,f(u_i)).$$

Proof: Let DNF$(u_j,f(u_j))$ be true for some $j \in \{1,\dots,m\}$. Then at least one conjunction must be true, for example, $(u_j = u_j) \wedge (f(u_j) = v_j)$. From that, it follows that $f(u_j) = v_j$ and, therefore, also the implication $(u_j = u_j) \Rightarrow (f(u_j) = v_j)$ is true. Since f is a function, there is no $v_k \neq v_j$ such that $f(u_j) = v_k$, and so the given CNF contains no implication $(u_j = u_j) \Rightarrow (f(u_j) = v_k)$ which could be false (this would otherwise happen, because we already know from the first conjunction that $f(u_j) = v_j$). Because no other equality $u_j = u_i$, $j \neq i$ is true, all implications $(u_j = u_i) \Rightarrow (f(u_j) = v_i)$ are trivially true, and we conclude that the whole conjunctive normal form CNF$(u_j,f(u_j))$ is true.

Conversely, let CNF$(u_j,f(u_j))$ hold true for some $j \in \{1,\dots,m\}$. Then, all implications $(u_j = u_i) \Rightarrow (f(u_j) = v_i)$, $i = 1,\dots,m$ must be true, namely, either $u_i \neq u_j$ and then the implication $(u_j = u_i) \Rightarrow (f(u_j) = v_i)$ is trivially true or $i = j$, which means that $u_j = u_j$, and so the equality $f(u_j) = v_j$ is also true. This also means that there is no k such that $u_j = u_k$ and $v_k \neq v_j$ because then the implication $(u_j = u_k) \Rightarrow (f(u_j) = v_k)$ could not be true (we already know that $f(u_j) = v_j$). This confirms that f is a function. Finally, since the conjunction $(u_j = u_j) \wedge (f(u_j) = v_j)$ is true, it follows that also DNF$(u_j,f(u_j))$ is true. ■

If we consider fuzzy sets instead of elements u_i, v_i, $i = 1,\dots,m$, we obtain the so-called *fuzzy function* of type 2 (see [75]), which is a *classical function over sets of fuzzy sets.*[1] Strictly speaking, a fuzzy function (of type 2) is a classical function

$$F : \mathscr{F}(U) \longrightarrow \mathscr{F}(V),$$

[1]Let us remark that the concept of a fuzzy function of type 2 should not be confused with the notion of fuzzy set of type 2 [151]. The latter is a generalization of a fuzzy set whose membership degrees are replaced by fuzzy numbers.

where $\mathscr{F}(U)$ and $\mathscr{F}(V)$ are sets of all fuzzy sets on U and V. If a fuzzy function is finite, then we can represent it analogously as above using the table

$$F : \begin{array}{c||c|c|c|c} x & A_1 & A_2 & \dots & A_m \\ \hline y & B_1 & B_2 & \dots & B_m \end{array}, \tag{3.6}$$

where $A_1,\dots,A_m \subsetneq U$ and $B_1,\dots,B_m \subsetneq V$.

If we represent a fuzzy function of type 2 by a set of pairs of fuzzy sets $\{(A_i, B_i) \mid i = 1,\dots,m\}$, we obtain a *fuzzy graph* (see [151]). Because the universes U, V can be infinite, we can understand a finite fuzzy graph F (i.e., a finite fuzzy function of type 2) as a *finite characterization* of a classical (possibly infinite) function $f : U \longrightarrow V$. A schematic depiction of a fuzzy graph is in Figure 3.1.

Analogously as above, we can describe the function F given by (3.6) either by the set of rules

$$\begin{aligned} \mathscr{R}_1 &: X \text{ is } \mathscr{A}_1 \text{ AND } Y \text{ is } \mathscr{B}_1 \text{ OR} \\ \mathscr{R}_2 &: X \text{ is } \mathscr{A}_2 \text{ AND } Y \text{ is } \mathscr{B}_2 \text{ OR} \\ &\quad \dots\dots\dots\dots \text{ OR} \\ \mathscr{R}_m &: X \text{ is } \mathscr{A}_m \text{ AND } Y \text{ is } \mathscr{B}_m \end{aligned} \tag{3.7}$$

or by the set of IF-THEN rules

$$\begin{aligned} \mathscr{R}_1 &: \text{IF } X \text{ is } \mathscr{A}_1 \text{ THEN } Y \text{ is } \mathscr{B}_1 \text{ AND} \\ \mathscr{R}_2 &: \text{IF } X \text{ is } \mathscr{A}_2 \text{ THEN } Y \text{ is } \mathscr{B}_2 \text{ AND} \\ &\quad \dots\dots\dots\dots \text{ AND} \\ \mathscr{R}_m &: \text{IF } X \text{ is } \mathscr{A}_m \text{ THEN } Y \text{ is } \mathscr{B}_m. \end{aligned} \tag{3.8}$$

If we take (3.7) or (3.8) on surface level as certain symbolic expressions, then we can again apply Proposition 3.1 and conclude that both descriptions describe the fuzzy function F equivalently.

Recall that the above sets of rules are called *linguistic descriptions*. For people, the IF-THEN form (3.8) is more natural than (3.7). Therefore, when speaking about a linguistic description, they usually have only (3.8) in mind. We will take an advantage of such terminological license in this book, too.

The situation here, though, is different from (3.3) (or (3.2)). Actually, we are not primarily interested in the fuzzy function F but in some (classical) function f, which is *approximated* by F. Therefore, each rule in (3.8) (or (3.7)) characterizes the function f locally in a proper neighborhood, which depends on the shape of f. The neighborhoods are schematically depicted by dashed rectangles in Figure 3.1. The question is, how the neighborhoods should be constructed. Our goal is to construct them in a proper way enabling us to derive a function f^A *approximating* the given function f. The procedure is described below.

Remark 3.1 *From the logical point of view, however, there is a difference between both descriptions: in (3.7), we consider only positive cases when values of X* indeed have *one of the corresponding properties denoted by $\mathscr{A}_i$. In (3.8), we may also consider situations when values of X need not have any of the properties $\mathscr{A}_i$. Note that in (3.7), the domain of F is, in fact, the set $\{A_1,\ldots,A_m\}$. However, in (3.8), the domain of F can be understood as the set $\mathscr{F}(U)$ of all fuzzy sets on U. Implication is thus a very general way how to express that some phenomenon depends on some other one. Note that Proposition 3.1 says that the descriptions (3.7) and (3.8) are equivalent only when speaking about these positive cases.*

3.1.2 Relational Interpretation of Linguistic Descriptions

In this section, we follow three levels of analysis, as introduced in Section 2.3.4.

As mentioned above, we disregard the surface form (3.7) of the description of f. On the level of formal syntax as well as on the semantic level, though, we will keep this distinction and say that linguistic descriptions are interpreted in two possible ways. The reason is that these interpretations are not semantically equivalent, as we will see below.

Level of formal syntax. First, we will define sets of predicate symbols $\mathbf{A}_1(x)$, $\ldots$, $\mathbf{A}_m(x)$ and $\mathbf{B}_1(y),\ldots,\mathbf{B}_m(y)$. These are some arbitrary symbols that we understand as representations of certain properties of the variables x and y, respectively.

EXAMPLE 3.1 Let our task be to characterize a dependence between position of a control lever (in centimeters to the left or right) and position of a robotic arm (in millimeters to the left or right). The former is characterized by values from $U = [-10, 10]$ (cm) and the latter by values from $V = [-5, 5]$ (mm). This dependence is represented by a function $f : U \longrightarrow V$ which is not known to us precisely but we have a rough idea about its course. In both universes, we are able to specify seven positions. For better orientation, the positions are denoted by symbols NB (negatively big), NM (negatively medium), NS (negatively small), ZE (zero), PS (positively small), PM (positively medium), and PB (positively big); the situation is schematically depicted in Figure 5.1. Then we can introduce the following predicate symbols:

(i) $\mathbf{NB}_{CL}(x)$, $\mathbf{NM}_{CL}(x)$, $\mathbf{NS}_{CL}(x)$, $\mathbf{ZE}_{CL}(x)$, $\mathbf{PS}_{CL}(x)$, $\mathbf{PM}_{CL}(x)$, $\mathbf{PB}_{CL}(x)$ for a position of the control lever (CL),

(ii) $\mathbf{NB}_{RA}(y)$, $\mathbf{NM}_{RA}(y)$, $\mathbf{NS}_{RA}(y)$, $\mathbf{ZE}_{RA}(y)$, $\mathbf{PS}_{RA}(y)$, $\mathbf{PM}_{RA}(y)$, $\mathbf{PB}_{RA}(y)$ for a position of the robotic arm (RA).

□

Each rule $\mathscr{R}_i$ in the linguistic description (3.8) is assigned either the formula

$$\mathbf{A}_i(x) \mathbin{\&} \mathbf{B}_i(y) \tag{3.9}$$

(*conjunctive interpretation*), where & denotes a fuzzy logical conjunction, or the formula

$$\mathbf{A}_i(x) \Rightarrow \mathbf{B}_i(y) \tag{3.10}$$

(*implicative interpretation*), where $\Rightarrow$ denotes a fuzzy logical implication. The overall linguistic description (3.8) is then assigned either the *disjunctive normal form*

$$\mathrm{DNF}(x, y) = \bigvee_{i=1}^{m} (\mathbf{A}_i(x) \,\&\, \mathbf{B}_i(y)), \tag{3.11}$$

or the *conjunctive normal form*

$$\mathrm{CNF}(x, y) = \bigwedge_{i=1}^{m} (\mathbf{A}_i(x) \Rightarrow \mathbf{B}_i(y)). \tag{3.12}$$

Note that the formulas $\mathbf{A}_i(x)$, $i = 1, \ldots, m$ are assigned to the antecedents (i.e., expressions "X is $\mathscr{A}_i$") and $\mathbf{B}_i(y)$ to the consequents (i.e., expressions "Y is $\mathscr{B}_i$") of the corresponding rules in (3.8).

Semantic level. On this level, the forms (3.11) and (3.12) are assigned specific fuzzy relations. This is done step by step by assigning interpretations to each of their components.

First, a suitable fuzzy set $A_i \underset{\sim}{\subset} U$ is assigned to each formula $\mathbf{A}_i(x)$, and a suitable fuzzy set $B_i \underset{\sim}{\subset} V$ is assigned to each formula $\mathbf{B}_i(y)$, $i = 1, \ldots, m$. These fuzzy sets should be chosen properly to cover both the domain and the range of the function f. This means that we must guarantee the following: to every $u \in \mathrm{dom}(f)$, there is $i \in \{1, \ldots, m\}$ such that $u \in \mathrm{Supp}(A_i)$ and $f(u) \in \mathrm{Supp}(B_i)$. In practice, we usually suppose that shapes of these membership functions are triangles (Figure 2.5) or trapezoids (Figure 2.7) and that all the considered fuzzy sets form a fuzzy partition as in Figure 5.1.

Furthermore, we must choose interpretation of the conjunction & or of the implication $\Rightarrow$. We know from Section 2.2.2 that the conjunction is interpreted by some t-norm $\mathbf{T}$ and implication by the corresponding fuzzy implication (residuum) $\rightarrow$. Each rule $\mathscr{R}_i$ is then interpreted by a fuzzy relation $R_i \underset{\sim}{\subset} U \times V$ given by one of the following membership functions:

$$R_i(u, v) = A_i(u) \,\mathbf{T}\, B_i(v), \quad u \in U, v \in V, \tag{3.13}$$

which interprets formula (3.9), or

$$R_i(u, v) = A_i(u) \rightarrow B_i(v), \quad u \in U, v \in V, \tag{3.14}$$

which interprets formula (3.10).

How a suitable t-norm in (3.13) should be determined? This depends on the chosen structure of truth values—cf. Section 2.3. The most common possibility is to put $\mathbf{T} =$ min or $\mathbf{T} = \cdot$ or also $\mathbf{T} = \otimes$ (Łukasiewicz conjunction), where the first two choices seem to give the best results. According to practical experiences, the best choice of implication $\rightarrow$ in (3.14) is the Łukasiewicz one (2.24).

Taking fuzzy relations (3.13) and (3.14) as interpretations of the fuzzy IF-THEN rules from the linguistic description (3.8), we obtain the interpretation of the whole linguistic description on the semantic level as one of the fuzzy relations $R_{\mathrm{DNF}} \underset{\sim}{\subset}$

$U \times V$ or $R_{\mathrm{CNF}} \subsetneqq U \times V$ assigned to the disjunctive or the conjunctive normal form, respectively. Their membership functions are

$$R_{\mathrm{DNF}}(u, v) = \bigvee_{i=1}^{m} (A_i(u) \, \mathbf{T} \, B_i(v)) \tag{3.15}$$

or

$$R_{\mathrm{CNF}}(u, v) = \bigwedge_{i=1}^{m} (A_i(u) \to B_i(v)). \tag{3.16}$$

The fuzzy relations R_{DNF} and R_{CNF} can also be written in an aggregate form as

$$R_{\mathrm{DNF}} = \bigcup_{i=1}^{m} (A_i \overset{\mathbf{T}}{\times} B_i), \tag{3.17}$$

$$R_{\mathrm{CNF}} = \bigcap_{i=1}^{m} (A_i \Rrightarrow B_i). \tag{3.18}$$

The fuzzy relation R_{DNF} is schematically depicted in Figure 3.2 and R_{CNF} in Figure 3.3. It is important to notice that the normal forms (3.11) and (3.12) do not depend on concrete universes U, V. It is necessary to understand them as a result of formal analysis of a linguistic description, which can have different interpretations depending on the given universes and also on the function f to be approximated. The reason to introduce the level of formal syntax is the need to clearly separate the structure of the linguistic description from its concrete interpretation, since the latter can be influenced by additional conditions stemming from the given problem. This is one of the reasons why fuzzy logic and its methods are universal and robust.

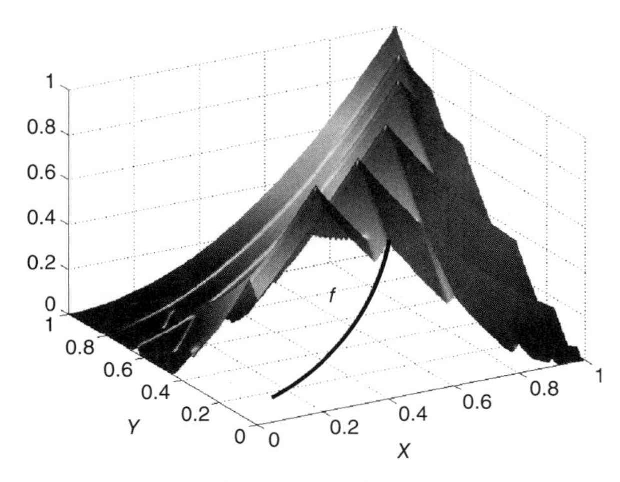

Figure 3.2 Schematic depiction of a fuzzy relation R_{DNF}, which interprets a disjunctive normal form approximating a function f.

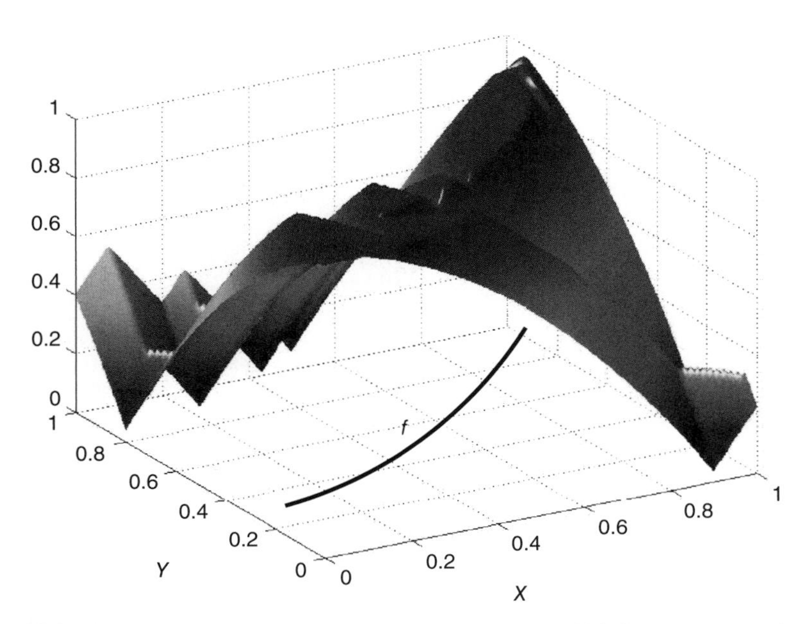

Figure 3.3 Schematic depiction of a fuzzy relation R_{CNF}, which interprets a conjunctive normal form approximating a function f.

EXAMPLE 3.2 We will also illustrate numerically the construction of normal forms interpreting a linguistic description. Let us consider a linguistic description formed on the surface level by two rules

$$\mathscr{R}_1 := \text{IF } X \text{ is } \textit{about 2} \text{ THEN } Y \text{ is } \textit{about 104},$$

$$\mathscr{R}_1 := \text{IF } X \text{ is } \textit{about 4} \text{ THEN } Y \text{ is } \textit{about 102},$$

where the expressions "about 2" and "about 4" are names of inexactly given regions in the universe $U = \{0, 1, \ldots, 5\}$ and "about 102" and "about 104" are similarly defined in the universe $V = \{100, 101, \ldots, 105\}$. This linguistic description characterizes a function $f : U \longrightarrow V$, whose exact course is not known. We only know that the value of f in the point 2 lies "somewhere around 104", and in the point 4, it lies "somewhere around 102".

First, we assign a formula $\mathbf{A}2(x)$ to the antecedent "X is *about 2*" and a formula $\mathbf{A}4(x)$ to "X is *about 4*". Similarly, we assign $\mathbf{B}102(y)$ and $\mathbf{B}104(y)$ to the respective consequents. Then, based on (3.11) and (3.12), we obtain the following interpretation of the given linguistic description on the level of formal syntax:

$$\text{DNF}(x, y) = (\mathbf{A}2(x) \,\&\, \mathbf{B}104(y)) \vee (\mathbf{A}4(x) \,\&\, \mathbf{B}102(y)), \tag{3.19}$$

$$\text{CNF}(x, y) = (\mathbf{A}2(x) \Rightarrow \mathbf{B}104(y)) \wedge (\mathbf{A}4(x) \Rightarrow \mathbf{B}102(y)). \tag{3.20}$$

Both these forms are the basis for relational interpretations of the linguistic description given above.

Let the fuzzy sets on the universes U, V assigned to the formulas above be the following:

$$\begin{aligned}
\mathbf{A}2(x) \ \textit{is assigned} \ A2 &= \left\{{}^{0.4}/1, {}^{1}/2, {}^{0.4}/3\right\}, \\
\mathbf{A}4(x) \ \textit{is assigned} \ A4 &= \left\{{}^{0.3}/3, {}^{1}/4, {}^{0.3}/5\right\}, \\
\mathbf{B}102(x) \ \textit{is assigned} \ B102 &= \left\{{}^{0.4}/101, {}^{1}/102, {}^{0.4}/103\right\}, \\
\mathbf{B}104(x) \ \textit{is assigned} \ B104 &= \left\{{}^{0.3}/103, {}^{1}/104, {}^{0.3}/105\right\}.
\end{aligned}$$

Now, let us set the t-norm $\mathbf{T} = \min$. Then, using (3.13), we obtain the following semantic interpretations of the rules $\mathscr{R}_1$ and $\mathscr{R}_2$:

$$\begin{aligned}
R_1 = \{&{}^{0.3}/\langle 1,103\rangle, {}^{0.4}/\langle 1,104\rangle, {}^{0.3}/\langle 1,105\rangle, {}^{0.3}/\langle 2,103\rangle, {}^{1}/\langle 2,104\rangle, \\
&{}^{0.3}/\langle 2,105\rangle, {}^{0.3}/\langle 3,103\rangle, {}^{0.4}/\langle 3,104\rangle, {}^{0.3}/\langle 3,105\rangle\}, \\
R_2 = \{&{}^{0.3}/\langle 3,101\rangle, {}^{0.3}/\langle 3,102\rangle, {}^{0.3}/\langle 3,103\rangle, {}^{0.4}/\langle 4,101\rangle, {}^{1}/\langle 4,102\rangle, \\
&{}^{0.4}/\langle 4,103\rangle, {}^{0.3}/\langle 5,101\rangle, {}^{0.3}/\langle 5,102\rangle, {}^{0.3}/\langle 5,103\rangle\}.
\end{aligned}$$

Finally, the fuzzy relation assigned to the disjunctive normal form (3.19) is obtained using (3.15):

$$\begin{aligned}
R_{\mathrm{DNF}} = R_1 \cup R_2 = \{&{}^{0.3}/\langle 1,103\rangle, {}^{0.4}/\langle 1,104\rangle, {}^{0.3}/\langle 1,105\rangle, {}^{0.3}/\langle 2,103\rangle, \\
&{}^{1}/\langle 2,104\rangle, {}^{0.3}/\langle 2,105\rangle, {}^{0.3}/\langle 3,101\rangle, {}^{0.3}/\langle 3,102\rangle, {}^{0.3}/\langle 3,103\rangle, \\
&{}^{0.4}/\langle 3,104\rangle, {}^{0.3}/\langle 3,105\rangle, {}^{0.4}/\langle 4,101\rangle, {}^{1}/\langle 4,102\rangle, \\
&{}^{0.4}/\langle 4,103\rangle, {}^{0.3}/\langle 5,101\rangle, {}^{0.3}/\langle 5,102\rangle, {}^{0.3}/\langle 5,103\rangle\}.
\end{aligned}$$

In matrix form, R_{DNF} can be rewritten as

$$R_{\mathrm{DNF}} = \begin{bmatrix}
0 & 0 & 0 & 0 & 0 & 0 \\
0 & 0 & 0 & 0.3 & 0.4 & 0.3 \\
0 & 0 & 0 & 0.3 & 1 & 0.3 \\
0 & 0.3 & 0.3 & 0.3 & 0.4 & 0.3 \\
0 & 0.4 & 1 & 0.4 & 0 & 0 \\
0 & 0.3 & 0.3 & 0.3 & 0 & 0
\end{bmatrix}.$$

If we construct conjunctive normal form using the Łukasiewicz implication, we obtain

$$R_{\mathrm{CNF}} = \begin{bmatrix}
1 & 1 & 1 & 1 & 1 & 1 \\
0.6 & 0.6 & 0.6 & 0.9 & 1 & 0.9 \\
0 & 0 & 0 & 0.3 & 1 & 0.3 \\
0.6 & 0.6 & 0.6 & 0.9 & 0.6 & 0.6 \\
0 & 0.4 & 1 & 0.4 & 0 & 0 \\
0.6 & 1 & 1 & 1 & 0.6 & 0.6
\end{bmatrix}.$$

□

The fuzzy relations R_{DNF} and R_{CNF} are not equal.[2] If the t-norm $\mathbf{T}$ is adjoint with the implication $\rightarrow$ (i.e., the equivalence (2.78) is fulfilled), then it is easy to show that

$$R_{\mathrm{DNF}} \subseteq R_{\mathrm{CNF}}. \tag{3.21}$$

Special case of DNF and CNF. Let us suppose that we know that the function f passes through (or, at least, is close enough to) a set of points

$$(u_1, v_1), \ldots, (u_m, v_m).$$

If we replace the crisp equality in (3.4) or (3.5) by a fuzzy one (cf. Section 2.2.6), then we can characterize the course of f by means of DNF or CNF in one of the following special forms:

$$\overline{\mathrm{DNF}}(x, y) = \bigvee_{i=1}^{m} ((x \approx u_i) \,\&\, (y \approx v_i)), \tag{3.22}$$

$$\overline{\mathrm{CNF}}(x, y) = \bigwedge_{i=1}^{m} ((x \approx u_i) \Rightarrow (y \approx v_i)). \tag{3.23}$$

The predicate $\approx$ is a *fuzzy equality* whose interpretation again depends on the chosen structure of truth values, because the transitivity of $\approx$ is determined by the t-norm used.

The predicate $x \approx u$ can be interpreted as "x is approximately equal to u". Then, a set of fuzzy IF-THEN rules, which is on the level of formal syntax interpreted by (3.22) or (3.23), looks on the surface level as follows:

IF X is *approximately equal to* u_1 THEN Y is *approximately equal to* v_1,

....

IF X is *approximately equal to* u_m THEN Y is *approximately equal to* v_m.

Interpretations of (3.22) or (3.23) on the semantic level are fuzzy relations with the following membership functions:

$$R_{\overline{\mathrm{DNF}}}(u, v) = \bigvee_{i=1}^{m} (\|u \approx u_i\| \;\mathbf{T}\; \|v \approx v_i\|), \tag{3.24}$$

or

$$R_{\overline{\mathrm{CNF}}}(u, v) = \bigwedge_{i=1}^{m} (\|u \approx u_i\| \rightarrow \|v \approx v_i\|). \tag{3.25}$$

If we suitably choose the points u_i, $i = 1, \ldots, m$, then these fuzzy sets form a fuzzy partition of the domain of function f (depicted, e.g., in Figure 5.1). Of course, the same holds also for the range of f, if we suitably choose the points v_i.

[2] Note that the third and fifth rows of the matrix forms are identical.

Remark 3.2 *The surface form of fuzzy IF-THEN rules in (3.8) as logical implications and careless analysis of formulas for the derivation of a conclusion led to an idea that there is always a* logical implication *present. The use of the minimum operation* $\mathbf{T} = \wedge$ *in (3.13) then led to incorrect name "Mamdani implication" (or "Larsen implication" if* $\mathbf{T}$ *is the product). These operations, however, are symmetrical and, therefore, they cannot be taken as implications.*

It must be emphasized that implication is by no means a symmetrical operation. Consider, for example, the implication

"If it is an evening then we go to the cinema".

Then, of course, from the fact that we go to the cinema, it does not follow *that it is necessarily an evening.*

The term "Mamdani implication" has been introduced by engineers. Correct use of terms need not be always that important in practical applications. But, it is important for the correct understanding of the essence of the solved problems and must not be neglected when developing well substantiated applications. Therefore, we will consistently avoid the terms "Mamdani implication" and "Larsen implication" in this book.

3.1.3 Managing More Variables

If we deal with more antecedent variables $X_1, \ldots, X_n$, then the fuzzy IF-THEN rules take the form (2.85). Consequently, the DNF and CNF in (3.11) and (3.12) take the form

$$\mathrm{DNF}(x_1, \ldots, x_n, y) = \bigvee_{i=1}^{m} ((\mathbf{A}_{i1}(x_1)\ \&'\ \ldots\ \&' \mathbf{A}_{in}(x_n))\ \&\ \mathbf{B}_i(y)), \tag{3.26}$$

$$\mathrm{CNF}(x_1, \ldots, x_n, y) = \bigwedge_{i=1}^{m} ((\mathbf{A}_{i1}(x_1)\ \&'\ \ldots\ \&' \mathbf{A}_{in}(x_n)) \Rightarrow \mathbf{B}_i(y)), \tag{3.27}$$

where $\mathbf{A}_{i1}(x_1)\ \&'\ \ldots\ \&'\mathbf{A}_{in}(x_n)$, $i = 1, \ldots, m$, is a formula interpreting the antecedent of (2.85). The $\&'$ is a conjunction which can be different from $\&$.

On semantic level, interpretations of (3.26) and (3.27) are fuzzy relations R_{DNF}, $R_{\mathrm{CNF}} \subsetneq U_1 \times \cdots \times U_n \times V$ given by membership functions

$$R_{\mathrm{DNF}}(u_1, \cdots, u_n, v) = \bigvee_{i=1}^{m} ((A_{i1}(u_1)\ \mathbf{T}' \cdots \mathbf{T}'\ A_{in}(u_n))\ \mathbf{T}\ B_i(v)), \tag{3.28}$$

$$R_{\mathrm{CNF}}(u_1, \cdots, u_n, v) = \bigwedge_{i=1}^{m} ((A_{i1}(u_1)\ \mathbf{T}' \cdots \mathbf{T}'\ A_{in}(u_n)) \rightarrow B_i(v)) \tag{3.29}$$

for all $u_j \in U_j$, $j = 1, \ldots, n$, and $v \in V$. The operation $\mathbf{T}'$ can be an arbitrary t-norm. However, recall the inequality (2.39) from which it follows that minimum gives the greatest value. Therefore, in practical applications, putting $\mathbf{T}' = \min$ seems to be the best choice. As stated above, $\rightarrow$ should be the Łukasiewicz implication (2.24).

3.2 APPROXIMATION OF FUNCTIONS USING FUZZY IF-THEN RULES

Before describing the way how approximating function f^A is constructed, we will first present a concept that plays an important role in the theory and practice of fuzzy modeling.

3.2.1 Defuzzification

Though we have shown in the introduction of this book that accuracy is only an illusion and absolutely precise values do not exist in practice, yet it is often the case that measured values are treated as if they were precise. In applications, it is important to have a possibility to work with precise values (or, at least to consider given values as precise). This is understandable, since, for example, in fuzzy control, the output must be a concrete signal (specific electrical impulse) to an actuator, which is followed by a concrete action. Therefore, we have to introduce the notion of *defuzzification*. It is a special operation that transforms a fuzzy set to one individual element. There are many ways how this operation can be implemented.

Defuzzification is an operation that assigns to a nonempty fuzzy set an element from its support, that is, it is a function $DEF : (\mathscr{F}(U) - \{\emptyset\}) \longrightarrow U$ such that

$$\mathrm{DEF}(A) \in \mathrm{Supp}(A)$$

holds for any nonempty fuzzy set $A \in \mathscr{F}(U)$.

In this section, we work only with fuzzy sets having a finite support, that is, the considered fuzzy sets have the form

$$A = \{{}^{a_1}/{}_{u_1}, \ldots, {}^{a_r}/{}_{u_r}\} \tag{3.30}$$

(we can also deal with them as if being defined on a finite universe U). This restriction is motivated by practical reasons. Note that most of the formulas given below can be defined for fuzzy sets on infinite universes as well.

Center of Gravity/Area defuzzification. This is the most often used defuzzification method. The result is obtained as a *center of gravity* (COG) of a fuzzy set A:

$$\mathrm{COG}(A) = \frac{\sum_{k=1}^{r} A(u_k) \cdot u_k}{\sum_{k=1}^{r} A(u_k)}. \tag{3.31}$$

This method is used in particular for fuzzy approximation problems. The disadvantage of it is its large computational complexity. The result of COG defuzzification is depicted in Figure 3.4 (for the sake of clarity, the fuzzy set is depicted by a continuous curve).

Mean of Maxima defuzzification. This method (MOM) is computationally simpler than COG:

$$\mathrm{MOM}(A) = \frac{1}{r_{\max}} \sum_{j=1}^{r_{\max}} u_j^{\max}, \tag{3.32}$$

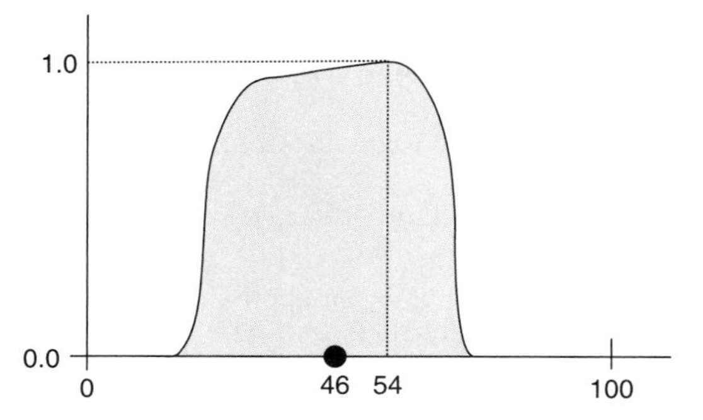

Figure 3.4 COG defuzzification—the resulting value (the center of gravity) is denoted by the thick dot.

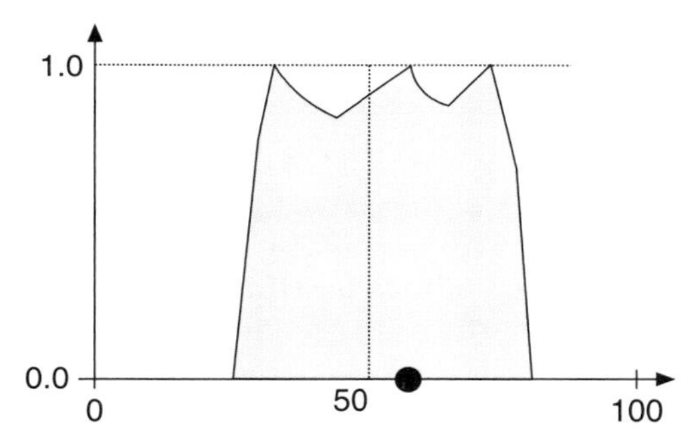

Figure 3.5 MOM defuzzification—the resulting value (the center of maximal values) is denoted by the thick dot.

where $u_j^{\max} = u_j$, if $A(u_j) = \max\{A(u_k) \mid k = 1,\ldots,r\}$, that is, $\{u_j^{\max} \mid j = 1,\ldots,r_{\max}\}$ are all the elements of the support of fuzzy set (3.30) whose membership degree $A(u_j^{\max})$ is maximal. This method gives identical result as the COG method if a defuzzified fuzzy set is symmetric and has one maximum only (e.g., a fuzzy number).

The result of the MOM defuzzification is depicted in Figure 3.5 (again, the fuzzy set is depicted by a continuous curve).

First of Maxima and Last of Maxima defuzzifications. Both these methods are the simplest defuzzification methods. Their practical significance is, however, marginal.

$$\mathrm{FOM}(A) = \min\{u_j^{\max} \mid j = 1,\ldots,r_{\max}\}, \tag{3.33}$$

where $\{u_j^{\max} \mid i = 1,\ldots,n_{\max}\}$ are all the elements of the support of fuzzy set (3.30) whose membership degree $A(u_j^{\max})$ is maximal. Obviously, the First of Maxima (FOM) method picks the first element $u_j^{\max}$ whose membership degree $A(u_j^{\max})$ is maximal. Analogously, the Last of Maxima (LOM) method

$$\mathrm{LOM}(A) = \max\{u_j^{\max} \mid j = 1,\ldots,r_{\max}\} \tag{3.34}$$

picks the last element $u_j^{\max}$ whose membership degree $A(u_j^{\max})$ is maximal.

Center of Sums defuzzification. This method (Center of Sums, COS) is a variant of the COG defuzzification. It is used quite often because its computational algorithm is fast enough.

Let us suppose that a fuzzy set A is a union of fuzzy sets

$$A = B_1 \cup \cdots \cup B_s.$$

The fuzzy sets B_i are usually results of an approximate deduction based on these rules from the corresponding linguistic description, which were fired according to the observation. Then, we can define the COS defuzzification as follows:

$$\mathrm{COS}(A) = \frac{\sum_{j=1}^{s} \left(\sum_{k=1}^{r} u_k \cdot B_j(u_k)\right)}{\sum_{j=1}^{s} \sum_{k=1}^{r} B_j(u_k)}. \tag{3.35}$$

The COS method counts in intersections of the fuzzy sets B_i repeatedly, contrary to the COG method.

DEE defuzzification This is a special kind of defuzzification convenient when we work with genuine linguistic expressions (see Chapter 5). It classifies a fuzzy set to be defuzzified into one of the three types—Z, M, and S. Shapes of the corresponding fuzzy sets are clear from Figure 3.6. Thus, the DEE defuzzification[3] works as follows:

$$\mathrm{DEE}(A) = \begin{cases} \mathrm{LOM}(A) & \text{if } A \text{ is of type Z,} \\ \mathrm{MOM}(A) & \text{if } A \text{ is of type M,} \\ \mathrm{FOM}(A) & \text{if } A \text{ is of type S.} \end{cases}$$

Alternatively, the MOM defuzzification for a fuzzy set of type M can be replaced by COG.

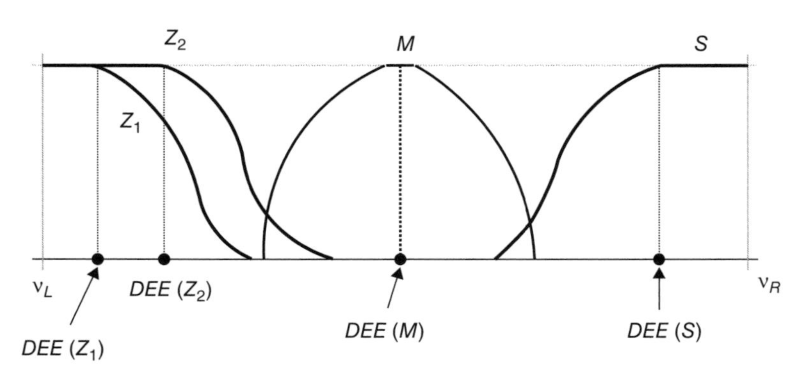

Figure 3.6 DEE defuzzification—the resulting value for each type of fuzzy set (Z, M, or S) is denoted by the thick dot. The universe is represented by an interval $[v_L, v_R]$.

[3]DEE means Defuzzification of Evaluative Expressions. The justification of this term is given in Chapter 5.

3.2.2 Fuzzy Approximation

3.2.2.1 Construction of approximating function

Fuzzy-relation-based approximating function. As discussed above, our goal is to find an approximation of the given function f whose precise course is unknown. The approximating function f^A is derived on the basis of a linguistic description given in the form of fuzzy IF-THEN rules. One may argue that we can never be sure that the found function f^A is the best possible one because we do not know f exactly. However, the fuzzy approximation theory provides means that guarantee (in the mathematical sense) that if the function f has certain reasonable properties, then there exists a suitable and sufficiently accurate approximating function f^A.

Definition 3.1 *Let $f : U \longrightarrow V$ be a given function, $\epsilon > 0$, and let $R \subsetneq U \times V$ be a fuzzy relation. We say that R* approximates the function f with the precision $\epsilon > 0$ *if for any $u \in U$ and $v \in V$,*

$$R(u, v) > 0 \text{ implies } \|f(u), v\| < \epsilon, \tag{3.36}$$

where $\|\cdot\|$ is a suitable metric defined on the set V. If $V = \mathbb{R}$ (real numbers), then we can rewrite (3.36) into

$$R(u, v) > 0 \text{ implies } |f(u) - v| < \epsilon. \tag{3.37}$$

If a measurement $u = u_0$ is given, then we can apply (2.66) (or (2.67)) to obtain the fuzzy set

$$B'_{u_0} = \{{}^r/v \mid r = R(u_0, v), v \in V\}. \tag{3.38}$$

The fuzzy set (3.38) can be taken as a result of a *derivation* on the basis of the given linguistic description following its relational interpretation.[4] Obviously, it is a projection of the fuzzy relation R to V when u_0 is fixed.

Remark 3.3 *If f is a function of n variables, that is, $f : U_1 \times \cdots \times U_n \longrightarrow V$, then using fuzzy IF-THEN rules, we construct the fuzzy relation R according to (3.28) or (3.29). Then, given values $u_{01} \in U_1, \ldots, u_{0n} \in U_n$, we compute the resulting fuzzy set by means of*

$$B'_{u_{01},..,u_{0n}} = \{{}^r/v \mid r = R(u_{01}, \ldots, u_{0n}, v), v \in V\}. \tag{3.39}$$

The function f^A is obtained using

$$f^A(u) = \text{DEF}(B'_u), \quad u \in U, \tag{3.40}$$

where DEF is a suitable defuzzification function. Because it always holds that

$$\text{DEF}(B'_u) \in \text{Supp}(B'_u),$$

[4]Some authors speak about *inference*. In the strict logical sense, however, such terminology is incorrect.

we immediately obtain that if R approximates the function f with the precision $\epsilon > 0$, then for any $u \in U$, it holds that

$$\|f^A(u), f(u)\| < \epsilon, \tag{3.41}$$

or, for $V = \mathbb{R}$,

$$|f^A(u) - f(u)| < \epsilon. \tag{3.42}$$

Formulas (3.41) or (3.42) tell us that the function f^A approximates the function f with the precision $\epsilon > 0$.

Note that the choice of the defuzzification function is irrelevant for this theoretical result. On the other hand, this choice is important if the quality of approximation is at play. In book [99], the reader can find one possible criterion, according to which the COG defuzzification is optimal. Thus, the most important question is how the fuzzy relation R should be created.

DNF/CNF-based approximating function. If we want to apply the results of Section 3.1.2, we have two possibilities: the fuzzy relation R can take either the form (3.15) or (3.16). Then the membership function of the fuzzy set (3.38) has one of the following two forms:

$$B'_{u_0}(v) = R_{\mathrm{DNF}}(u_0, v) = \bigvee_{i=1}^{m} (A_i(u_0)\,\mathbf{T}\,B_i(v)), \qquad v \in V, \tag{3.43}$$

or

$$B'_{u_0}(v) = R_{\mathrm{CNF}}(u_0, v) = \bigwedge_{i=1}^{m} (A_i(u_0) \rightarrow B_i(v)), \qquad v \in V. \tag{3.44}$$

After substituting $B'_{u_0}(v)$ to (3.40), we immediately obtain

$$f^A(u_0) = \mathrm{DEF}\left(\left\{ \bigvee_{i=1}^{m} (A_i(u_0)\,\mathbf{T}\,B_i(v)) \Big/ v \,\middle|\, v \in V \right\}\right), \tag{3.45}$$

or

$$f^A(u_0) = \mathrm{DEF}\left(\left\{ \bigwedge_{i=1}^{m} (A_i(u_0) \rightarrow B_i(v)) \Big/ v \,\middle|\, v \in V \right\}\right). \tag{3.46}$$

To obtain as good as possible approximation of the function f, it is necessary to tune shapes of fuzzy sets A_i and B_i, $i = 1, \ldots, m$.

EXAMPLE 3.3 Let us approximate the function $f(x) = x^3$ on the interval $[-1, 1]$, first using uniform partition of input and output variables by 17 triangular membership functions.[5] Then we tune membership functions according to the course of the approximated function.

[5]This example was prepared by Martin Dyba.

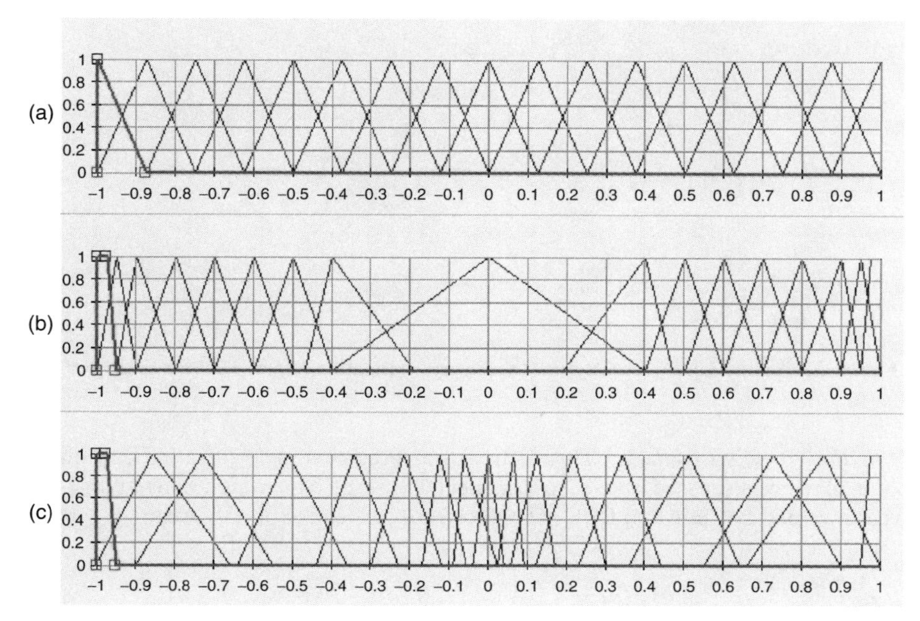

Figure 3.7 (a) Fuzzy partition of input and output variables of the original fuzzy approximation. (b) Fuzzy partition of the input variable of the tuned fuzzy approximation. (c) Fuzzy partition of the output variable of the tuned approximation for Example 3.3.

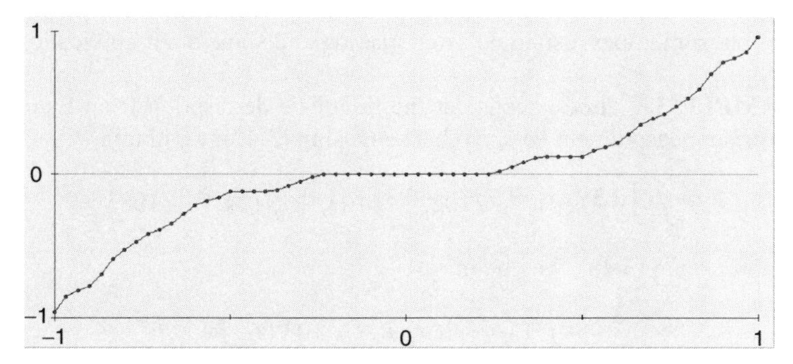

Figure 3.8 Function f^A_{orig} for the original fuzzy approximation of $f(x) = x^3$ for Example 3.3.

Fuzzy partitions of the original and tuned fuzzy approximations are depicted in Figure 3.7. Resulting functions f^A_{orig} for the original fuzzy approximation and f^A_{new} for the tuned one are depicted in Figures 3.8 and 3.9, respectively.

We can see much smoother course of the function f^A_{new} compared to the original f^A_{orig}. Figures in this example were prepared using software system LFL Controller© developed in the University of Ostrava.[6] □

[6]See the companion website http://www.wiley.com/go/novak/fuzzy/modeling.

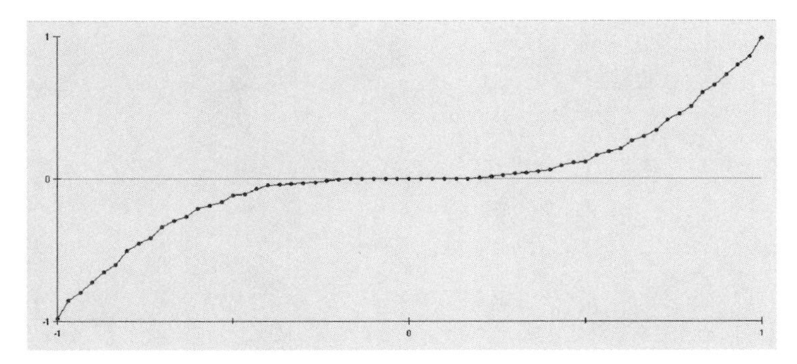

Figure 3.9 Function f^A_{new} for the tuned fuzzy approximation of $f(x) = x^3$ for Example 3.3.

Mamdani–Assilian method. This method was proposed in 1974 in [67]. From the point of view discussed above, it is a special case of equations (3.43) and (3.45), where we put $\mathbf{T} = \min$ and DEF = COG. Hence, we can explicitly write

$$B'_{u_0}(v) = \bigvee_{i=1}^{m} (A_i(u_0) \wedge B_i(v)), \quad v \in V, \tag{3.47}$$

and

$$f^A(u) = \text{COG}(B'_u). \tag{3.48}$$

This method is the most often used method in various kinds of practical applications in control but sometimes also in decision making and some other applications.

EXAMPLE 3.4 Let us consider the linguistic description from Example 3.2 and let a crisp measurement be $u_0 = 3$. Then, using (3.47), we obtain

$$B'_3 = \left\{ {}^{0.3}/101, {}^{0.3}/102, {}^{0.3}/103, {}^{0.4}/104, {}^{0.3}/105 \right\}.$$

Furthermore, using (3.48), we obtain

$$f^A(3) = \text{COG}(B'_3) = 103.06.$$

We can see nice approximation properties of this method. □

The Mamdani–Assilian fuzzy approximation method given by equations (3.47) and (3.48) is schematically depicted in Figure 3.10. The fuzzy sets A_1, A_2 are depicted under the x-axis, and the fuzzy sets B_1, B_2 to the left of the y-axis. The interpretation of fuzzy rules $R_1 = A_1 \times B_1$, $R_2 = A_2 \times B_2$ constructed using A_1, A_2 and B_1, B_2 is depicted by rectangles through which the approximated function f passes. If an element u_0 is given, then we construct the fuzzy set B'_{u_0} as follows: we find membership degrees $A_1(u_0)$ and $A_2(u_0)$ and restrict by them fuzzy sets B_1 and B_2. Using union of these restricted fuzzy sets we obtain the resulting fuzzy set B'_{u_0}, which is depicted by shading left to the y axis. The result $v_0 = f^A(u_0)$, which is approximately equal to

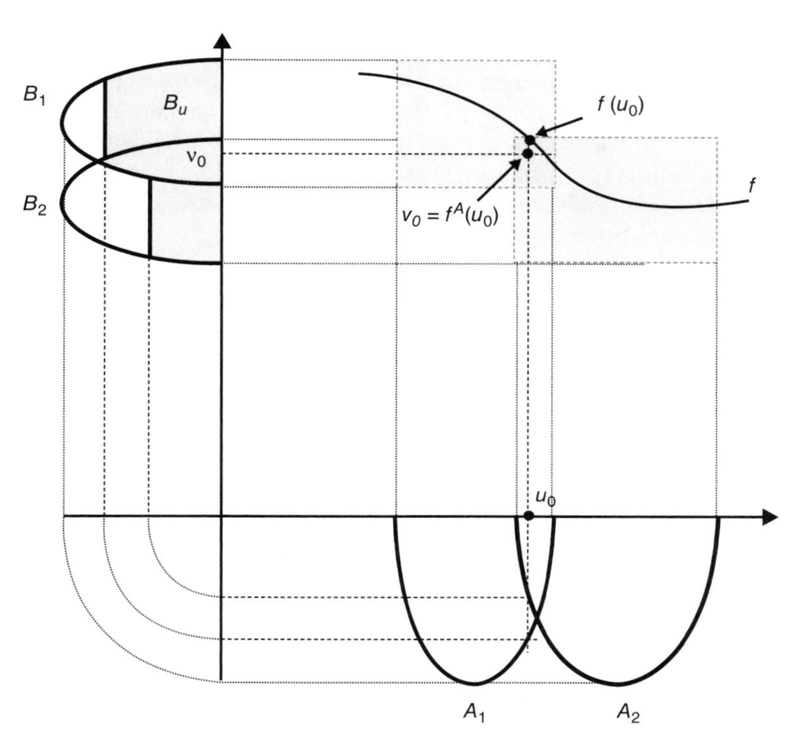

Figure 3.10 A schematic depiction of fuzzy approximation of function f at point u_0 ($v_0 = f^A(u_0)$).

the value $f(u_0)$, is obtained by means of the defuzzification of fuzzy set B'_{u_0} using the COG method.

Basic properties of fuzzy approximation. An important property of fuzzy approximation is the so-called *functional completeness* (we can also meet the term "universal approximator" in the literature). Roughly speaking, it says that any bounded continuous function can be approximated with arbitrary precision. This means that we can approximate any practically reasonable function with arbitrary precision by some set of fuzzy IF-THEN rules. Consequently, the method in question is sufficiently universal. Exact formulation of this result is stated in the following theorem, namely, for the Mamdani–Assilian method and the CNF. A more general formulation for the DNF can be found in [99, Chapter 5] (cf. also [17, 22]).

Theorem 3.2 *Let $U = [a, b] \subseteq \mathbb{R}$ be an interval of reals (a compact set). Then to any bounded continuous function $f : U \longrightarrow \mathbb{R}$ and $\epsilon > 0$, there exist fuzzy sets $A_i \underset{\sim}{\subset} U$, $B_i \underset{\sim}{\subset} \mathbb{R}$, $i = 1, \ldots, m$, and a defuzzification function* DEF $: (\mathscr{F}(\mathbb{R}) - \{\emptyset\}) \longrightarrow \mathbb{R}$ *such that*

$$|f(u_0) - \mathrm{DEF}(B'_{u_0})| < \epsilon, \quad u_0 \in U, \tag{3.49}$$

where B'_{u_0} is the fuzzy set computed either using DNF (3.47) or using CNF (3.44).

Proof: We give here the proof for the DNF only, that is, for B'_{u_0} computed using (3.47). The proof of the second part of this theorem can be found in [60].

On the basis of assumptions of the theorem, we can consider a function $f : [a, b] \longrightarrow [c, d]$, where a, b, c, d are real numbers, and it holds that $a < b$ and $c < d$ (i.e., $U = [a, b]$). Moreover, because f is, according to our assumptions, continuous, it is uniformly continuous as well. Let us choose $\epsilon > 0$ and consider $\epsilon/2$. Then there exists $\delta > 0$ such that for any u, u',

$$|u - u'| < \delta \quad \text{implies} \quad |f(u) - f(u')| < \frac{\epsilon}{2}. \tag{3.50}$$

Therefore, we can divide the interval $[a, b]$ to m open subintervals U_i of length smaller than δ in such a way that $\bigcup_{i=1}^{m} U_i = [a, b]$. Then, $f(U_i)$ is a subinterval of $[c, d]$, and from property (3.50), it follows that its length is smaller than $\epsilon/2$. To any interval $f(U_i)$, $i = 1, \ldots, m$, we find its center y_i and consider the open interval

$$V_i = \left(y_i - \frac{\epsilon}{2}, y_i + \frac{\epsilon}{2}\right).$$

It is clear that $f(U_i) \subseteq V_i$.

Now we define fuzzy sets $A_i \underset{\sim}{\subset} [a, b]$ and $B_i \underset{\sim}{\subset} [c, d]$ as follows:

$$A_i(u) > 0 \text{ iff } u \in U_i, \tag{3.51}$$

$$B_i(v) > 0 \text{ iff } v \in V_i. \tag{3.52}$$

We want to show that the fuzzy relation R_{DNF} defined by the membership function

$$R_{\text{DNF}}(u, v) = \bigvee_{i=1}^{m} (A_i(u) \wedge B_i(v))$$

approximates f with the precision ϵ.

Indeed, let $u_0 \in [a, b]$. Then $u_0 \in U_{j_1}, \ldots, u_0 \in U_{j_k}$ for some $1 \le k \le m$ (in practice, usually k is equal to 2). From the definition of the fuzzy set A_i (3.51), it follows that $A_{j_1}(u_0) > 0, \ldots, A_{j_k}(u_0) > 0$. Further, $f(u_0) \in f(U_{j_1}) \subseteq V_{j_1}, \ldots, f(u_0) \in f(U_{j_k}) \subseteq V_{j_k}$, hence $B_{j_1}(f(u_0)) > 0, \ldots, B_{j_k}(f(u_0)) > 0$ as well. It follows that

$$R_{\text{DNF}}(u_0, f(u_0)) = \bigvee_{i=j_1}^{j_k} (A_i(u_0) \wedge B_i(f(u_0))) > 0.$$

Let $R_{\text{DNF}}(u_0, v)$ is greater than 0 for some $v \in [c, d]$. We are going to show that $|f(u_0) - v| < \epsilon$. Recall that the length $|V_j| < \epsilon$ for all V_j, and so $|\bigcap_{j=j_1}^{j_k} V_j| < \epsilon$ as well.

It follows from the construction above that $B_{j_1}(v) > 0, \ldots, B_{j_k}(v) > 0$. This means that $v \in \bigcap_{j=j_1}^{j_k} V_j$; hence, for any $j, j = j_1, \ldots, j_k$,

$$|f(u_0) - v| \le |f(u_0) - y_j| + |y_j - v| < \frac{\epsilon}{2} + \frac{\epsilon}{2} = \epsilon,$$

because both $f(u_0) \in f(U_j)$ and $y_j \in f(U_j)$. The y_j is the center of the interval V_j, and v belongs either to its right or to its left half.

From the definition of the defuzzification DEF and (3.47), we obtain that also

$$\mathrm{DEF}(B'_{u_0}) \in \bigcap_{j=j_1}^{j_k} V_j,$$

which implies the claim (3.49). ∎

This result is very important because it explains why fuzzy approximation gives good results. Let us remark that this theorem can be generalized for (3.43), where **T** is an arbitrary continuous t-norm (the proof can be found in [99]). Note also that the shapes of the fuzzy sets A_i and B_i do not play a role in the proof of this theorem.

3.2.3 Choosing between DNF and CNF

From the point of view of functionality, both forms give similar results. The DNF interpretation of linguistic description is more popular because it is easier to understand. However, we cannot say that it gives (in some sense) better results than the CNF interpretation. Quite often, we face the contrary (cf. [11]).

Consistency and coherence. This is a property of linguistic description that has fundamental importance for obtaining correct results. Roughly speaking, the consistency certifies that a given linguistic description contains no conflicts, that is, there are no rules with the same (or similar) antecedents and significantly different (contradictory) consequents. A typical example of contradictory rules is the following:

IF *obstacle* is *left* OR *front* THEN *bypass* is *right*,

IF *obstacle* is *right* OR *front* THEN *bypass* is *left*.

If an obstacle lies in front, then this linguistic description leads to ill behavior because both antecedents overlap. Then in the DNF interpretation, the consequents are considered as possible conclusions aggregated disjunctively: "left or right bypass is possible" and the COG defuzzification leads to an unavoidable accident with the obstacle.

The CNF interpretation, on the other hand, gives both "left and right" and leads to an empty output, that is, B'_{u_0} is the empty fuzzy set. The problem consists in the fact that there are two different conclusions for one input. Consequently, we cannot avoid the idea of dealing with a function even in the fuzzy setting. This means that the linguistic description should lead to a fuzzy function understood as a mapping from $\mathscr{F}(U)$ to $\mathscr{F}(V)$.

Checking consistency of the DNF interpretation is surprisingly more complicated than that of the CNF interpretation because it requires study of fuzzy functions based on graded equalities [59, 60]. For the CNF interpretation, we can effectively use the concept of *coherence* introduced in [33].

A fuzzy relation $R \subsetneq U \times V$ interpreting a linguistic description is called *coherent* if for each $u \in U$, there exists $v \in V$ such that $R(u, v) = 1$. In other words, the coherence corresponds with the requirement that $\mathrm{Ker}(R(u,\cdot))$ is nonempty for every $u \in U$. For the CNF interpretation R_{CNF}, we obtain

$$\mathrm{Ker}(R_{\mathrm{CNF}}(u,\cdot)) = \bigcap_{i=1}^{m} \{v \mid A_i(u) \le B_i(v)\}. \tag{3.53}$$

Hence, the coherence of the $\mathbb{R}_{\mathrm{CNF}}$ is equivalent to

$$\bigcap_{i=1}^{m} \{v \mid A_i(u) \le B_i(v)\} \neq \emptyset$$

for all $u \in U$. This means that, for every $u \in U$, the kernels of the outputs of all the fired rules must overlap in at least one point. Therefore, fundamentally different outputs for the same inputs are impossible. Thus, the coherence of the CNF interpretation ensures that the linguistic description does not contain conflicting rules.

EXAMPLE 3.5 Let A_1, A_2, A_3 and B_1, B_2, B_3 be two triangular fuzzy partitions on $U = [1, 5]$ and $\mathrm{V} = [1, 3]$, respectively (Figure 3.11).

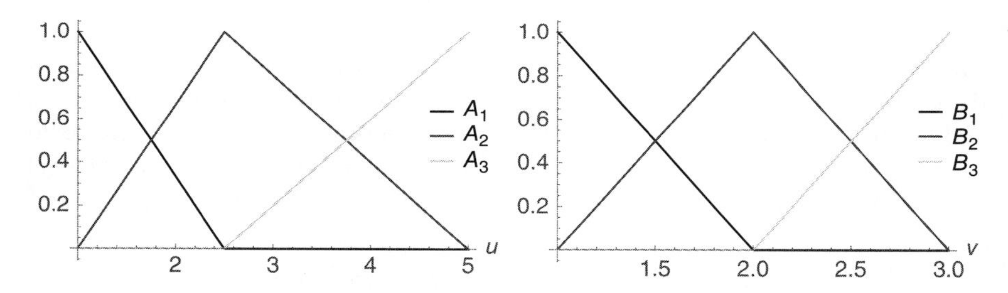

Figure 3.11 Triangular fuzzy partitions for Example 3.5.

Let the following two sets of fuzzy IF-THEN rules be given:

$$\mathscr{R}_1 := \begin{array}{l} \text{IF } X \text{ is } \mathscr{A}_1 \text{ THEN } Y \text{ is } \mathscr{B}_2, \\ \text{IF } X \text{ is } \mathscr{A}_2 \text{ THEN } Y \text{ is } \mathscr{B}_3, \\ \text{IF } X \text{ is } \mathscr{A}_3 \text{ THEN } Y \text{ is } \mathscr{B}_2, \end{array}$$

and

$$\mathscr{R}_2 := \begin{array}{l} \text{IF } X \text{ is } \mathscr{A}_1 \text{ THEN } Y \text{ is } \mathscr{B}_2, \\ \text{IF } X \text{ is } \mathscr{A}_2 \text{ THEN } Y \text{ is } \mathscr{B}_3, \\ \text{IF } X \text{ is } \mathscr{A}_3 \text{ THEN } Y \text{ is } \mathscr{B}_1. \end{array}$$

Note that $\mathscr{R}_1$ and $\mathscr{R}_2$ differ only in the consequent of their third IF-THEN rule. $\mathscr{R}_1$ is coherent, but $\mathscr{R}_2$ is not. For example, for $u_0 = 3.75$, the set Ker $(R_{1,\mathrm{CNF}}(u_0,\cdot))$ from (3.53) is equal to $[1,3] \cap [1.5, 2.5] \cap [2.5, 3] = \{2.5\}$ for $\mathscr{R}_1$, but Ker $(R_{2,\mathrm{CNF}}(u_0,\cdot))$

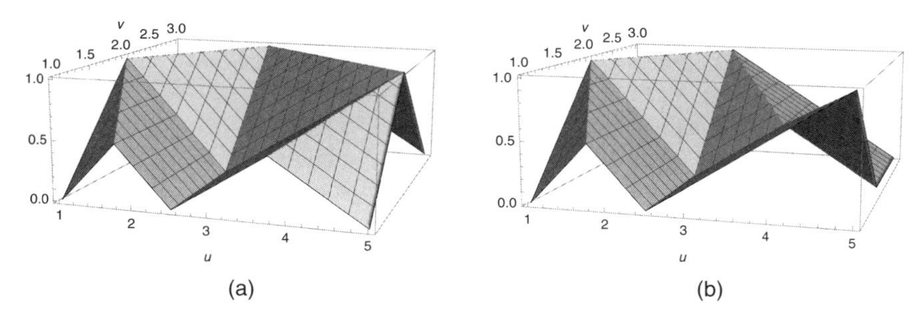

Figure 3.12 The fuzzy relations $R_{1,\mathrm{CNF}}$ (a) and $R_{2,\mathrm{CNF}}$ (b) from Example 3.5.

is equal to $[1,3] \cap [1,1.5] \cap [2.5,3] = \emptyset$ for $\mathscr{R}_2$. The fuzzy relations $R_{1,\mathrm{CNF}}$ and $R_{2,\mathrm{CNF}}$ (cf. (3.16), $\rightarrow$ is the Łukasiewicz implication) are visualized in Figure 3.12. Notice the difference between $R_{1,\mathrm{CNF}}$ and $R_{2,\mathrm{CNF}}$ around the point $u = 3.75$. □

For the DNF interpretation, however, coherence is not helpful because

$$\begin{aligned}\mathrm{Ker}(R_{\mathrm{DNF}}(u,\cdot)) &= \bigcup_{i=1}^{m} \{v \mid A_i(u)\,\mathbf{T}\,B_i(v) = 1\} \\ &= \bigcup_{i=1}^{m} \{v \mid A_i(u) = 1 \ \text{ and } \ B_i(v) = 1\}\end{aligned}$$

holds, that is, coherence of the DNF interpretation requires that for all $u \in U$, there must exist A_i such that $A_i(u) = 1$, and the corresponding output fuzzy set B_i must be normal. But this has little to do with the consistency.

Let us remark that an analogous principle based on the concept of convexity instead of normality has been proposed for the DNF interpretation by Coufal [29]. Checking this condition, however, is more complex, while checking coherence of the CNF interpretation is comparably simpler. For details, see [27, 28, 33].

Continuity. One of the arguments in favor of the DNF interpretation is its continuity. The following theorems demonstrate that CNF has the same power.

Theorem 3.3 *Let $U, V \subset \mathbb{R}$ be intervals of real numbers and all the membership functions as well as the used t-norm be continuous. Then the approximating function f^A in (3.45) with the COG defuzzification is continuous in every input value u for which at least one rule fires. The latter happens if the membership degree $A_i(u)$ of u in the antecedent of the firing rule $\mathscr{R}_i$ is nonzero.*

Theorem 3.4 *Let $U, V \subset \mathbb{R}$ be intervals of real numbers and the fuzzy relation R_{CNF} be coherent. Furthermore, let all fuzzy sets A_i and B_i, $i = 1,\ldots,m$, be normal, strictly convex and have a continuous membership function and, either* $\mathrm{Supp}(A_i) = U$ *or* $\mathrm{Supp}(B_i) = V$ *(or both) for all $i = 1,\ldots,m$. Then the approximating function f^A in*

(3.46) with MOM defuzzification is continuous in every input for which at least one rule fires (i.e., whose antecedent is fulfilled in a nonzero degree for at least one rule).

For the proof of these theorems, see [135].

It follows from these theorems that fuzzy sets with Gaussian membership functions (cf. Figure 2.5) or with other kind of membership function having unbounded support either in the antecedent or in the consequent *ensure continuity* of the CNF interpretation R_{CNF} with the MOM defuzzification. We see that *there is no significant difference between both kinds of the relational interpretation of fuzzy IF-THEN rules.*

3.3 GENERALIZED MODUS PONENS AND FUZZY FUNCTIONS

When applying relational fuzzy IF-THEN rules, we should not neglect the problem of solvability of the associated fuzzy relation equations. The fact that fuzzy IF-THEN rules resemble logical formulas suggests that the relational interpretation of them can also provide a computational method for a certain kind of modus ponens (2.84). This situation can be on surface level described as follows:

$$\begin{array}{rl} & \mathscr{R}_1 := \text{ IF } X \text{ is } \mathscr{A}_1 \text{ THEN } Y \text{ is } \mathscr{B}_1, \\ \text{Condition:} & \cdots\cdots\cdots\cdots\cdots\cdots \\ & \mathscr{R} := \text{ IF } X \text{ is } \mathscr{A}_m \text{ THEN } Y \text{ is } \mathscr{B}_m, \\ \text{Observation:} & X \text{ is } \mathscr{A}', \\ \hline \text{Conclusion:} & Y \text{ is } \mathscr{B}'. \end{array} \tag{3.54}$$

This is the well-known scheme of *generalized modus ponens*, which is the basis of the theory of approximate reasoning. The condition is formed by a linguistic description. The expression $\mathscr{A}'$ represents a property that can be slightly different from any of $\mathscr{A}_1, \ldots, \mathscr{A}_m$. Naturally, it follows that a conclusion $\mathscr{B}'$ can be slightly different from any of $\mathscr{B}_1, \ldots, \mathscr{B}_m$.

We will describe analysis of (3.54) on the semantic level only. Its formal description on syntactic level can be found, for example, in [44, 99, 112].

On the semantic level, (3.54) can be interpreted as follows. The expression "X is $\mathscr{A}'$" is assigned a fuzzy set $A' \underset{\sim}{\subset} U$ and $R \underset{\sim}{\subset} U \times V$ is one of the fuzzy relations R_{DNF} or R_{CNF} from (3.15) or (3.16), respectively. Then (3.54) leads to the computation of an image $B' \underset{\sim}{\subset} V$ of A' in R, that is,

$$B' = A' \circ R, \tag{3.55}$$

where we used the notation as in (2.69) (alternatively, we can write $B' = R''A'$). If we rewrite (3.55) using the membership functions (3.15) or (3.16), we obtain

$$B'(v) = \bigvee_{u \in U} \left(A'(u) \mathbf{T}_1 \bigvee_{i=1}^{m} (A_i(u) \mathbf{T}_2 B_i(v)) \right) \tag{3.56}$$

or

$$B'(v) = \bigvee_{u \in U} \left(A'(u) \, \mathbf{T}_1 \, \bigwedge_{i=1}^{m} (A_i(u) \to B_i(v)) \right) \tag{3.57}$$

for all $v \in V$. The scheme (3.55) (or, in more detail, (3.56) and (3.57)) is in the literature called *computational form of the generalized rule of modus ponens*, or simply, *computational rule of modus ponens*. The t-norms $\mathbf{T}_1, \mathbf{T}_2$ in (3.56) are different, in general, but quite often we put $\mathbf{T}_1 = \mathbf{T}_2 = \min$.

Remark 3.4 *Note that if we set $A' = \{^1/u_0\}$, then the fuzzy set $B'(v)$ in (3.56) or (3.57) reduces to the fuzzy set $B'_{u_0}(v)$ in (3.43) or (3.44), respectively.*

Computational modus ponens and fuzzy relation equations. It can be proved that the fuzzy relations R_{DNF} (3.15) and R_{CNF} (3.16) give rise to a fuzzy function (of type 2; cf. [75])

$$F_R : \mathscr{F}(U) \longrightarrow \mathscr{F}(V), \tag{3.58}$$

which is a classical function between sets of fuzzy sets. Then (3.55) corresponds to a computation of a functional value of F_R in a given argument A'. Moreover, it is required that $F_R(A_i) = B_i$ for all $i \in \{1,\ldots,m\}$. Another argument in favor of the latter requirement follows from the idea of modus ponens in (3.54): we again expect that for each antecedent A_i, $i \in \{1,\ldots,m\}$, the result of modus ponens should be the corresponding consequent B_i. In both cases, this means that

$$B_i = A_i \circ R \tag{3.59}$$

must hold true for each $i = 1,\ldots,m$.

Finding a suitable fuzzy relation R in (3.59) is the well-known problem of solving fuzzy relation equations. This problem was first raised by E. Sanchez in [129]. Unfortunately, there may be no solution of (3.59) if arbitrary fuzzy sets A_i and B_i, $i = 1,\ldots,m$, are given. It turned out that these fuzzy sets must necessarily fulfill certain conditions. There are many papers dealing with this problem (e.g., [40, 41, 58, 110, 120]). The simplest conditions have been proposed by I. Perfilieva in [109].

First, all the fuzzy sets $A_i \subsetapprox U, B_i \subsetapprox V$, $i = 1,\ldots,m$ must be *normal*, that is, there must exist elements $u_i \in U, v_i \in V$ such that $A_i(u_i) = 1$ as well as $B_i(v_i) = 1$. Furthermore, a fuzzy equality $\approx_1$ must be given such that $A_i(u) = \|u \approx_1 u_i\|$ for all $u \in U$ and a fuzzy equality $\approx_2$ such that $B_i(v) = \|v \approx_2 v_i\|$ for all $v \in V$ (all fuzzy sets are fuzzy points in the sense of (2.71)).

Let us remark that the latter fuzzy equalities can always be constructed if the fuzzy sets A_i, B_i are normal (cf. Section 2.2.6) and both families fulfill condition (2.73). Then, the system (3.59) is solvable if and only if

$$A_i(u_k) \leq B_i(v_k) \tag{3.60}$$

holds true for all $i, k = 1,\ldots,m$. The meaning of this condition becomes clear from Figure 3.13. One can see that the condition assures a weaker form of functionality: the values B_1, B_2 of the fuzzy function (3.58) must be closer to each other than the arguments A_1, A_2. Let us also remark that the coherence condition discussed above implies (3.60).

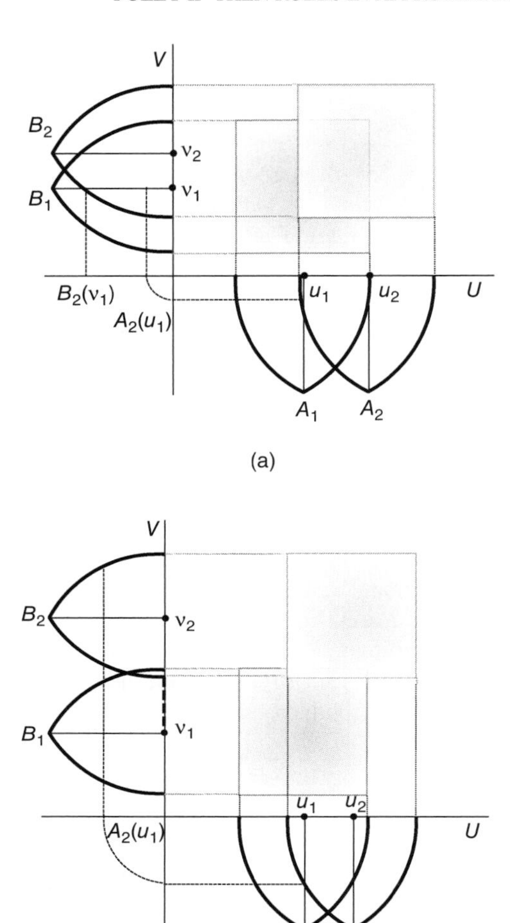

Figure 3.13 A necessary position of fuzzy sets for correct generalized modus ponens. (a) The fuzzy sets A_1, A_2 and B_1, B_2 fulfill condition (3.60). (b) The fuzzy sets A_1, A_2 and B_1, B_2 violate condition (3.60) (because $B_2(v_1) = 0$).

It follows from this discussion that if the fuzzy sets in (3.15) or (3.16) do not fulfill the solvability condition (3.60), then the linguistic description (3.7) does not work correctly, which means that (3.59) does not hold and we should speak neither about fuzzy function (3.58) nor about generalized modus ponens (3.54).

3.4 TAKAGI–SUGENO RULES

3.4.1 Basic Concepts

In this section, we describe a special type of fuzzy IF-THEN rules for approximation of functions. Up to now, we analyzed rules that characterize a relationship between inexactly given values of independent and dependent variables. In 1985, T. Takagi and M. Sugeno [138] came up with an idea to introduce conditional rules, where values of the dependent variable Y are determined by exact linear expressions confined to certain areas characterized imprecisely. This means that the antecedent of each rule has the form introduced above, and the consequent is a linear expression. Thus, we work with a set of rules of the following form:

$$\begin{aligned}
\mathscr{R}_1 := \ & \textsf{IF } X_1 \text{ is } \mathscr{A}_{11} \textsf{ AND } \ldots \textsf{ AND } X_n \text{ is } \mathscr{A}_{1n} \\
& \qquad \textsf{THEN } Y = b_{10} + b_{11}X_1 + \ldots + b_{1n}X_n, \\
& \cdots\cdots\cdots\cdots\cdots\cdots\cdots\cdots\cdots\cdots \\
\mathscr{R}_m := \ & \textsf{IF } X_1 \text{ is } \mathscr{A}_{m1} \textsf{ AND } \ldots \textsf{ AND } X_n \text{ is } \mathscr{A}_{mn} \\
& \qquad \textsf{THEN } Y = b_{m0} + b_{m1}X_1 + \cdots + b_{mn}X_n.
\end{aligned} \tag{3.61}$$

As a special case, it is possible to consider a situation when the consequent of each rule is just a crisp number z, that is,

$$\begin{aligned}
\mathscr{R}_1 := \ & \textsf{IF } X_1 \text{ is } \mathscr{A}_{11} \textsf{ AND } \cdots \textsf{ AND } X_n \text{ is } \mathscr{A}_{1n} \textsf{ THEN } Y = z_1, \\
& \cdots\cdots\cdots\cdots\cdots\cdots\cdots\cdots\cdots\cdots \\
\mathscr{R}_m := \ & \textsf{IF } X_1 \text{ is } \mathscr{A}_{m1} \textsf{ AND } \cdots \textsf{ AND } X_n \text{ is } \mathscr{A}_{mn} \textsf{ THEN } Y = z_m.
\end{aligned} \tag{3.62}$$

The rules of the form (3.61) (or (3.62)) are called *Takagi–Sugeno fuzzy rules*, or shortly TS rules. The meaning of each rule $\mathscr{R}_i$, $i \in \{1,\ldots,m\}$ in (3.61) can be expressed as follows: if we confine values of independent variables X_j, $j = 1,\ldots,n$, to an area that can be characterized by expression "$\mathscr{A}_{i1}$ and $\cdots$ and $\mathscr{A}_{in}$", then a dependence of Y on $X_1,\ldots,X_n$ is linear and can be described by the formula $Y = b_{i0} + b_{i1}X_1 + \cdots + b_{in}X_n$. The expression "$\mathscr{A}_{i1}$ and $\cdots$ and $\mathscr{A}_{in}$" is interpreted by a fuzzy relation—see below.

3.4.2 Fuzzy Approximation Using TS Rules

We will interpret TS rules on the semantic level only. Let us suppose that universes $U_1,\ldots,U_n$, V and a function $f : U_1 \times \cdots \times U_n \longrightarrow V$ are given. Each U_i is a universe of values of the corresponding variable X_i, $i = 1,\ldots,n$.

First, the connective " AND " in (3.61) (or (3.62)) is assigned a certain t-norm $\mathbf{T}$. We usually take $\mathbf{T} = \min$. Furthermore, the expressions "X_i is $\mathscr{A}_{ji}$" are assigned fuzzy sets in the corresponding universes, namely, "X_1 is $\mathscr{A}_{j1}$" is assigned a fuzzy set $A_{j1} \underset{\sim}{\subset} U_1$, ..., "$X_n$ is $\mathscr{A}_{jn}$" is assigned a fuzzy set $A_{jn} \underset{\sim}{\subset} U_n$ for all rules $\mathscr{R}_j$, $j = 1,\ldots,m$. Then each rule $\mathscr{R}_j$ in (3.61) characterizes a linear dependence of the variable Y on the independent variables $X_1,\ldots,X_n$. The dependence is aggregated over inexactly delineated areas, each of which is given by the Cartesian product of fuzzy sets

$$A_{j1} \overset{\mathbf{T}}{\times} \cdots \overset{\mathbf{T}}{\times} A_{jn} \underset{\sim}{\subset} U_1 \times \cdots \times U_n, \qquad j = 1,\ldots,m.$$

The whole set of rules (3.61) characterizes a *nonlinear* dependence $f^A : U_1 \times \cdots \times U_n \to V$ over the area

$$\bigcup_{j=1}^{m} (A_{j1} \overset{\mathbf{T}}{\times} \cdots \overset{\mathbf{T}}{\times} A_{jn}) \subsetneq U_1 \times \cdots \times U_n. \tag{3.63}$$

The function f^A is supposed to approximate f in the sense of the minimal root-mean-square error, provided that the domain of the function f lies inside the support of (3.63).

The function f^A is obtained using the formula

$$f^A(x_1, \ldots, x_n) =$$

$$= \frac{\sum_{j=1}^{m} (A_{j1}(x_1)\, \mathbf{T} \cdots \mathbf{T}\, A_{jn}(x_n)) \cdot (b_{j0} + b_{j1}x_1 + \cdots + b_{jn}x_n)}{\sum_{j=1}^{m} (A_{j1}(x_1)\, \mathbf{T} \cdots \mathbf{T}\, A_{jn}(x_n))}, \tag{3.64}$$

where $x_1 \in U_1, \ldots, x_n \in U_n$. If we put

$$\beta_j(x_1, \ldots, x_n) = \frac{A_{j1}(x_1)\, \mathbf{T} \cdots \mathbf{T}\, A_{jn}(x_n)}{\sum_{j=1}^{m} (A_{j1}(x_1)\, \mathbf{T} \cdots \mathbf{T}\, A_{jn}(x_n))}, \tag{3.65}$$

we can rewrite (3.64) into

$$\begin{aligned} f^A(x_1, \ldots, x_n) &= \beta_1(x_1, \ldots, x_n)(b_{10} + b_{11}x_1 + \cdots + b_{1n}x_n) + \cdots \\ &\quad + \beta_m(x_1, \ldots, x_n)(b_{m0} + b_{m1}x_1 + \cdots + b_{mn}x_n) = \\ &= \sum_{j=1}^{m} \beta_j(x_1, \ldots, x_n)(b_{j0} + b_{j1}x_1 + \cdots + b_{jn}x_n). \end{aligned} \tag{3.66}$$

Note that there is no defuzzification function to be considered here. Note also that the function (3.66) has $m(n + 1)$ parameters b_{ji} to be computed.

If we work with special rules of the form (3.62), then the computation of $f^A(x_1, \ldots, x_n)$ takes the following simpler form:

$$f^A(x_1, \ldots, x_n) = \sum_{j=1}^{m} \beta_j(x_1, \ldots, x_n)\, z_j. \tag{3.67}$$

Now, for simplicity, let us assume that there is only one independent variable X, that is, the TS rules have the form

$$\begin{aligned} \mathscr{R}_1 &:= \text{IF } X \text{ is } \mathscr{A}_1 \text{ THEN } Y = b_{10} + b_{11}\, X, \\ &\cdots\cdots\cdots\cdots\cdots\cdots\cdots\cdots\cdots\cdots \\ \mathscr{R}_m &:= \text{IF } X \text{ is } \mathscr{A}_m \text{ THEN } Y = b_{m0} + b_{m1}\, X. \end{aligned} \tag{3.68}$$

Then the approximating function f^A (3.66) reduces to

$$f^A(x) = \sum_{j=1}^{m} \beta_j(x)(b_{j0} + b_{j1}x), \tag{3.69}$$

where

$$\beta_i(x) = \frac{A_i(x)}{\sum_{j=1}^{m} A_j(x)}, \qquad i = 1, \ldots, m. \tag{3.70}$$

EXAMPLE 3.6 Let A_1, A_2, and A_3 be fuzzy sets forming a trapezoidal fuzzy partition on $U = [1, 12]$ (Figure 3.14). Let the following TS rules be given:

$$\mathscr{R}_1 := \text{IF } X \text{ is } \mathscr{A}_1 \text{ THEN } Y = 1.25,$$
$$\mathscr{R}_2 := \text{IF } X \text{ is } \mathscr{A}_2 \text{ THEN } Y = 0.5 + 0.2\ X,$$
$$\mathscr{R}_3 := \text{IF } X \text{ is } \mathscr{A}_3 \text{ THEN } Y = 3 - 0.1\ X.$$

Note that the coefficient b_{11} from (3.68) is equal to 0. The function f^A computed by (3.69) along with fuzzy partition A_1, A_2, A_3 is depicted in Figure 3.14.

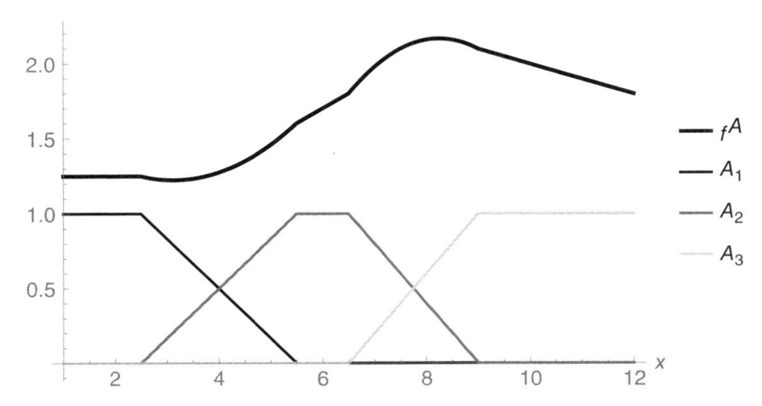

Figure 3.14 Trapezoidal fuzzy partition A_1, A_2, A_3 on $U = [1, 12]$ and the approximating function f_A from Example 3.6.

Note that the function f^A is just the linear function given by coefficients b_{i0}, b_{i1} in regions where a kernel of a fuzzy set A_i is located and, at the same time, membership functions of other fuzzy sets $A_j, j \neq i$, are equal to 0. For example, for $x \in [5.5, 6.5]$, $f^A(x) = 0.5 + 2\ x$. Between these regions (e.g., in [2.5, 5.5]), Takagi–Sugeno method uses both linear functions (in this case, $f_1(x) = 1.25$ and $f_2(x) = 0.5 + 2\ x$) and computes their weighted average with weights equal to β_1 and β_2 from (3.70). We can see that the function f^A does not look particularly smooth; possibilities of how it can be amended are discussed in [4]. □

Similarly as for the fuzzy approximation methods described above, approximation using Takagi–Sugeno rules also possesses the property of the functional completeness.

Theorem 3.5 *Let $U = [a, b] \subseteq \mathbb{R}$ be an interval of reals (a compact set). Then to any bounded continuous function $f : U \longrightarrow \mathbb{R}$ and $\epsilon > 0$, there exists a system of TS rules (3.68) such that*

$$|f(u) - f^A(u)| < \epsilon,$$

where $f^A(u)$ is function (3.69).

Details and the proof can be found, for example, in [15].

The following theorem provides estimation of the approximation error of TS rules under special conditions. Let us assume that fuzzy rules are formulated in such a way that membership functions of the fuzzy sets on the left side of rules (3.68) are isosceles triangles with vertices in equidistant points $u_1, \ldots, u_m$ covering a (compact) set $U \subset \mathbb{R}$.

Theorem 3.6 *([106]) Let $f(u)$ be a two times differentiable function on $U = [a, b]$ and*

$$q_2 = \max_{x \in U} \{|f''(x)|\}.$$

Furthermore, let $\epsilon > 0$ be a given precision and m be given in such a way that the value $h = \frac{(b-a)}{m}$ is not greater than $\sqrt{\frac{8\epsilon}{q_2}}$. For any $u_j = a + jh$, $0 \le j \le m$, let us choose constants b_{j0}, b_{j1} in (3.64) in such a way that

$$b_{j0} = f'(u_j), \quad b_{j1} = f(u_j) - u_j f'(u_j).$$

Then the function f^A (3.69) approximates the function f with precision given by the right side of the inequality

$$|f(x) - f^A(x)| \le \epsilon + h \max_{0 \le j \le m-1} \left\{ \frac{|f'(u_{j+1}) - f'(u_j)|}{4} \right\}, \qquad x \in U.$$

More details about fuzzy approximation and further results can be found, for example, in [99, 106, 107].

3.4.3 Identification of TS Rules

In this section, we describe how the system of TS rules can be identified, that is, how to determine

(i) the fuzzy sets $A_{j1}, \ldots, A_{jn}$ for all $j = 1, \ldots, m$.
(ii) the parameters $b_{j0}, \ldots, b_{jn}$, $j = 1, \ldots, m$.

The identification is realized on the basis of data that have the following structure:

$$\begin{aligned}&\langle u_{11},\ldots,u_{1n},v_1\rangle,\\&\ldots\ldots\ldots\ldots\ldots\ldots\\&\langle u_{r1},\ldots,u_{rn},v_r\rangle.\end{aligned} \tag{3.71}$$

The values $u_{k1} \in U_1, \ldots, u_{kn} \in U_n$, $k = 1,\ldots,r$, are values of the independent variables $X_1,\ldots,X_n$ and $v_k \in V$ are values of the dependent variable Y. The r is the number of objects (lines). To obtain reliable results, it should be a sufficiently large number (from hundreds to thousands or more).

The algorithm for generation of TS rules can be described as follows:

1. For each variable $X_1,\ldots,X_n$, we must construct the fuzzy sets $A_{j1} \lessapprox U_1, \ldots, A_{jn} \lessapprox U_n$, $j = 1,\ldots,m$. There are several ways of how to do it:
   (a) Determine them by an expert.
   (b) Form a certain kind of fuzzy partition of each universe U_i.
   (c) Find these fuzzy sets using fuzzy cluster analysis whose methods will be discussed in Chapter 6.
2. To determine the parameters $b_{j0},\ldots,b_{jn}$, $j = 1,\ldots,m$, minimize the expression

$$S = \sum_{k=1}^{r}\left(v_k - \sum_{j=1}^{m}\beta_j(u_{k1},\ldots,u_{kn})\cdot(b_{j0} + b_{j1}u_{k1} + \cdots + b_{jn}u_{kn})\right)^2. \tag{3.72}$$

In case of one independent variable only (i.e., $n = 1$), this formula is simplified to

$$S = \sum_{k=1}^{r}\left(v_k - \sum_{j=1}^{m}\beta_j(u_k)(b_{j0} + b_{j1}u_k)\right)^2. \tag{3.73}$$

The $\langle u_{k1},\ldots,u_{kn},v_k\rangle$, $k = 1,\ldots,r$ are data (3.71) and $\beta_j(u_{k1},\ldots,u_{kn})$ are coefficients (3.65) (or (3.70)). One can see that it is a slight modification of the classical least squares method, where $\beta_j(u_{k1},\ldots,u_{kn})$ play the role of weights. If more complicated model with several variables and large number m of rules is given, then the precise solution of (3.72) is impossible and we must find the coefficients $\beta_j(u_{k1},\ldots,u_{kn})$ using iterative methods.

In case when n equals to 1, the result of solving (3.73) is the following:

$$b_{j1} = \frac{\sum_{k=1}^{r}\beta_j(u_k)^2\cdot\sum_{k=1}^{r}\beta_j(u_k)u_kv_k - \sum_{k=1}^{r}\beta_j(u_k)^2u_k\cdot\sum_{k=1}^{r}\beta_j(u_k)v_k}{\sum_{k=1}^{r}\beta_j(u_k)^2\ \sum_{k=1}^{r}(\beta_j(u_k)u_k)^2 - (\sum_{k=1}^{r}\beta_j(u_k)^2u_k)^2}, \tag{3.74}$$

$$b_{j0} = \frac{\sum_{k=1}^{r}\beta_j(u_k)v_k - b_{j1}\sum_{k=1}^{r}\beta_j(u_k)^2u_k}{\sum_{k=1}^{r}\beta_j(u_k)^2}. \tag{3.75}$$

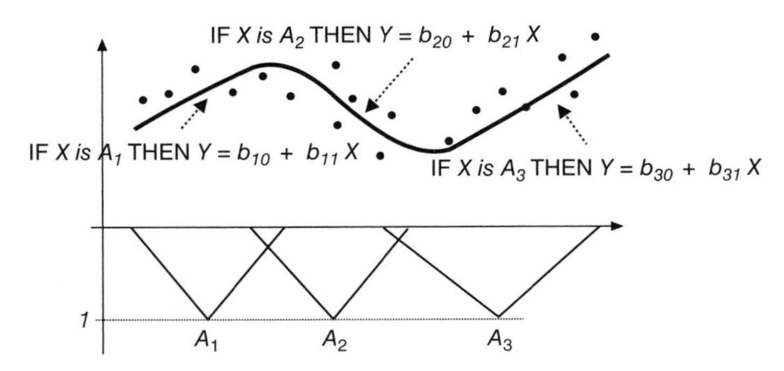

Figure 3.15 A schematic depiction of the fuzzy approximating function f^A using TS rules with $n = 1$.

The approximating function f^A for $n = 1$ is schematically depicted in Figure 3.15.

Example of derivation of Takagi–Sugeno rules and the approximating function f^A can be found in Example 6.3.

Takagi and Sugeno used a linear dependence in their original proposal. It is clear that this can be generalized, that is, rules (3.61) can generally have the following form:

$$\mathscr{R} := \text{IF } X_1 \text{ is } \mathscr{A}_1 \text{ AND } \cdots \text{ AND } X_n \text{ is } \mathscr{A}_n \text{ THEN}$$

$$Y = f(X_1, \ldots, X_n, b_0, \ldots, b_p), \tag{3.76}$$

where $b_0, \ldots, b_p$ are parameters. It is questionable, however, whether a nonlinear form of the functions f can be justified, because the resulting dependence characterized by means of fuzzy TS rules is nonlinear, too. TS rules can be generalized further, we can, for example, consider coefficients $b_0, \ldots, b_p$ to be fuzzy numbers, etc.

The Takagi–Sugeno rules were for the first time used in solving the problem of a car parking. From that time, TS rules have been widespread and are used in special problems of fuzzy modeling, when it is necessary to construct a model of the studied process. It is sometimes called *fuzzy regression analysis*. TS rules are useful if we have measurements and partial knowledge of the process in certain inexactly specified regions at disposal. The TS rules can also be effectively used in the predictive fuzzy control. This means that we model the controlled process by a fuzzy model. Then we derive control action on the basis of reaction on the future behavior of the controlled process. The future behavior is computed using a fuzzy model.

4

FUZZY TRANSFORM

In this chapter, we describe a very powerful and relatively simple technique called *fuzzy transform* (*F-transform* for short). Its author is Irina Perfilieva who published it in detail in [111].

The F-transform encompasses both classical (usually, integral) transforms as well as fuzzy approximation models (usually based on the technique of fuzzy IF-THEN rules). Recall from Chapter 1 that the core idea is to transform an infinitary object that is a real bounded function $f\colon [a,b] \longrightarrow [c,d]$ into a finite vector of components. This phase is called *direct F-transform*. The *inverse F-transform* transforms the latter finite vector into a function $\hat{f}$ that approximates the original function f. By a proper setting of parameters, we can obtain the function $\hat{f}$ with desired properties.

The F-transform has a lot of various interesting applications. It is applied in time series analysis, signal processing, numerical solution of differential equations, image processing, computer vision, and elsewhere. We will present some of these applications in Chapters 8 and 9.

4.1 FUZZY PARTITION

The starting point of the F-transform is the definition of a *fuzzy partition*. Recall that this concept has been introduced already in Section 2.2.4. We will redefine it in this section from the point of view of the F-transform because it plays an essential role in this technique. Recall that a fuzzy partition is a finite collection of fuzzy subsets of the universe that determines a discrete kernel and thus also the corresponding transform. Therefore, we have as many F-transforms as fuzzy partitions. Throughout this section, we deal with an interval $[a,b] \subset \mathbb{R}$ of real numbers.

Insight into Fuzzy Modeling, First Edition. Vilém Novák, Irina Perfilieva, and Antonín Dvořák.

Companion Website: www.wiley.com/go/novak/fuzzy/modeling

The following concept of fuzzy partition was introduced in [111].

Definition 4.1 *Let $c_0 < \cdots < c_n$ be fixed nodes within $[a, b]$ such that $c_0 = a$, $c_n = b$ and $n \geq 2$. We say that fuzzy sets $A_0, \dots, A_n \subsetneq [a, b]$ establish a fuzzy partition of $[a, b]$ if they fulfill the following conditions for $k = 0, \dots, n$:*

1. *$A_k\colon [a, b] \longrightarrow [0, 1]$, $A_k(c_k) = 1$;*
2. *$A_k(x) = 0$ if $x \notin (c_{k-1}, c_{k+1})$ (for uniformity of notation we set $c_{-1} = a$ and $c_{n+1} = b$);*
3. *A_k is continuous;*
4. *For $k = 1, \dots, n$, A_k strictly increases on $[c_{k-1}, c_k]$, and for $k = 0, \dots, n-1$, A_k strictly decreases on $[c_k, c_{k+1}]$;*
5. *For all $x \in [a, b]$,*

$$\sum_{k=0}^{n} A_k(x) = 1. \tag{4.1}$$

The fuzzy sets $A_0, \dots, A_n$ are often called *basic functions*. We will equivalently speak either about fuzzy sets or about basic functions.

Condition (4.1) is called the *Ruspini condition*. It is quite strong because it forces the basic functions to overlap in a specific way. We can understand it also as a certain kind of orthogonality between them.

We say that a fuzzy partition of $[a, b]$ is *h-uniform* if its nodes $c_0, \dots, c_n$, where $n \geq 2$, are equidistant. This means that $c_k = a + hk$, where $h = (b - a)/n$, $k = 0, \dots, n$, and the two additional properties are fulfilled:

6. $A_k(c_k - x) = A_k(c_k + x)$, for all $x \in [0, h]$, $k = 1, \dots, n-1$,
7. For all $k = 1, \dots, n-1$ and $x \in [c_k, c_{k+1}]$,

$$A_k(x) = A_{k-1}(x - h),$$
$$A_{k+1}(x) = A_k(x - h).$$

A point $x \in [a, b]$ is *covered* by the fuzzy set A_k if $A_k(x) > 0$. Note that the shape of the basic functions (fuzzy sets) is not predetermined. Therefore, it can be chosen according to additional requirements (e.g., smoothness).

EXAMPLE 4.1 The triangular nonuniform fuzzy partition is determined by the following formulas:

$$A_0(x) = \begin{cases} 1 - \dfrac{(x - c_0)}{h_0}, & x \in [c_0, c_1], \\ 0, & \text{otherwise}, \end{cases} \tag{4.2}$$

$$A_k(x) = \begin{cases} \dfrac{(x - c_{k-1})}{h_{k-1}}, & x \in [c_{k-1}, c_k], \\ 1 - \dfrac{(x - c_k)}{h_k}, & x \in [c_k, c_{k+1}], \\ 0, & \text{otherwise}, \end{cases} \tag{4.3}$$

$$A_n(x) = \begin{cases} \frac{(x - c_{n-1})}{h_{n-1}}, & x \in [c_{n-1}, c_n], \\ 0, & \text{otherwise}, \end{cases} \tag{4.4}$$

where $k = 1, \dots, n-1$ and $h_k = c_{k+1} - c_k$. To make this fuzzy partition h-uniform, it is sufficient to put $h_0 = \cdots = h_{n-1} = h$. This fuzzy partition is depicted in Figure 4.1.

The cosine h-uniform fuzzy partition is determined by the following formulas:

$$A_0(x) = \begin{cases} 0.5\left(\cos\frac{\pi}{h}(x - c_0) + 1\right), & x \in [c_0, c_1], \\ 0, & \text{otherwise}, \end{cases} \tag{4.5}$$

$$A_k(x) = \begin{cases} 0.5\left(\cos\frac{\pi}{h}(x - c_k) + 1\right), & x \in [c_{k-1}, c_{k+1}], \\ 0, & \text{otherwise}, \end{cases} \tag{4.6}$$

$$A_n(x) = \begin{cases} 0.5\left(\cos\frac{\pi}{h}(x - c_n) + 1\right), & x \in [c_{n-1}, c_n], \\ 0, & \text{otherwise}, \end{cases} \tag{4.7}$$

where $k = 1, \dots n-1$. This fuzzy partition is depicted in Figure 4.2. □

Fuzzy partitions can be generalized in two ways: first, we can replace the number 1 in the Ruspini condition (4.1) by an arbitrary integer number. This leads to denser fuzzy partitions, as can be seen in Figure 4.3. Further generalization was introduced in [114] in connection with the notion of the higher-degree F-transform. Besides relaxing some of the conditions above, a generalized fuzzy partition is constructed using a generating function A that is shifted along the universe. More details can be found in [51, 52, 133].

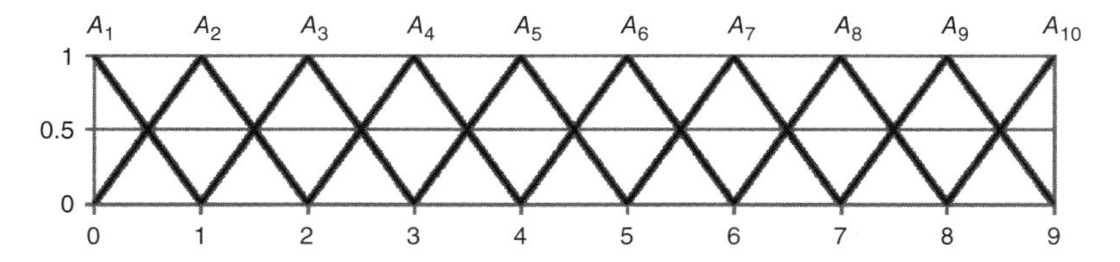

Figure 4.1 Example of the 1-uniform triangular fuzzy partition.

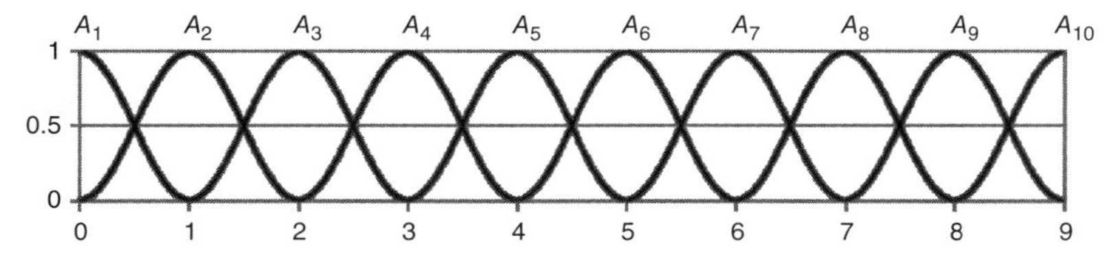

Figure 4.2 Example of 1-uniform cosine fuzzy partition.

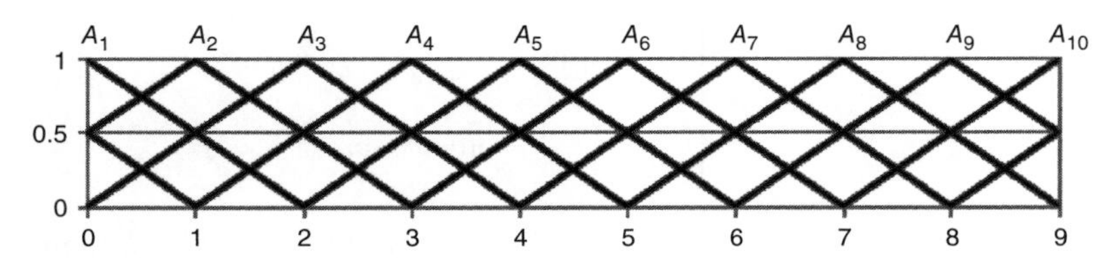

Figure 4.3 Example of a fuzzy 2-partition with triangular basic functions.

4.2 THE CONCEPT OF F-TRANSFORM

As mentioned above, the F-transform establishes a correspondence between a set of continuous functions on an interval of real numbers and a set of $(n + 1)$-dimensional vectors, where n is a finite number. Each component of the resulting vector is obtained as a weighted local mean of a given function over an area covered by a corresponding basic function. The vector of the F-transform components can be construed as a simplified representation of the original function. Because of its simplicity, we can often use it in applications more effectively than the original function. The F-transform has two phases: direct and inverse. In this section, we explain the basic principles and properties of the F-transform. Proofs and more results can be found in [111].

4.2.1 Direct F-Transform

We assume that $c_0 < \cdots < c_n$ are fixed nodes from $[a, b]$ such that $c_0 = a$, $c_n = b$, and $n \geq 2$. We also formally extend the set of nodes by $c_{-1} = a$ and $c_{n+1} = b$.

Definition 4.2 *Let $A_0, \ldots, A_n$ be a fuzzy partition of $[a, b]$ and $f : [a, b] \longrightarrow [c, d]$ be a continuous function on $[a, b]$. We say that the $(n + 1)$-tuple of real numbers $\mathbf{F}[f] = (F_0[f], \ldots, F_n[f])$ given by*

$$F_k[f] = \frac{\int_a^b f(x)\, A_k(x)dx}{\int_a^b A_k(x)dx}, \qquad k = 0, \ldots, n, \tag{4.8}$$

is the (integral) F-transform of f with respect to $A_0, \ldots, A_n$.

The elements $F_0[f], \ldots, F_n[f]$ are called *components of the F-transform.*

If $A_0, \ldots, A_n$ forms an h-uniform fuzzy partition, then the expression (4.8) can be simplified as follows:

$$\begin{aligned} F_0[f] &= \frac{2}{h} \int_{c_0}^{c_1} f(x)\, A_1(x)dx, \\ F_n[f] &= \frac{2}{h} \int_{c_{n-1}}^{c_n} f(x)\, A_n(x)dx, \\ F_k[f] &= \frac{1}{h} \int_{c_{k-1}}^{c_{k+1}} f(x)\, A_k(x)dx, \quad k = 1, \ldots, n-1. \end{aligned} \tag{4.9}$$

Remark 4.1 *By direct F-transform, we will understand both the technique as well as the vector* $\mathbf{F}[f]$ *for the given function* f.

Properties of direct F-transform:

(a) If for all $x \in [a,b], f(x) = C$, where $C \in \mathbb{R}$ (i.e., f is a constant function), then $F_k[f] = C$ for all $k = 0, \ldots, n$;

(b) If $f = \alpha_1 g_1 + \alpha_2 g_2$, then $\mathbf{F}[f] = \alpha_1 \mathbf{F}[g_1] + \alpha_2 \mathbf{F}[g_2]$ (linearity);

(c) Each k-th component, $k = 0, \ldots, n$, minimizes the function

$$\Phi(y) = \int_a^b (f(x) - y)^2 A_k(x) dx$$

defined on $[f(a), f(b)]$;

(d) For each $k = 1, \ldots, n-1$ and $t \in [c_k, c_{k+1}]$,

$$|f(t) - F_k[f]| \leq 2\omega(h,f) \quad \text{and} \quad |f(t) - F_{k+1}[f]| \leq 2\omega(h,f),$$

where

$$\omega(h,f) = \max_{|\delta| \leq h} \max_{x \in [a,b-\delta]} |f(x+\delta) - f(x)| \tag{4.10}$$

is the modulus of continuity of f;

(e) If $A_0, \ldots, A_n$ form an h-uniform fuzzy partition on $[a,b]$, then

$$\int_a^b f(x)dx = h\left(\frac{F_0[f]}{2} + \frac{F_n[f]}{2} + \sum_{k=1}^{n-1} F_k[f]\right).$$

4.2.2 Inverse F-Transform

The inverse F-transform is a procedure how to reconstruct the original function f from the vector $\mathbf{F}[f]$. It is clear, however, that if f is nonconstant, then it cannot be reconstructed precisely because we lose information when passing from f to $\mathbf{F}[f]$. Despite this, the inverse F-transform $\hat{f}$ approximates f in such a way that a universal convergence result can be established.

Definition 4.3 *Let* $A_0, \ldots, A_n$ *be a fuzzy partition of* $[a,b]$ *and* $\mathbf{F}[f] = (F_0, \ldots, F_n)$ *be the F-transform of* f *with respect to* $A_0, \ldots, A_n$. *Then, the function* $\hat{f} : [a,b] \longrightarrow \mathbb{R}$ *given by*

$$\hat{f}(x) = \sum_{k=0}^{n} F_k[f] \, A_k(x) \tag{4.11}$$

is called the inverse F-transform.

Both direct as well as inverse F-transform of a given function f are schematically depicted in Figure 4.4.

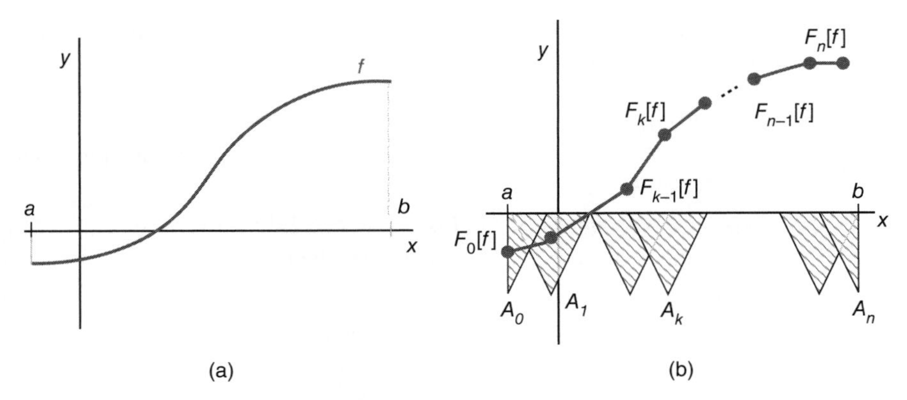

Figure 4.4 Scheme of the F-transform. (a) The original function f. (b) Components of the direct F-transform $\mathbf{F}[f]$ depicted as big full circles and also the inverse F-transform $\hat{f}$, which is depicted by a broken line passing through the components. The fuzzy partition is depicted under the x-axis.

Remark 4.2 *Similarly as above, by inverse F-transform, we will understand both the technique as well as the function* $\hat{f}$.

The most important property of the inverse F-transform $\hat{f}$ is its ability to approximate any continuous function f with an arbitrary precision. This is formulated in the following theorem, which we present including its proof.

Theorem 4.4 *Let f be a continuous function on $[a, b]$. Then, for any $\epsilon > 0$, there exist n_ϵ and a fuzzy partition $A_0, \ldots, A_{n_\epsilon}$ of $[a, b]$ such that for all $x \in [a, b]$,*

$$|f(x) - \hat{f}_\epsilon(x)| \leq \epsilon, \tag{4.12}$$

where $\hat{f}_\epsilon$ is the inverse F-transform of f with respect to the fuzzy partition $A_0, \ldots, A_{n_\epsilon}$.

Proof: Note that the function f is uniformly continuous on $[a, b]$, that is, for each $\epsilon > 0$, there exists $\delta = \delta(\epsilon) > 0$ such that for all $x', x'' \in [a, b]$, $|x' - x''| < \delta$ implies $|f(x') - f(x'')| < \epsilon$. To prove our theorem, we choose some $\epsilon > 0$ and find nodes $c_0, \ldots, c_n \in [a, b]$ such that $a = c_0 < \cdots < c_n = b$ and $|f(x') - f(x'')| < \epsilon$ whenever $x', x'' \in [c_{k-1}, c_{k+1}], k = 1, \ldots, n-1$. Let us put $n = n_\epsilon$ and take any fuzzy partition determined by the chosen nodes and basic functions $A_0, \ldots, A_n$. To complete the proof, we must verify (4.12). Let $F_0[f], \ldots, F_n[f]$ be components of the F-transform of f with respect to the basic functions $A_0, \ldots, A_n$. Then for each $t \in [c_k, c_{k+1}]$, $k = 0, \ldots, n-1$, we evaluate

$$|f(t) - F_k[f]| = \left| f(t) - \frac{\int_{c_{k-1}}^{c_{k+1}} f(x)A_k(x)dx}{\int_{c_{k-1}}^{c_{k+1}} A_k(x)dx} \right| \leq$$

$$\leq \frac{\int_{c_{k-1}}^{c_{k+1}} |f(t) - f(x)|A_k(x)dx}{\int_{c_{k-1}}^{c_{k+1}} A_k(x)dx} \leq \epsilon$$

and, analogously, we also obtain $|f(t) - F_{k+1}[f]| \leq \epsilon$, where for the uniformity of notation, we put $c_{-1} = a$ and $c_{n+1} = b$. Therefore, having in mind (4.1), we obtain

$$\left|f(t) - \sum_{i=0}^{n} F_i[f]\, A_i(t)\right| = \left|f(t) \sum_{i=0}^{n} A_i(t) - \sum_{i=0}^{n} F_i[f]\, A_i(t)\right| \leq$$

$$\leq \sum_{i=0}^{n} A_i(t)|f(t) - F_i[f]| = \sum_{i=k}^{k+1} A_i(t)|f(t) - F_i[f]| \leq \epsilon \sum_{i=k}^{k+1} A_i(t) =$$

$$= \epsilon \sum_{i=0}^{n} A_i(t) = \epsilon.$$

Because the argument t has been chosen arbitrarily within the interval $[a, b]$, this proves the inequality (4.12). ■

EXAMPLE 4.2 In Figure 4.5, we illustrate how the inverse F-transform approximates the function $f = \sin(15x)\, x^2$. The approximating function $\hat{f}$ is piecewise linear because the fuzzy partition is formed by triangular basic functions. The figure illustrates the uniform approximation property depending on the number of basic functions. □

If we do not require special properties of $\hat{f}$, then it is sufficient to compute the F-transform with respect to the simplest triangular fuzzy partition. The precise statement is in the following theorem.

Theorem 4.5 *Let f be a continuous function on $[a, b]$, and let $A'_0, \dots, A'_n$ and $A''_0, \dots, A''_n$ $(n \geq 3)$ be two sets of basic functions forming two h-uniform fuzzy partitions of $[a, b]$. Let $\hat{f}'$ and $\hat{f}''$ be the two corresponding inverse F-transforms of f, respectively. Then, for arbitrary $x \in [a, b]$,*

$$|\hat{f}'(x) - \hat{f}''(x)| \leq 2\omega(h, f),$$

where $h = \frac{b-a}{n}$ and $\omega(h, f)$ is the modulus of continuity (4.10) of f on the interval $[a, b]$.

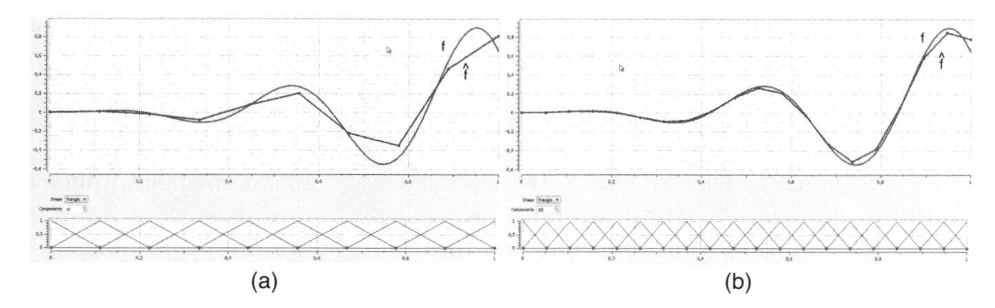

Figure 4.5 The function $f = \sin(15x)\, x^2$ defined on the interval $[0, 1]$ and its inverse F-transform. The fuzzy partition is uniform and formed by triangular-shaped basic functions. (a) contains 10 and (b) contains 20 basic functions. The F-transform components are marked by small circles.

Properties of inverse F-transform:

(a) If $f(x)$ is equal to C for all $x \in [a, b]$ (i.e., f is a constant function), then the inverse F-transform $\hat{f}(x)$ is equal to C.

(b) If $f = \alpha_1 g_1 + \alpha_2 g_2$, then $\hat{f} = \alpha_1 \hat{g_1} + \alpha_2 \hat{g_2}$ (linearity).

(c)

$$\int_a^b f(x)dx = \int_a^b \hat{f}(x)dx.$$

(d) If the basic functions $A_0, \ldots, A_n$ are triangular, then $\hat{f}$ is piecewise linear. If they are cosine or have other nonlinear shape, then $\hat{f}$ is piecewise nonlinear.

The property (b) says that the inverse F-transform is linear. This means that if we transform a function that is obtained as a sum of two other functions (possibly multiplied by some constants), then its inverse F-transform is equal to the sum of inverse F-transform of the respective summands (possibly multiplied by some constants). The property (c) says that the integral of the original function is equal to the integral of its inverse F-transform. Both properties are very important for applications. The details and proofs of the theorems and the above listed properties can be found in [111].

Concerning property (d), in practice, the triangular shape of the basic functions is sufficient. If the approximation of f by $\hat{f}$ is sufficiently tight, then we hardly see any difference between piecewise linear and nonlinear course of the latter.

4.3 DISCRETE F-TRANSFORM

The original function f is in practice usually defined on a finite set

$$P = \{x_1, \ldots, x_p\} \subseteq [a, b]$$

only. Therefore, in computation of the F-transform, we must replace integrals by sums and we speak about *discrete* F-transform. The basic properties, of course, remain the same as in the case of the integral one.

A special condition not necessary for the integral transform is the following.

Definition 4.6 *The domain P of the function f is* sufficiently dense *with respect to the fixed partition if for each fuzzy set A_k, $k = 0, \ldots, n$, from the fuzzy partition, there is an element $x_j \in P$ belonging to the support of A_k; formally, we write this condition as follows:*

$$(\forall k)(\exists j)A_k(x_j) > 0.$$

This definition means that each basic function covers at least one point from the domain P.

The following two definitions introduce the discrete F-transform. The formulas given below can be directly used in applications and easily implemented in computer program.

Definition 4.7 *Let $A_0, \ldots, A_n$ be a fuzzy partition of $[a, b]$ and a function f be defined on the set $P = \{x_1, \ldots, x_p\} \subseteq [a, b]$ that is sufficiently dense with respect to $A_0, \ldots, A_n$.*

We say that the $(n+1)$*-tuple of real numbers* $\mathbf{F}[f] = (F_0[f], \ldots, F_n[f])$ *is a discrete F-transform of* f *with respect to* $A_0, \ldots, A_n$ *if*

$$F_k[f] = \frac{\sum_{j=1}^{p} f(x_j)A_k(x_j)}{\sum_{j=1}^{p} A_k(x_j)}. \tag{4.13}$$

It is not difficult to demonstrate that components of the discrete F-transform have similar properties as those listed on page 85.

The inverse discrete F-transform is defined in the same way as the integral one.

Definition 4.8 *Let* $A_0, \ldots, A_n$ *be a fuzzy partition of* $[a, b]$ *and a function* f *be defined on the set* $P = \{x_1, \ldots, x_p\} \subseteq [a, b]$ *that is sufficiently dense with respect to the above partition. Moreover, let* $\mathbf{F}[f] = (F_0[f], \ldots, F_n[f])$ *be the discrete F-transform of* f *with respect to* $A_0, \ldots, A_n$*. Then, the function* $\hat{f} : P \longrightarrow \mathbb{R}$ *given by*

$$\hat{f}(x_j) = \sum_{k=0}^{n} F_k[f]\, A_k(x_j), \qquad j = 1, \ldots, p \tag{4.14}$$

is the inverse discrete F-transform *of* f.

Analogously to Theorem 4.4, we can show that the inverse discrete F-transform $\hat{f}$ can approximate the original discrete function f on P with an arbitrary precision.

Remark 4.3 *An interesting comparison between the discrete F-transform and the least-square approximation was presented in paper [105]. It was demonstrated there that the discrete F-transform is invariant with respect to interpolating and least-squares approximation of the set* $\{(x_j, f(x_j)) \mid j = 1, \ldots, p\}$*. This means that the best approximation of* f *on* P *in the form of* $\sum_{i=0}^{n} \alpha_i A_i$*, where* $n \leq p$*, has the same direct discrete F-transform as the original* f.

4.4 F-TRANSFORM OF FUNCTIONS OF TWO VARIABLES

The direct and inverse F-transform of a function of two (and more) variables is a direct generalization of the case of one variable. We introduce it briefly, and refer to [111] for more details.

Continuous F-transform. Suppose that the universe is a rectangle $[a, b] \times [c, d] \subseteq \mathbb{R} \times \mathbb{R}$ and that $c_0 < \cdots < c_n$ are fixed nodes of $[a, b]$ and $e_0 < \cdots < e_m$ are fixed nodes of $[c, d]$ such that $c_0 = a$, $c_n = b$, $e_0 = c$, $e_m = d$, and $n, m \geq 2$. Let us formally extend the set of nodes by setting $c_{-1} = a$, $e_{-1} = c$, $c_{n+1} = b$, and $e_{m+1} = d$. Assume that $A_0, \ldots, A_n$ are basic functions that form a fuzzy partition of $[a, b]$ and $B_0, \ldots, B_m$ are basic functions that form a fuzzy partition of $[c, d]$. Then, the rectangle $[a, b] \times [c, d]$ is partitioned into fuzzy sets $(A_k \dot{\times} B_l)$ with the membership functions

$$\left(A_k \dot{\times} B_l\right)(x, y) = A_k(x) \cdot B_l(y),$$

$k = 0, \ldots, n$, $l = 0, \ldots, m$.

Definition 4.9 *Let $A_0, \ldots, A_n \subsetneqq [a,b]$ and $B_0, \ldots, B_m \subsetneqq [c,d]$ be fuzzy sets using which a fuzzy partition of $[a,b] \times [c,d]$ introduced above is formed. Let $f : ([a,b] \times [c,d]) \longrightarrow \mathbb{R}$ be a continuous function. We say that the $((n+1) \times (m+1))$-matrix of real numbers*

$$\mathbf{F}[f] = \begin{pmatrix} F_{00}[f] & \cdots & F_{0m}[f] \\ \vdots & \vdots & \vdots \\ F_{n0}[f] & \cdots & F_{nm}[f] \end{pmatrix}$$

is the (integral) F-transform of f with respect to $A_0, \ldots, A_n$ and $B_0, \ldots, B_m$ if for each $k = 0, \ldots, n$, $l = 0, \ldots, m$,

$$F_{kl}[f] = \frac{\int_c^d \int_a^b f(x,y)\, A_k(x)\, B_l(y) dx dy}{\int_c^d \int_a^b A_k(x)\, B_l(y) dx dy}. \tag{4.15}$$

In Figure 4.6, the F-transform of two variables is schematically depicted. Figure 4.6(a) depicts a two-dimensional fuzzy partition, and Figure 4.6(b) schematically depicts one of the components.

The components $F_{kl}[f]$ (4.15) have properties (adapted to the case of two variables) similar to those listed on page 85. For example, the property (e) has the following form (we assume that $A_0, \ldots, A_n$ form an h_1-uniform fuzzy partition of $[a,b]$ and $B_0, \ldots, B_m$ form an h_2-uniform fuzzy partition of $[c,d]$):

$$\begin{aligned}\int_c^d \int_a^b f(x,y) dx dy &= \frac{h_1 h_2}{4}(F_{00}[f] + F_{0m}[f] + F_{n0}[f] + F_{nm}[f]) + \\ &+ \frac{h_1 h_2}{2}\left(\sum_{k=1}^{n-1} F_{k1}[f] + \sum_{k=1}^{n-1} F_{km}[f] + \sum_{l=1}^{m-1} F_{1l}[f] + \sum_{l=1}^{m-1} F_{nl}[f]\right) + \\ &+ h_1 h_2 \sum_{k=1}^{n-1} \sum_{l=1}^{m-1} F_{kl}[f].\end{aligned}$$

Discrete F-transform. The discrete case of the F-transform of two variables is analogous to that of one variable. Let us consider two sets $P = \{x_1, \ldots, x_N\} \subset [a,b]$ and $Q = \{y_1, \ldots, y_M\} \subset [c,d]$ and a function $f : P \times Q \longrightarrow \mathbb{R}$. Note that this case is important in applications of the F-transform to image processing (see Chapter 8).

Definition 4.10 *Let $A_0, \ldots, A_n \subsetneqq [a,b]$ and $B_0, \ldots, B_m \subsetneqq [c,d]$ be fuzzy partitions using which a fuzzy partition of $[a,b] \times [c,d]$ is constructed and sets P and Q be sufficiently dense with respect to them. Let a function $f : P \times Q \longrightarrow \mathbb{R}$ be given. We say that the $((n+1) \times (m+1))$-matrix of real numbers $\mathbf{F}[f] = (F_{kl}[f])_{nm}$ is the discrete F-transform of f with respect to $A_0, \ldots, A_n$ and $B_0, \ldots, B_m$ if*

$$F_{kl}[f] = \frac{\sum_{j=1}^{M} \sum_{i=1}^{N} f(x_i, y_j) A_k(x_i) B_l(y_j)}{\sum_{j=1}^{M} \sum_{i=1}^{N} A_k(x_i) B_l(y_j)} \tag{4.16}$$

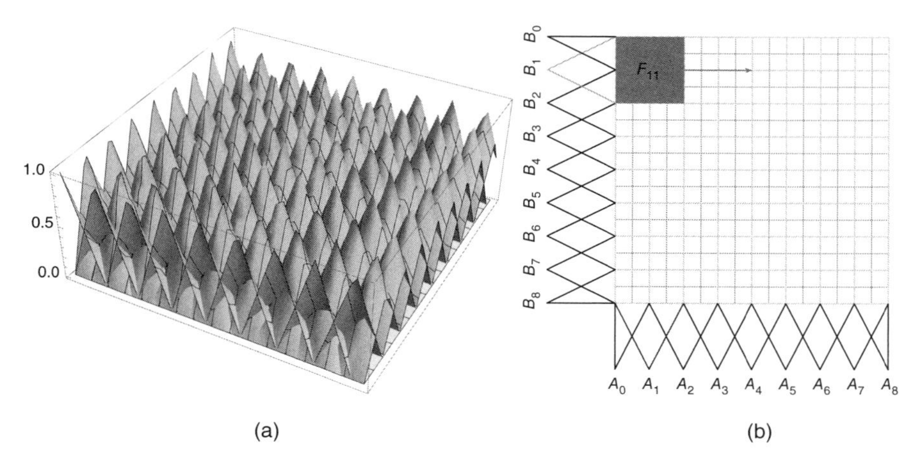

Figure 4.6 Scheme of the F-transform of two variables. (a) A fuzzy partition and (b) one of the components with respect to the fuzzy partition constructed over both variables.

holds for all $k = 0, \ldots, n$, $l = 0, \ldots, m$.

The inverse F-transform of a function of two variables is a simple extension of (4.11). It will be given below for the continuous version of F-transform.

Definition 4.11 *Let* $A_0, \ldots, A_n \subsetneq [a,b]$ *and* $B_0, \ldots, B_m \subsetneq [c,d]$ *be fuzzy partitions. Let* $\mathbf{F}[f]$ *be the F-transform of* $f : [a,b] \times [c,d] \longrightarrow \mathbb{R}$ *with respect to* $A_0, \ldots, A_n$ *and* $B_0, \ldots, B_m$. *Then, the function* $\hat{f} : [a,b] \times [c,d] \longrightarrow \mathbb{R}$ *defined as*

$$\hat{f}(x,y) = \sum_{k=0}^{n} \sum_{l=0}^{m} F_{kl}[f]\, A_k(x) B_l(y) \tag{4.17}$$

is called the inverse F-transform *of* f.

Analogously to the case of a function of one variable, we can prove that the inverse F-transform $\hat{f}$ can approximate the original continuous function f with an arbitrary precision, and the (adapted) properties (a)–(c), which are listed on page 88, are fulfilled.

4.5 F¹-TRANSFORM

The F-transform presented above can be further generalized, namely, we can introduce higher-degree F-transform (F^m-transform for $m \geq 0$). Its components are polynomials of degree m. Hence, the original F-transform described in the previous sections coincides with the F^0-transform.

In this section, we describe only the F^1-transform. Detailed definitions, theorems, and their proofs for higher-degree F-transforms can be found in [113].

In the same way as above, we will fix an interval $[a,b]$ of reals and consider a continuous function $f : [a,b] \longrightarrow \mathbb{R}$. Moreover, we will consider a set of nodes $\{a =$

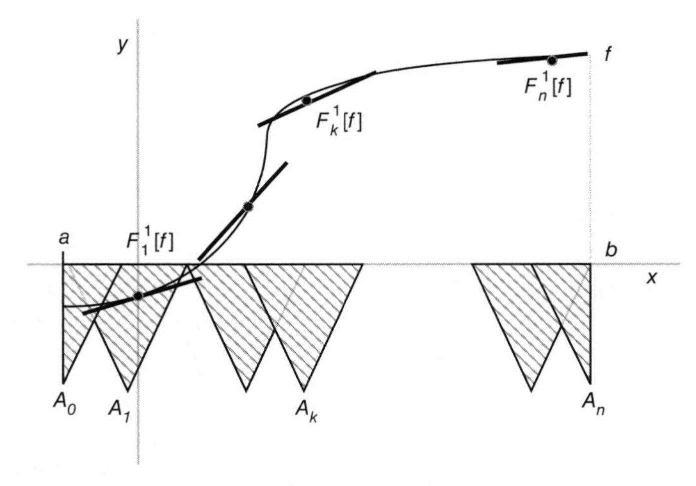

Figure 4.7 Scheme of the direct F^1-transform $\mathbf{F}^1[f]$. The components of the direct F-transform are schematically depicted as the corresponding lines. The fuzzy partition is depicted under the x-axis.

$c_0, \ldots, c_n = b\} \subset [a, b]$, $n \geq 2$, and an h-uniform fuzzy partition defined over these nodes.

It is important to emphasize that in the sequel, we will deal with "full" fuzzy sets $A_1, \ldots, A_{n-1} \subsetneqq [a, b]$ only. This means that since the fuzzy sets A_0 and A_n are only halves in comparison to the other ones, they are omitted from our considerations.

Definition 4.12 *The vector of linear functions*

$$\mathbf{F}^1[f] = (\beta_1^0[f] + \beta_1^1[f] \cdot (x - c_1), \ldots, \beta_{n-1}^0[f] + \beta_{n-1}^1[f] \cdot (x - c_{n-1})) \tag{4.18}$$

is the F^1-transform of f with respect to the h-uniform fuzzy partition $A_1, \ldots, A_{n-1}$, where

$$\beta_k^0[f] = \frac{\int_{c_{k-1}}^{c_{k+1}} f(x) A_k(x) dx}{h}, \tag{4.19}$$

$$\beta_k^1[f] = \frac{\int_{c_{k-1}}^{c_{k+1}} f(x)(x - c_k) A_k(x) dx}{\int_{c_{k-1}}^{c_{k+1}} (x - c_k)^2 A_k(x) dx} \tag{4.20}$$

for every $k = 1, \ldots, n - 1$.

Note that $\beta_k^0[f] = F_k^0[f]$ is, in fact, the component (4.8). Note also that each component $F_k^1[f](x)$ in (4.18) is a function of x. The direct F^1-transform of a given function f is schematically depicted in Figure 4.7.

It can be proved that if the partition $A_1, \ldots, A_{n-1}$ is uniform and the basic functions A_k have triangular shape, then (4.20) becomes

$$\beta_k^1[f] = \frac{12 \int_{c_{k-1}}^{c_{k+1}} f(x)(x - c_k)A_k(x)dx}{h^3}. \tag{4.21}$$

The inverse F^1-transform of a function f is, similarly as the inverse F^0-transform, defined as a linear combination of basic functions with "coefficients" given by the F^1-transform components.

Definition 4.13 *Let $f : [a, b] \to \mathbb{R}$ be a given function and*

$$\mathbf{F}^1[f] = (F_1^1[f], \ldots, F_{n-1}^1[f])$$

be the direct F^1-transform of f with components (4.18) computed with respect to the fuzzy partition $A_1, \ldots, A_{n-1}$. Then the function $\hat{f}^1 : [a, b] \longrightarrow \mathbb{R}$ defined by

$$\hat{f}^1(x) = \sum_{k=1}^{n-1} F_k^1[f](x) \cdot A_k(x) \tag{4.22}$$

is called the inverse F^1-transform *of f with respect to $\mathbf{F}^1[f]$ and $A_1, \ldots, A_{n-1}$.*

The F^1-transform has similar approximation properties as the basic (zero degree) one (described in the previous sections), namely, if $n \to \infty$, where n is the number of basic functions $A_1, \ldots, A_{n-1}$, then the obtained sequence of inverse F^1-transforms $\hat{f}^1_{(n)}$ of f uniformly converges to f (see [113] for the details). This property is demonstrated in Figure 4.8. One of the important advantages of the inverse F^1-transform is that $\hat{f}^1$ *provides better approximation of f than $\hat{f}^0$.*

Another important property of the F^1-transform is its ability to approximate derivatives of the function f by the coefficient $\beta_k^1[f]$. More precisely, the following theorem can be proved.

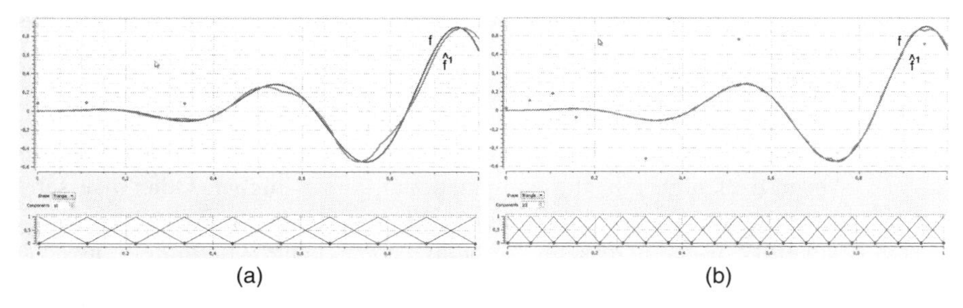

Figure 4.8 The function $f = x^2 \sin(15\,x)$ defined on the interval $[0, 1]$ and its inverse F^1-transform $\hat{f}^1$. The fuzzy partition is uniform and formed by triangular-shaped basic functions. (a) contains 10 and (b) contains 20 basic functions.

Theorem 4.14 ([62]) *Let $A_1, \ldots, A_{n-1}$ be an h-uniform partition of $[a, b]$, and let functions f and A_k, $k = 1, \ldots, n-1$, be four times continuously differentiable on $[a, b]$. Finally, let $\mathbf{F}^1[f]$ be the F^1-transform (4.18) of f. Then, for all $k = 1, \ldots, n-1$,*

$$|\beta_k^1[f] - f'(c_k)| \leq Mh^2 \tag{4.23}$$

for some $M > 0$.

According to Theorem 4.14, each coefficient $\beta_k^1[f]$ in (4.20) or (4.21) provides a reasonable estimation of an average value of the first derivative of the function f in the interval $[c_{k-1}, c_{k+1}]$. Thus we have a strong tool for a rough characterization of a function. This is useful especially when the function has very volatile course, which is often the case of time series or noisy signals. Let us also remark that Theorem 4.14 can be extended even to higher-order F-transform, which means that we can obtain average estimations of arbitrary derivatives of a function (for details, see [62, 113]).

4.6 METHODOLOGICAL REMARKS TO APPLICATIONS OF THE F-TRANSFORM

In this section, we describe necessary steps to apply the F-transform. We suppose that the function $f : [a, b] \longrightarrow [c, d]$ is given in the form of data:

$$\begin{bmatrix} x_1 & f(x_1) \\ x_2 & f(x_2) \\ \vdots & \vdots \\ x_p & f(x_p) \end{bmatrix}, \tag{4.24}$$

where $p > 3$, $x_1, \ldots, x_p \in [a, b] \subset \mathbb{R}$ and $f(x_1), \ldots, f(x_p) \in [c, d] \subset \mathbb{R}$.

1. **Construction of the fuzzy partition:** This is the first necessary step that must be done. It depends on the goal of our model, that is, if we want to find as precise approximation of f as possible or we want the inverse F-transform $\hat{f}$ of f to have special properties. Though the fuzzy partition can be nonuniform, in most practical situations, an h-uniform fuzzy partition is fully sufficient, and so we will suppose the latter in the sequel.
   (a) Specify the number n of components and set the step $h = (b - a)/n$.
   The width of h depends on the goal of our model. If we want to obtain as best approximation of f as possible, then h should be small. Of course, because of higher n, the computational cost is higher. Otherwise set h according to further requirements. To obtain smooth $\hat{f}$ (i.e., to remove volatility of f), increase h. To remove components with high frequencies (e.g., in time series), set $h = d\,T$ for $d > 0$ and an appropriate periodicity T—see the explanation in Chapter 9.
   (b) Construct the nodes $c_0 < \cdots < c_n$, where $c_k = a + hk$, $k = 0, \ldots, n$, and $c_n = b$.

Be careful to keep the fuzzy partition sufficiently dense, that is, each interval $[c_{k-1}, c_{k+1}]$ must contain at least one value x_j. Note that the nodes c_k, $k = 0, \ldots, n$, need not coincide with any $x_j, j = 1, \ldots, p$.

(c) Decide whether the basic functions should be triangular or they should have another shape. In practice, we mostly use triangular or cosine basic functions defined in Example 4.1. Recall that shape of the basic functions determines the course of $\hat{f}$, that is, whether it is piecewise linear or nonlinear.

(d) Construct the fuzzy partition using formulas from Example 4.1. Note that each basic function A_k spans over three nodes c_{k-1}, c_k, and c_{k+1}, $k = 1, \ldots, n-1$, so that $A_k(c_{k-1}) = A_k(c_{k+1}) = 0$ and $A_k(c_k) = 1$.

2. **Computation of the direct F-transform:** Using the fuzzy partition $A_0, \ldots, A_n$ constructed in Step 1, compute the F-transform components $F_0[f], \ldots, F_n[f]$ using one of the following formulas:
   (a) Formula (4.13) if we apply F^0-transform to the function f of one variable.
   (b) Formula (4.16) if we apply F^0-transform to the function f of two variables.
   (c) Formula (4.13) and

$$\beta_k^1[f] = \frac{\sum_{j=1}^{p} f(x_j)(x_j - c_k)A_k(x_j)}{\sum_{j=1}^{p} (x_j - c_k)^2 A_k(x_j)} \tag{4.25}$$

   if we apply F^1-transform. Then compute the components

$$F_k^1[f](x) = \beta_k^0[f] + \beta_k^1[f] \cdot (x - c_k)$$

   for all $k = 1, \ldots, n-1$.

3. **Computation of the inverse F-transform:** Compute the function $\hat{f}$ using one of the following formulas:
   (a) Formula (4.14) if we apply F^0-transform to the function f of one variable.
   (b) Formula (4.17) if we apply F^0-transform to the function f of two variables.
   (c) Formula (4.22) if we apply F^1-transform.

It should be emphasized that the inverse F-transform provides *explicit formula* for $\hat{f}$. Therefore, if needed, it can be taken as an explicit formula determining a function originally known only in the form of data (4.24).

5

FUZZY NATURAL LOGIC AND APPROXIMATE REASONING

5.1 LINGUISTIC SEMANTICS AND LINGUISTIC VARIABLE

In our exposition presented in the previous chapters, we argued that fuzzy sets are a reasonable mathematical tool using which semantics of certain classes of words and linguistic expressions can be modeled. Let us emphasize, however, that elaboration of natural language semantics in full, which means to capture semantics of sentences and to cope with many exceptions and finenesses, is extremely difficult and still quite far from being solved.[1]

In this chapter, we present parts of the so-called *Fuzzy Natural Logic* (FNL), which will be applied in the subsequent chapters.[2] FNL is a group of mathematical theories that extend mathematical fuzzy logic in narrow sense (cf. Section 2.3 and also [86]). Its goal is to develop a mathematical model of special human reasoning schemes that employ natural language but are independent on a concrete one. Therefore, FNL also includes a model of the semantics of some parts of natural language.

The main constituents of FNL are at present the following:

(i) Theory of evaluative linguistic expressions.

[1]Such a model must in the first place characterize semantics of verbs, which are the most complicated constituents of sentences. The advanced mathematical model of linguistic semantics, which also covers its vagueness, is presented in the book [76]. The model is based on nonstandard mathematical theory—the so-called Alternative Set Theory [140]. It should be noted that fuzzy sets take in this model the role of a reasonable approximation tool.

[2]FNL continues the program initiated by the concept of *fuzzy logic in broader sense* (FLb) introduced in [77].

Insight into Fuzzy Modeling, First Edition. Vilém Novák, Irina Perfilieva, and Antonín Dvořák.

Companion Website: www.wiley.com/go/novak/fuzzy/modeling

(ii) Theory of fuzzy/linguistic IF-THEN rules and logical inference based on them.
(iii) Theory of fuzzy generalized quantifiers with emphasis on intermediate quantifiers, generalized Aristotle syllogisms, and square of opposition.

In this chapter, we focus on the theories (i) and (ii) because they are currently the most elaborated and play an essential role in fuzzy modeling and its applications (cf. [87]). More about the theory (iii) can be found in [36, 84, 71, 73].

5.1.1 Linguistic Variable

A special concept, introduced by L. A. Zadeh in his famous paper [151], is that of *linguistic variable*. It is a variable whose values can be words or some expressions of natural language, for example, "small or medium", "very big", and "around 10". In linguistics, such expressions are called *syntagms*. We will prefer to call values of linguistic variables simply *linguistic expressions*.

According to the original Zadeh's definition, a *linguistic variable* is a quintuple

$$\langle \mathscr{X}, T(\mathscr{X}), U, G, M \rangle,$$

where $\mathscr{X}$ is a *name* of the variable, $T(\mathscr{X})$ is its *term set*, that is, a *set of possible values* (linguistic expressions) of $\mathscr{X}$; U is a *universe*, G is a *syntactic rule*using which linguistic expressions $\mathscr{A}, \mathscr{B}, \ldots$ from the set $T(\mathscr{X})$ are formed, and M is a *semantic rule* using which each linguistic expression $\mathscr{A} \in T(\mathscr{X})$ is assigned its meaning, which is a fuzzy set

$$A = M(\mathscr{A}) \underset{\sim}{\subset} U.$$

EXAMPLE 5.1 A nice example of linguistic variable, often cited in literature, is $\mathscr{X}$ = *height* (of a person). Its term set $T(\mathscr{X})$ consists of linguistic expressions such as *small*, *very big*, *not small and not big*, and *roughly medium*. These linguistic expressions are generated by a special grammar G. The meaning of any of them is modeled by a fuzzy set on the universe $U = [0, 250]$ (cm). Possible shapes of their membership functions follow from the analysis of the meaning of the so-called evaluative linguistic expressions, which will be discussed below. □

5.1.2 Intension, Context, Extension

When developing a model of semantics of natural language expressions, we cannot avoid three essential concepts: intension, context, and extension. Loosely speaking, *intension* of a linguistic expression (or a concept) is a *representation of a property* that is denoted by the former. The property remains the same notwithstanding the context in which it is used. For example, let us consider the concept *low temperature*. Intension of this concept is the *property of linearly ordered degrees of Celsius*, namely, "to be low". This property can occur in any place and time. Indeed, we can speak about low temperature in Africa, which can be about, say, 10 °C; in Antarctica, which can be about −60 °C; in Central Europe in the summer, which can be about 15 °C, or in the winter, which can be about −10 °C; etc. However, the objects (degrees) themselves are not contained in the intension. It is meaningful to speak about them only when an individual context of speech is specified. For example, it can be the temperature of water, body, and air, possibly the temperature of air in the summer, winter, etc.

The following definition introduces a model of intension and extension due to R. Carnap [21]. First, we introduce a set W of *contexts*.[3] A *context* is the state of "all things in the given place and time". From a formal point of view, we can take context as some parameter whose character is unimportant.

Definition 5.1 *Let a set W of contexts and a universal set U of some elements be given. Let $\mathscr{A}$ be a natural language expression. Then* intension *of $\mathscr{A}$ is a function*

$$\mathrm{Int}(\mathscr{A}) : W \longrightarrow \mathscr{F}(U), \tag{5.1}$$

which assigns to any context $w \in W$ a fuzzy set of objects from U. This fuzzy set is called extension *of the expression $\mathscr{A}$ in the context w, that is,*

$$\mathrm{Ext}_w(\mathscr{A}) = \mathrm{Int}(\mathscr{A})(w) \subsetneq U. \tag{5.2}$$

We can speak about individual objects only when a concrete context is given. This gives us extension that contains all objects that possess the given property in the given context. From it follows that one intension leads to a class of extensions, each of which is a class of elements. Formally, the extension $\mathrm{Ext}_w(\mathscr{A})$ in a context $w \in W$ is a value of intension $\mathrm{Int}(\mathscr{A})$ at the point w. According to (5.1), this is the fuzzy set (5.2).

EXAMPLE 5.2 Let us consider the linguistic expression "height is *small*" and the context w to be *heights of people in Central Europe*. This means that people smaller than 165 cm are surely small, people 175 cm tall are typically medium tall, while those taller than 185 cm are surely tall. Then we obtain the following fuzzy set being mathematical model of the extension $\mathrm{Ext}_w(\mathit{small})$:

$$\mathrm{Ext}_w(\mathit{small})(x) = \begin{cases} 1 & \text{if } x \le 165, \\ 0 & \text{if } x > 185, \\ 1 - \dfrac{1}{2}\left(\dfrac{x-165}{10}\right)^2 & \text{if } 165 < x \le 175, \\ \dfrac{1}{2}\left(\dfrac{185-x}{10}\right)^2 & \text{if } 175 < x \le 185. \end{cases} \tag{5.3}$$

We can see that this model is the same fuzzy set as in Example 2.2. The difference is that we obtain another fuzzy set (i.e., extension) if we consider another context. All

[3] In the general theory of linguistic semantics, one can meet the notion of *possible world*. The latter can be intuitively understood as a maximal consistent set of all possible affairs (this means that it does not lead to contradiction and cannot be extended). In this book, instead of possible world we prefer to use the term *context* which is more convenient for our purpose. We can take it as a special case of a possible world.

the fuzzy sets defined in all the possible contexts form an intension that represents the *meaning* of the expression "height is *small*". □

Let us emphasize that intensions *must* be distinguished from extensions. In Example 5.2, we characterized the fuzzy set of heights of small people. However, it was only an example in one possible context. In China, Africa, North Europe, or Lilliput, we must modify this definition though the *shapes* of the considered fuzzy sets remain the same. Moreover, we can use the concept *small* also for other objects than people.

EXAMPLE 5.3 Let us follow Example 5.2 and consider a set of contexts for heights of people $W = \{w_1 :=$ "China", $w_2 :=$ "Africa", $w_3 :=$ "North Europe", $w_4 :=$ "Lilliput", $\dots\}$. If we assign to each $w_i \in W$ a fuzzy set of the form (5.3) in which we modify the parameters represented by numbers 165, 175, and 185 accordingly, then we have just constructed the intension Int("height is small"), that is, a function from W to $\mathscr{F}(\mathbb{R})$. □

5.1.3 Refined Definition of Linguistic Variable

We see from the discussion in the previous subsection that the original concept of linguistic variable does not take into account the concept of intension.

Let a linguistic expression $\mathscr{A}$ and a set of contexts W be given. Then the *meaning* of $\mathscr{A}$ is a couple

$$\text{Mean}(\mathscr{A}) = \langle \text{Int}(\mathscr{A}), \{\text{Ext}_w(\mathscr{A}) \mid w \in W\}\rangle, \tag{5.4}$$

that is, the meaning of the linguistic expression $\mathscr{A}$ is characterized both by its intension and by its extensions in all (considered) contexts.

The *linguistic variable* is a tuple

$$\langle \mathscr{X}, T(\mathscr{X}), G, U, W, \mathscr{M}\rangle,$$

where $\mathscr{X}$ is its name, $T(\mathscr{X})$ is its term set (a set of linguistic expressions), G is a *syntactic rule* using which linguistic expressions $\mathscr{A}, \mathscr{B}, \dots \in T(\mathscr{X})$ are formed, U is a *universe*, and $\mathscr{M}$ is a *semantic rule* using which each linguistic expression $\mathscr{A} \in T(\mathscr{X})$ is assigned its meaning Mean($\mathscr{A}$) (5.4).

The concept of linguistic variable allows the set of linguistic expressions $T(\mathscr{X})$ to be infinite, generated by some formal grammar. In practice, however, we quite often work with simplified variables whose sets of values (linguistic expressions) contain only a limited number of elements (usually several tens).

Remark 5.1 *In various practical applications, one can see the following expressions:* negative big *(NB),* negative medium *(NM),* negative small *(NS),* zero *(ZE),* positive small *(PS),* positive medium *(PM), and* positive big *(PB). These are* word labels *taken from a* rating scale, *for example, a questionnaire prepared for some kind of investigation in sociology, psychology, marketing, etc. These labels cannot be taken as expressions of natural language. Their meaning is derived on the basis of some chosen* prototype *picked up from the rating scale and the labels are often represented by triangular fuzzy sets (see Figure 5.1). These fuzzy sets cannot represent extensions*

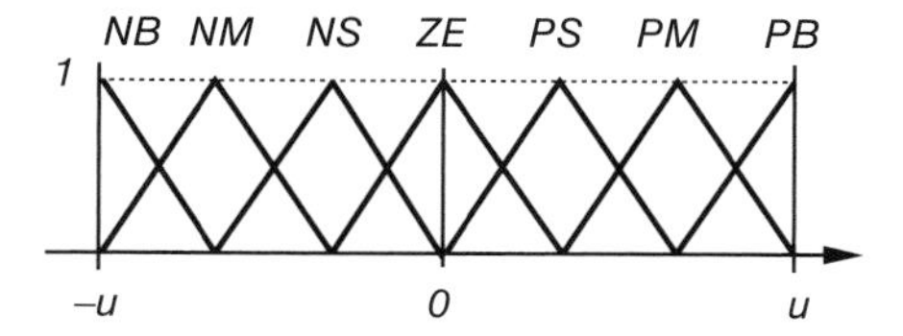

Figure 5.1 Fuzzy sets characterizing extensions of vague categories.

of the underlying linguistic expressions, otherwise we arrive at a contradiction with results of the logical analysis of the latter (cf. [83]).

For example, let us assume that in Figure 5.1, $u = 90$. *If we took PS as the extension of the expression "positively small", then it would imply that, for example, the membership degrees* $PS(1), PS(2), PS(3), \ldots$ *would be close to 0 and* $PS(0) = 0$. *This means that the numbers 1, 2, 3 are practically not small and 0 is not small at all. The membership degree 1 would be assigned only to the number* 30, *that is,* $PS(30) = 1$. *But this contradicts with our understanding of what does it mean "small", because 1, 2, 3, etc. are small without doubts. Picking up 30 as just one example of typically "small" clearly shows that triangles represent only selected prototypical examples with respect to a given context.*

5.2 THEORY OF EVALUATIVE LINGUISTIC EXPRESSIONS

In this section, we present the basic constituent of fuzzy natural logic, which is used in the theory of fuzzy modeling and in various applications of it.

5.2.1 The Concept and Structure of Evaluative Expressions

A very important class of natural language expressions is formed by the so-called *evaluative linguistic expressions*. In the sequel, we will often omit the adjective "linguistic" and use only the term "evaluative expressions" . These expressions are used by people whenever they need to evaluate some situation, phenomenon, and result; to decide on the basis of complex information; to classify something; and also in technical applications, such as fuzzy control. For example, when choosing a mobile phone, one can use the following words:

> This phone is *light* and *very thin* but with *quite large* display, *extremely long* battery life and has *more or less attractive* design.

By italics, we marked the evaluative expressions. Generally speaking, these are special expressions of natural language that represent either some value on an ordered scale (usually a certain number) or they characterize an approximate position on it.

The class of evaluative expressions includes, in particular, *simple evaluative expressions*, for example, *small*, *medium*, *big*, and *fuzzy numbers*. Expressions from these categories can be supplemented by the so-called *linguistic hedges*. Simple evaluative expressions can be connected by the connectives "and" and "or".

Note that evaluative expressions are typical values of linguistic variables introduced in the previous section and are also considered in majority of applications of fuzzy modeling. More about this field including a detailed formal theory can be found in paper [83].

5.2.1.1 ***Grammatical structure of evaluative expressions.*** This can be characterized in a simplified form as follows:

(i) *Simple evaluative expressions*:

(a) ⟨pure evaluative expression⟩ := ⟨linguistic hedge⟩⟨TE-adjective⟩

(b) ⟨fuzzy number⟩ := ⟨linguistic hedge⟩⟨numeral⟩

⟨linguistic hedge⟩ := ⟨empty⟩ | ⟨narrowing adverb⟩ | ⟨widening adverb⟩ | ⟨specifying adverb⟩

(ii) *Negative evaluative expressions*:

not ⟨pure evaluative expression⟩

(iii) *Compound evaluative expressions*:

(a) ⟨pure evaluative expression⟩ or ⟨pure evaluative expression⟩

(b) ⟨pure evaluative expression⟩ and/but ⟨negative evaluative expression⟩

The linguistic hedge ⟨empty⟩ has a special role described below.

TE-adjectives. These adjectives[4] play a central role when people need to judge sizes, measures, volumes, heights, etc. Typical examples are the adjectives "small", "medium", and "big" and also similar triples such as "strong", "medium strong", "weak" and "soft", "medium hard", "hard".[5] The basic principle of their meaning is connected with a specification of a position on a linearly ordered scale, namely, it is natural to distinguish three basic positions—*left*, *middle*, and *right*—that are, depending on the context, replaced by the words introduced above. The three canonical TE-adjectives form a *basic evaluative trichotomy*. The words "small" and "big" are *antonyms*. The middle term is an expression characterizing the so-called *extensional gap*. However, note that there also exist *complementary adjectives* that do not have middle member, for example, "complete–incomplete", but their existence does not deny our theory below.

We argue that all TE-adjectives behave semantically in a very similar way. This enables us to develop a unified logical theory of their semantics. Hence, in the sequel we will limit ourselves to expressions "small", "medium" and "big" only. We will take them as *canonical TE-adjectives*.

Linguistic hedges. These are special linguistic expressions standing in the position of adverbs that make the meaning of the adjacent linguistic expressions more or less

[4]TE stands for *trichotomous evaluative*.

[5]From the point of view of linguistic theory, the class of TE-adjectives includes gradable adjectives (e.g., big, high, small) as well as evaluative ones (e.g., good, nice).

specific. Hence, they enlarge the repertoire of evaluative expressions. The most elaborated is the case when linguistic hedges are realized using special adverbs (they fall into the class of the so-called *intensifying adverbs*). We distinguish three subclasses of these adverbs, namely, adverbs with *narrowing effect*, those with *widening effect*, and *specifying adverbs*. We will also consider a special linguistic hedge *empty*. This enables us to develop a unified theory of the semantics of simple evaluative expressions (see Section 5.2.3).

Numerals and fuzzy numbers. These are names of numbers, for example, "one million", "fifty five hundred", and "three and a half thousand".

Fuzzy numbers are imprecise numbers.[6] Moreover, we can supplement numerals by some hedge. The hedges are mostly widening ones. We will consider here the hedges *roughly, quite roughly* and *very roughly*, and also *about* and *approximately* (the last two hedges are specific for fuzzy numbers). It should be emphasized, however, that in practice, every number should be taken as imprecise. For example, if we say "three million", we usually do not have in mind "precisely three million" but only "about three million". From it also follows that there is a narrowing hedge for fuzzy numbers, namely, *precisely*.

Negative and compound evaluative expressions. The particle *not* denotes negation. Since negation is a fairly complicated concept causing problems both to logicians and to linguists, we should be very careful using it. Luckily, evaluative expressions are not that much complicated, but even here, the situation is not so simple.

First of all, it is necessary to distinguish the concepts of negative evaluative expression and the *antonym*. Evaluative expressions *small* and *big* are antonyms and it holds that semantically

$$big \neq not\ small \quad \text{and} \quad small \neq not\ big.$$

When analyzing other negative evaluative expressions, for example, "not very big", we face a more complicated situation. They have two possible readings: either the linguistic hedge "very" is negated or the whole expression "very big" is negated.[7] The first case is more common, but the theory of negation of linguistic hedges has not been developed so far. Therefore, in this book, we will consider the second case only.

Simple evaluative expressions enable us to form also compound ones using the connectives "and" and "or". This means that we can form expressions like "small or roughly medium", "X is small and Y is very big",[8] etc. We must be careful, however, because natural language is complicated and does not make it possible to form arbitrary Boolean combinations. For example, the expression "very small or big and rather medium or extremely big" has no meaning and cannot be taken as an expression of natural language. A more detailed model of compound evaluative expressions has still not been developed.

[6]See also discussion in Section 2.2.3.

[7]This is an example of an important linguistic phenomenon called *topic–focus* articulation. Negation applies always to focus. In the first case, the topic is "big" and the focus is "very"; in the second case, the whole expression is in the focus.

[8]Note that this is in fact an evaluative predication, see the next paragraph.

5.2.1.2 Pure evaluative linguistic expressions. Examples of pure evaluative expressions are "very small", "extremely strong", "more or less tall", "very roughly medium", etc. In this book, we will consider only few selected hedges, namely, *extremely, significantly, very* (narrowing), *more or less, roughly, quite roughly, very roughly, very very roughly* (widening), and *rather* (specifying). While widening hedges can be used with all TE-adjectives, narrowing and specifying ones can be in most cases used only with *small* and *big*. Indeed, for example, "very medium" has no sense. A special narrowing hedge used with "medium" is *typically*.

Note that considering the hedge ⟨empty⟩ enables us to take the TE-adjectives also as simple (or pure) evaluative expressions of the form (i)(a) above. For example, "big" is taken as the simple evaluative expression "⟨empty⟩ big".

It follows from the logical analysis that pure evaluative linguistic expressions can be naturally ordered. There are two possibilities.

First ordering of pure evaluative expressions. This ordering (denoted by $\prec$) is based on the position of extensions on the scale within a given context and, further, on comparison of widths of their supports. At the same time, while the expressions "⟨linguistic hedge⟩small " are ordered from a narrower extension to a wider one, the expressions "⟨linguistic hedge⟩big" are ordered in the opposite way.

For the few selected evaluative expressions, the ordering is as follows:

zero $\prec$ extremely small $\prec$ significantly small $\prec$ very small $\prec$ **small** $\prec$
$\prec$ more or less small $\prec$ roughly small $\prec$ quite roughly small $\prec$
$\prec$ very roughly small $\prec$ **medium** $\prec$ more or less medium $\prec$ roughly medium $\prec$
$\prec$ quite roughly medium $\prec$ very roughly medium $\prec$
$\prec$ very roughly big $\prec$ quite roughly big $\prec$ roughly big $\prec$
$\prec$ more or less big $\prec$ **big** $\prec$ very big $\prec$ significantly big $\prec$ extremely big. (5.5)

Second ordering of pure evaluative expressions. This ordering (denoted by $\lll$) is obtained as a lexicographic ordering (cf. (2.2)) based on the natural ordering of the canonical TE-adjectives together with the numeral *zero*:

$$\text{Ze (zero)} \lll \text{Sm (small)} \lll \text{Me (medium)} \lll \text{Bi (big)} \tag{5.6}$$

and the following ordering of hedges:

$$\begin{aligned} &\text{Ex (extremely)} \lll \text{Si (significantly)} \lll \text{Ve (very)} \lll \langle\text{empty hedge}\rangle \lll \\ &\lll \text{ML (more or less)} \lll \text{Ro (roughly)} \lll \text{QR (quite roughly)} \lll \\ &\lll \text{VR (very roughly)} \lll \text{VV (very very roughly)}. \end{aligned} \tag{5.7}$$

The linear lexicographic ordering $\lll$ of simple evaluative expressions can be interpreted as "to be more specific". This means that the TE-adjective laying to the left of $\lll$ in (5.6) is more specific than that on the right and, at the same time, expressions with the same TE-adjective and linguistic hedge laying to the left of $\lll$ (e.g., "extremely") are more specific than any other expression to the right of $\lll$ (e.g., "more or less"). The following are examples of the lexicographic ordering $\lll$:

(*a*) *extremely small* $\lll$ *very small* $\lll$ *small* $\lll$ *roughly small*;

(*b*) *more or less small* $\lll$ *very roughly small* $\lll$ *medium* $\lll$ *roughly medium*;

(*c*) *typically medium* $\lll$ *medium* $\lll$ *roughly medium* $\lll$ *big*.

In words, it means that "extremely small" is more specific than "very small", which is more specific than "small", and this is more specific than "roughly small"; similarly for the other cases. Of course, we must not forget that some combinations of hedges and TE-adjectives have no meaning, and so they do not form evaluative expressions.

5.2.2 Evaluative Linguistic Predications

The evaluative expressions discussed above can be taken as *abstract*, which means that they alone do not address any specific objects. If we consider objects addressed by them, then we arrive at a special linguistic notion called *evaluative linguistic predication*. This is a linguistic expression of the form

$$\langle \text{noun} \rangle \text{ is } \mathscr{A}, \tag{5.8}$$

where $\mathscr{A}$ is an evaluative linguistic expression. Typical examples of evaluative predications are expressions "the temperature is big", "the pressure is very small", "the velocity is more or less small", etc.

In applications of fuzzy modeling, we are usually not interested in the character of specific objects. In particular, it does not matter whether we deal with temperatures, height of people, or other specific positions on some scale. Therefore, we often replace $\langle$noun$\rangle$ by some abstract variable. Hence, evaluative predications will be always in this book taken as possessing the form

$$X \text{ is } \mathscr{A}, \tag{5.9}$$

where X is a variable representing objects.

In some cases, we can write evaluative predications also as

$$\mathscr{A}\ X, \tag{5.10}$$

for example, "big temperature" and "very small pressure". There are two readings of (5.9). Either it is a fuzzy proposition having a truth value (laying usually in [0, 1]) or it is a vague expression whose extension is a fuzzy set. In the latter case, meanings of (5.9) and (5.10) are taken as identical.

We will call the predication pure, negative, or compound depending on the form of the evaluative expression $\mathscr{A}$. Then both orderings $\prec$ and $\lll$ can also be extended to the evaluative predications as follows:

$$X \text{ is } \mathscr{A} \prec X \text{ is } \mathscr{B} \text{ iff } \mathscr{A} \prec \mathscr{B},$$

$$X \text{ is } \mathscr{A} \lll X \text{ is } \mathscr{B} \text{ iff } \mathscr{A} \lll \mathscr{B}.$$

We can also form a second kind of *compound evaluative predications*, in which the variables can be different:

$$X \text{ is } \mathscr{A} \text{ AND } Y \text{ is } \mathscr{B},$$

for example, "*distance* is *very small* AND *speed* is *high*".

Definition 5.2 *The set of evaluative (linguistic) expressions $EvExpr$ is the set of all abstract evaluative expressions as well as of evaluative predications (5.8) (or, alternatively, (5.9)). The orderings $\prec$ and $\lll$ enable us to consider two partially ordered sets, namely, $(EvExpr, \prec)$ and $(EvExpr, \lll)$.*

We will write $Ev \in EvExpr$ for an arbitrary element from $EvExpr$. The abstract evaluative expressions from $EvExpr$ will be denoted by script letters $\mathscr{A}, \mathscr{B}, \ldots$.

5.2.3 Mathematical Model of the Semantics of Evaluative Linguistic Expressions

In this section, we present a mathematical model that captures well the semantics of simple evaluative expressions. Details and justification of this model can be found in [83].

5.2.3.1 Linguistic context. This concept has already been mentioned in Section 5.1.2. It should be emphasized that the possibility to consider various contexts immensely expands the expressive power of natural language because the same word can be used in various situations, though its meaning (extension) seems to vary considerably. For example, *long distance* can mean 1000 km if we travel by car but only 15 km if we walk.

From the point of view of general linguistics, the concept of context is quite complicated and reaches deeply into issues of knowledge representation and modeling of the world properties. Luckily, the situation is much simpler in the case of evaluative expressions, so we can introduce the following definition.

Definition 5.3 *Let $v_L, v_S, v_R \in \mathbb{R}$ be distinguished points such that $v_L < v_S < v_R$. Then the* linguistic context *is a strictly increasing bijection*

$$w : [0, 1] \longrightarrow [v_L, v_S] \cup [v_S, v_R], \tag{5.11}$$

where $w(0) = v_L$, $w(0.5) = v_S$, and $w(1) = v_R$.

We may alternatively take the context as a triple of numbers

$$w = \langle v_L, v_S, v_R \rangle$$

with the following meaning: $v_L = w(0)$ denotes *the least* value that makes sense in a given situation (left bound); $v_R = w(1)$ is the opposite—*the greatest* value that makes sense (right bound). Finally, the value v_S is *the common middle* value, which is neither small nor big (typical center). We will often use the context $\langle 0, 0.5, 1\rangle$ and call it *standard*.

EXAMPLE 5.4 Let us consider again the body temperature. Then *extremely low body temperature* can be 35 °C and *extremely high body temperature* is 42 °C. It is clear that temperature higher than 42 °C is unrealistic, because it leads to death. The standard middle value is 36.6 °C. Therefore, the linguistic context is $w = \langle v_L, v_S, v_R\rangle = \langle 35, 36.6, 42\rangle$. Note that the value v_S does not lie in the center of the interval $[35, 42]$. Such situations are quite common. □

EXAMPLE 5.5 The context in Example 5.4 can be defined as the following function:

$$w(x) = \begin{cases} 3.2x + 35 & \text{if } x \in [0, 0.5], \\ 10.8x + 31.2 & \text{if } x \in (0.5, 1]. \end{cases}$$

Obviously, $w(0) = 35$, $w(0.5) = 36.6$, and $w(1) = 42$. □

The *set of all linguistic contexts* for evaluative expressions is

$$W = \{w \mid w \text{ is a context due to Definition 5.3}\}. \tag{5.12}$$

We will formally write that x *belongs to (is a member of) the given context* w as

$$x \in w.$$

This is tantamount to $x \in [v_L, v_S] \cup [v_S, v_R]$.[9] Moreover, because the linguistic context characterizes a real situation, it may happen that though the context is determined by the interval $[v_L, v_R]$, in real life, there may occur values that fall outside this interval and still we need to evaluate them using some evaluative expression. For example, we may consider heights of people determined by the context $\langle 155, 175, 190\rangle$ but, exceptionally, we can meet people 140 cm tall or 200 cm tall. To cope with such situations, we introduce the *extended inverse* of w in (5.11) as a function $w^{(-1)} : \mathbb{R} \to [0, 1]$ as follows:

$$w^{(-1)}(x) = \begin{cases} w^{-1}(x) & \text{if } x \in [v_L, v_R], \\ 0 & \text{if } x < v_L, \\ 1 & \text{if } x > v_R. \end{cases} \tag{5.13}$$

Remark 5.2 *The role of linguistic context is very important. In particular, it comes along in connection with linguistic (fuzzy) control, which we will present in Chapter 7. The concept of context offers there intriguing possibilities, for example, to increase precision of control, in learning, and in adaptivity. From the technical point of view, the setting of the linguistic context means specific scaling. For example, if the variable represents the control error with respect to the set-point value, then to set the linguistic context means to set the magnitude of gain.*

[9] Of course, this union is equal to $[v_L, v_R]$. We write it as a union explicitly to emphasize the role of the middle point v_S.

5.2.3.2 Semantics of pure evaluative expressions. In this section, we describe mathematical model of the semantics of evaluative expressions. We will confine only to pure ones, because they have the best elaborated applications.

Horizon and its role in semantics of evaluative expressions. Evaluative expressions can be understood as linguistic characterizations of the abstract concept of quantity which is taken from an ordered scale. According to *empirical observation*, scales considered in the linguistic meaning of evaluative expressions are *linearly ordered and bounded*. Since they can be very extensive, the requirement to name every quantity would lead to very large number of necessary linguistic expressions; in limit case, even infinite. The power of natural language enables people to use only small (finite) number of expressions that, surprisingly, may characterize any element of any ordered set. The price we must pay is vagueness of the meaning of the used expressions.

The question that arises is, what is the source of vagueness of extensions of the evaluative expressions. The justification, based on ideas of P. Vopěnka (see [140]), leads us to the concept of horizon. This is a sharp line determining imprecisely a part of the world laying before it. Moreover, it can be shifted “along the world” nearer or further.

Each context for evaluative expressions is represented by an ordered scale bounded by two limit points: a *left bound* and a *right bound*. These points are the “most typical” small value and the “most typical” big value, respectively. The properties of being “small” and “big” are naturally vague because, though we can always point out some concrete small (big) value, there does not exist the last small (the first big) one. The only fact we know is that small values run somewhere toward a certain point that is the *horizon of our seeing of all small values*. Everything beyond this point is *surely not small*. Note that this reasoning embraces the sorites paradox.[10]

The way how the sorites paradox is resolved in fuzzy logic (cf. [45]) consists in introduction of *degrees of truth* expressing that “we find ourselves still before the horizon”.

Quite similarly, starting from the right bound and going in the opposite direction, we find a *horizon of big values* such that everything beyond it is surely not big. As a consequence, there exists a certain point that lies somewhere between the left and right bound, and such that both horizons vanish at it. This point will be called the *central limit point*.[11]

Human mind (and, consequently, natural language) enables us to distinguish parts of the horizon more subtly by modifying it. In other words, we may say that our mind *shifts horizon* along the world. We obtain new either more or less specific horizons that determine extensions of the evaluative expressions. Consequently, if an element of the scale falls in the extension of a more specific (i.e., “narrower”) evaluative expression, then it falls in the extensions of all less specific (i.e., “wider”) ones (provided that they exist). For example, very small values cease to be “very small” sooner than to be “small”. This means that each “very small” value is at the same time “small”, but there exist small values that are not “very small”. Analogous reasoning

[10]One grain does not form a heap. Adding one grain to what is not yet a heap does not make a heap. Consequently, there are no heaps.

[11]This is a distinguished inner element of the scale. If some metric is defined on the scale, then, however, this point does not necessarily need to lie in its center.

can be made for the pair of evaluative expressions "small" and "roughly small" and for the other ones. These arguments lie in the background of the mathematical model of the semantics of evaluative expressions.

Mathematical model of horizon. The "world" of an evaluative expression is represented by the context introduced above. Because the horizon is always determined with respect to a position of an observer, we place the observer in one of the three distinguished points of the context: v_L, v_S, v_R. Depending on that, we obtain three kinds of horizon, namely, *left horizon (LH), middle horizon (MH)*, and *right horizon (RH)*. Thus, starting with the smallest value v_L, we encounter that all small values lie inside the left horizon (*LH*). Similarly, all big values lie inside *RH* (the big values vanish from right to left) and medium values lie inside *MH*. All three horizons are schematically depicted in Figure 5.2.

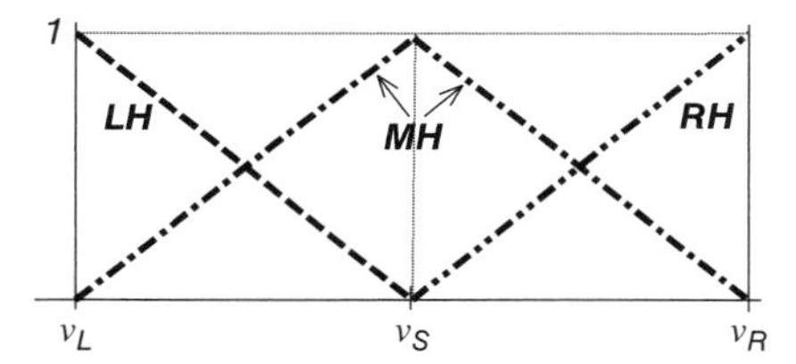

Figure 5.2 Left horizon (*LH*) given by the observer v_L, middle horizon (*MH*) by the observer v_S, and right horizon (*RH*) by the observer v_R (in a given context w). Note that the middle horizon spreads out to the left as well as to the right.

The following simple model of all three horizons in the standard context $w = \langle 0, 0.5, 1\rangle$ is obtained using the formal apparatus of higher-order fuzzy logic:[12]

$$LH(x) = \left(\frac{0.5 - x}{0.5}\right)^*, \qquad \text{(left horizon)} \tag{5.14}$$

$$MH(x) = \left(\frac{x}{0.5}\right)^* \wedge \left(\frac{1 - x}{0.5}\right)^*, \qquad \text{(medium horizon)} \tag{5.15}$$

$$RH(x) = \left(\frac{x - 0.5}{0.5}\right)^*, \qquad \text{(right horizon)} \tag{5.16}$$

where $x \in [0, 1]$ and the asterisk means that the given function is "cut" to the interval $[0, 1]$, that is, $f^*(x) = 0$ if $f(x) < 0$ and $f^*(x) = 1$ if $f(x) > 1$; otherwise $f^*(x) = f(x)$.

For example, the truth degree of the statement "the value 0.3 lies inside the left horizon" is $LH\ (0.3) = 0.4$. Similarly, we obtain the truth degree $MH\ (0.3) = 0.6$ for the medium horizon and $RH\ (0.3) = 0$ for the right horizon. The value 0.5 lies outside both left and right horizons, that is, $LH\ (0.5) = RH\ (0.5) = 0$, but with the truth degree 1, it lies inside medium horizon. Of course, for a different context, the formulas (5.14)–(5.16) must be modified accordingly.

[12]Equations (5.14)–(5.16) are obtained by interpreting three formulas of the form $\bot \sim x$, $\dagger \sim x$, $\top \sim x$ of the higher-order fuzzy logic in a model, where interpretation of $\sim$ is $\leftrightarrow^2$. This formally well captures the idea of a horizon "seen from the three distinguished points", namely, the leftmost $\bot$, middle $\dagger$, and rightmost $\top$.

Mathematical model of linguistic hedges. In [147], L. A. Zadeh suggested to model hedges by means of certain operations using which extensions of more complex linguistic expressions can be computed from simpler ones. In his original theory (see also [75, 150, 151]), he specified several elementary operations applied to membership functions of fuzzy sets. The first and most often cited is the hedge *very*. Zadeh proposed to interpret this hedge by the operation $CON(a) = a^2$, $a \in [0, 1]$. According to his proposal, the meaning of, for example, the expression *very small* is the fuzzy set $M(\textit{very small}) = (M(\textit{small}))^2$, where M(small) is the fuzzy set assigned to the word *small*. Similarly, the linguistic hedge *more or less* is assigned the function $DIL(a) = a^{0.5}$, $a \in [0, 1]$.

Later on, G. Lakoff in his thorough analysis [63] pointed out that this model is unsatisfactory. He clearly demonstrated that linguistic hedges enforce the following modifications of extensions of evaluative expressions:

(a) The kernel of the membership function is made narrower or wider depending on the type of the hedge.

(b) The membership function is further modified to be steeper for linguistic hedges with a narrowing effect (e.g., *very* and *significantly*), or less steep for linguistic hedges with a widening effect (e.g., *more or less* and *roughly*). It can be a combination of both modes for specifying hedges.

Typical action of narrowing and widening hedges on the extension of the TE-adjective *small* is depicted in Figure 5.3.

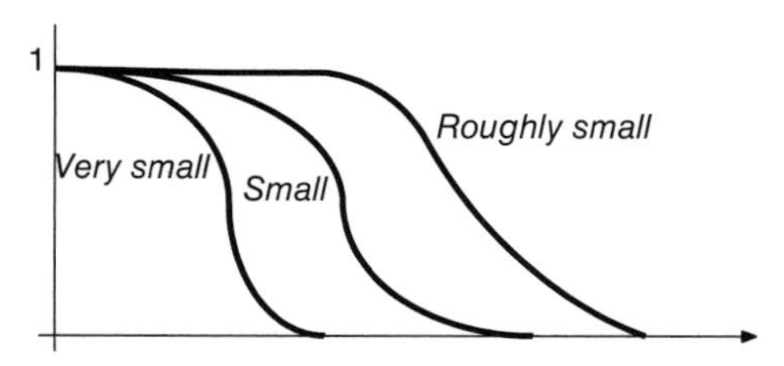

Figure 5.3 Modification of the shape of extensions of simple evaluative expression "⟨linguistic hedge⟩*small*" for typical hedges *very* and *roughly*.

Note that both $CON(1) = 1$ and $DIL(1) = 1$. Hence, these operations cannot change the kernel of a fuzzy set. Consequently, they do not enable the proper shifting of membership functions.

The first modification of Zadeh's theory fulfilling the above requirements was proposed by V. Novák in book [75]. It was later followed by other authors, for example, Bouchon-Meunier and Jia [12] and others.

The model of hedges described below fits the linguistic meaning well and is fairly simple. It is based on the results of linguistic analysis, which reveals that the effect of linguistic hedges consists in a contraction or a prolongation of horizon. Consequently, the meaning of any pure evaluative expression is determined by a specific horizon obtained by the modification of the corresponding horizon LH, MH, or RH.

For example, let us think of the meaning of the expression "very small numbers". The linguistic hedge (adverb) *very* specifies more closely the meaning of *small numbers*. Very small numbers are small, but there exist small numbers that are not very small. On the other hand, in the same way as for small numbers, there does not exist the last very small number, though we can point out a number that is not very small. Consequently, the horizon of very small numbers must lie closer to zero than that of small numbers.

There are various other hedges behaving similarly, for example, *significantly* or *extremely*. On the other hand, there exist linguistic hedges with the opposite, that is, widening effect. Typical examples are *more or less*, *roughly*, etc. We also argue that the semantics of the basic evaluative expressions "small", "medium", and "big" is based on the same principles as semantics of the discusses expressions "very small, more or less big", etc. Consequently, these expressions are construed as containing the *empty* linguistic hedge. This approach allows us to built a unified and elegant model of the semantics of all evaluative expressions.

The mathematical model of hedges is in our theory realized by means of a class of nondecreasing functions $\nu_{a,b,c} : [0,1] \to [0,1]$ determined by three parameters a, b, c, $a < b < c$, with the following meaning:

(i) $\nu_{a,b,c}(x) = 0$ for all $x \leq a$.
(ii) $\nu_{a,b,c}(b) = b$.
(iii) $\nu_{a,b,c}(x) = 1$ for all $c \leq x$.

The functions $\nu_{a,b,c}$ enable to realize *shifting* (and/or deformation) of the horizon. Therefore, each ⟨linguistic hedge⟩ will be assigned some function $\nu_{a,b,c}$ for specific values of a, b, c.

The following is a possible definition of the functions $\nu_{a,b,c}$:

$$\nu_{a,b,c}(y) = \begin{cases} 1, & c \leq y, \\ 1 - \dfrac{(c-y)^2}{(c-b)(c-a)}, & b \leq y < c, \\ \dfrac{(y-a)^2}{(b-a)(c-a)}, & a \leq y < b, \\ 0, & y < a. \end{cases} \tag{5.17}$$

Shapes of these functions are depicted in Figure 5.4. The quadratic shapes can be replaced by simpler linear ones.

5.2.3.3 Semantics of pure evaluative predications. Pure evaluative expressions are abstract, that is, they do not characterize sizes of concrete objects. Therefore, it has a good sense to consider only standard context $w = \langle 0, 0.5, 1 \rangle$ for them. Consequently, their extension can be identified with their intension and we can put

$$\text{Int}(\langle\text{linguistic hedge}\rangle\ \textit{small})(x) = \nu_{a,b,c}(LH(x)),$$

$$\text{Int}(\langle\text{linguistic hedge}\rangle\ \textit{medium})(x) = \nu_{a,b,c}(MH(x)),$$

$$\text{Int}(\langle\text{linguistic hedge}\rangle\ \textit{big})(x) = \nu_{a,b,c}(RH(x)),$$

where $x \in [0,1]$ and $\nu_{a,b,c}$ is the function (5.17) assigned to ⟨linguistic hedge⟩.

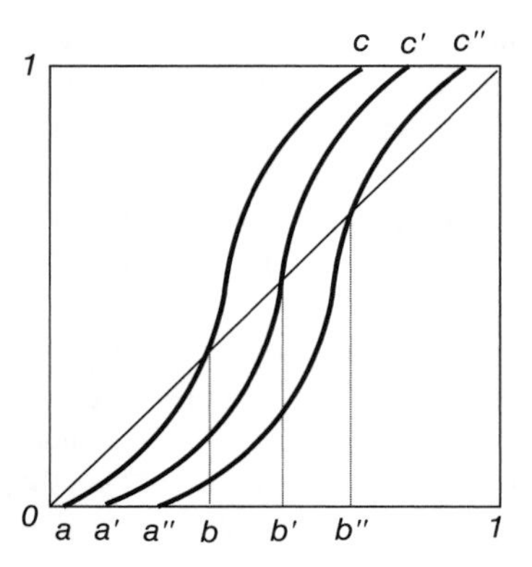

Figure 5.4 Shapes of the functions $v_{a,b,c}$ realizing horizon modification for various values of the parameters a, b, c.

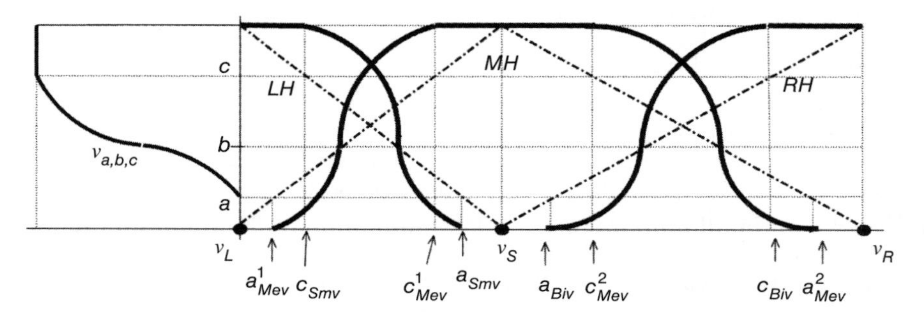

Figure 5.5 Scheme of the construction of extensions of pure evaluative predications.

Recall that pure evaluative predications have the form

$$X \text{ is } \langle\text{linguistic hedge}\rangle\langle\text{TE-adjective}\rangle, \tag{5.18}$$

where X is a variable whose values are various features of objects, for example, size, depth, velocity, pressure, and temperature. Values of these features depend on an individual context. Therefore, the essential role in the meaning of evaluative predications is played by the set of contexts W defined in (5.12). Intension of (5.18) has the form (5.1), and extensions have the form (5.2). In our case, it means that if $w = \langle v_L, v_S, v_R\rangle$ is a context, then $\text{Int}(X \text{ is } \mathscr{A})(w) = \text{Ext}_w(X \text{ is } \mathscr{A})$ is a fuzzy set

$$\text{Ext}_w(X \text{ is } \mathscr{A}) \subsetneq [v_L, v_S] \cup [v_S, v_R].$$

On the basis of the considerations above and the fact that a context is taken as a bijection $w : [0,1] \to [v_L, v_S] \cup [v_S, v_R]$, we get

$$\mathrm{Ext}_w(X \text{ is } \langle\text{linguistic hedge}\rangle \textit{ small})(x) = \nu_{a,b,c}(LH(w^{(-1)}(x))), \tag{5.19}$$

$$\mathrm{Ext}_w(X \text{ is } \langle\text{linguistic hedge}\rangle \textit{ medium})(x) = \nu_{a,b,c}(MH(w^{(-1)}(x))), \tag{5.20}$$

$$\mathrm{Ext}_w(X \text{ is } \langle\text{linguistic hedge}\rangle \textit{ big })(x) = \nu_{a,b,c}(RH(w^{(-1)}(x))), \tag{5.21}$$

where $w \in W$ and $x \in w$ (i.e., $x \in [v_L, v_S] \cup [v_S, v_R]$) and $w^{(-1)}$ is the extended inverse defined in (5.13). Construction of these extensions is schematically depicted in Figure 5.5.

Formulas for extensions of evaluative predications. If we put formulas (5.14)–(5.16) and (5.17) together and use (5.19)–(5.21), we obtain the following formulas for the computation of extensions of pure evaluative predications in the context $w = \langle v_L, v_S, v_R\rangle$.

Let us denote $K_1 = (c-b)(c-a)$ and $K_2 = (b-a)(c-a)$. Then,

$$\nu_{a,b,c}(LH(w^{(-1)}x)) = \begin{cases} 1, & x \le c_{Smv}, \\ & c_{Smv} = cv_L + (1-c)v_S, \\ 1 - \dfrac{(x - c_{Smv})^2}{K_1(v_S - v_L)^2}, & x \in (c_{Smv}, b_{Smv}], \\ & b_{Smv} = bv_L + (1-b)v_S, \\ \dfrac{(a_{Smv} - x)^2}{K_2(v_S - v_L)^2}, & x \in (b_{Smv}, a_{Smv}), \\ & a_{Smv} = av_L + (1-a)v_S, \\ 0, & x \ge a_{Smv}, \end{cases}$$

$$\nu_{a,b,c}(MH(w^{(-1)}x)) = \begin{cases} 1, \quad x \in [c^1_{Mev}, c^2_{Mev}], & \\ \quad c^1_{Mev} = cv_S + (1-c)v_L,\ c^2_{Mev} = cv_S + (1-c)v_R, & \\ 1 - \dfrac{(c^1_{Mev} - x)^2}{K_1(v_S - v_L)^2}, & x \in [b^1_{Mev}, c^1_{Mev}), \\ & b^1_{Mev} = bv_S + (1-b)v_L, \\ 1 - \dfrac{(x - c^2_{Mev})^2}{K_1(v_R - v_S)^2}, & x \in (c^2_{Mev}, b^2_{Mev}], \\ & b^2_{Mev} = bv_S + (1-b)v_R, \\ \dfrac{(x - a^1_{Mev})^2}{K_2(v_S - v_L)^2}, & x \in (a^1_{Mev}, b^1_{Mev}), \\ & a^1_{Mev} = av_S + (1-a)v_L, \\ \dfrac{(a^2_{Mev} - x)^2}{K_2(v_R - v_S)^2}, & x \in (b^2_{Mev}, a^2_{Mev}), \\ & a^2_{Mev} = av_S + (1-a)v_R, \\ 0, & x \le a^1_{Mev} \text{ or } x \ge a^2_{Mev}, \end{cases}$$

$$\nu_{a,b,c}(RH(w^{(-1)}x)) = \begin{cases} 1, & x \geq c_{Biv}, \\ & c_{Biv} = cv_R + (1-c)v_S, \\ 1 - \dfrac{(c_{Biv} - x)^2}{K_1(v_R - v_S)^2}, & x \in [b_{Biv}, c_{Biv}), \\ & b_{Biv} = bv_R + (1-b)v_S, \\ \dfrac{(x - a_{Biv})^2}{K_2(v_R - v_S)^2}, & x \in (a_{Biv}, b_{Biv}), \\ & a_{Biv} = av_R + (1-a)v_S, \\ 0, & x \leq a_{Biv}. \end{cases}$$

The meaning of the parameters a_{Smv} and c_{Smv} in extensions of the evaluative expressions "⟨linguistic hedge⟩ *small*" is apparent from Figure 5.5 as well as the meaning of the corresponding parameters for evaluative expressions "⟨linguistic hedge⟩ *medium*" and "⟨linguistic hedge⟩ *big*".

Intensions $\text{Int}(X \text{ is } \mathscr{A}) : W \longrightarrow \mathscr{F}(\mathbb{R})$ of the several pure evaluative expressions are schematically depicted in Figure 5.6. Note that because contexts are changing, extensions are modified in size, but their shapes remain the same.

It should be emphasized that shapes of membership functions of extensions of pure evaluative predications are in all contexts constructed in the same way. Recall that as a special case, the evaluative expressions "small", "medium", and "big" are also modified by a linguistic hedge—the *empty hedge*—which is also assigned a specific function $\nu_{a,b,c}$.

Table 5.1 contains experimentally estimated values of the parameters a, b, c in $\nu_{a,b,c}$ of few selected linguistic hedges. They follow from the natural ordering of

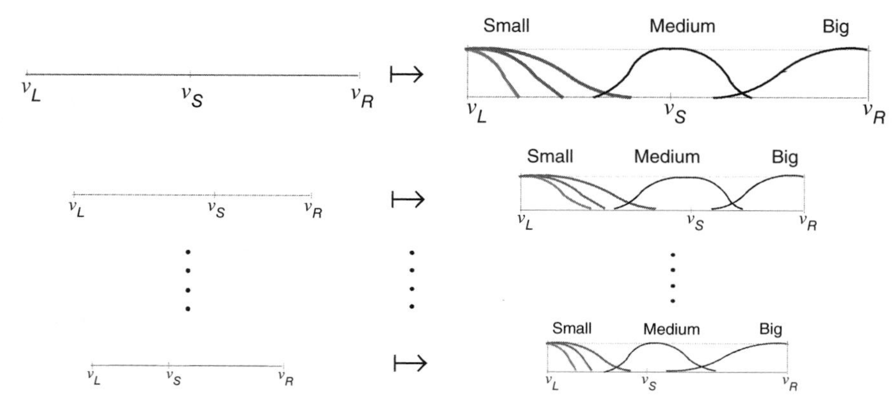

Figure 5.6 Intensions of the pure evaluative expressions "very small", "small", "roughly small", "medium", and "big". Because intension is a function, each context $w = \langle v_L, v_S, v_r \rangle \in W$ on the left side of the arrow are assigned extensions of the corresponding linguistic expressions (right side of the arrow).

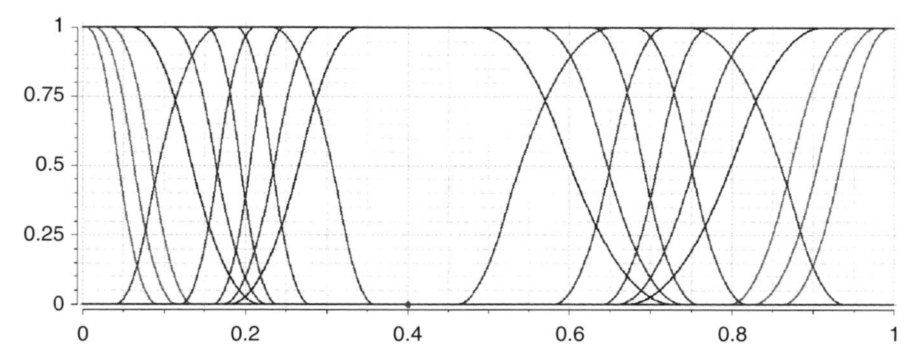

Figure 5.7 Experimentally estimated extensions of selected pure evaluative predications in the context $\langle 0, 0.4, 1\rangle$. The curves depict membership functions of extensions of "⟨linguistic hedge⟩ *small*" from the left and "⟨linguistic hedge⟩ *big*" from the right where ⟨linguistic hedge⟩:= extremely | significantly | very | ⟨empty hedge⟩ | more or less | roughly | quite roughly | very roughly. Finally, in the midsection, there are membership functions of the extensions "⟨linguistic hedge⟩ *medium*" for hedges with widening effect.

TABLE 5.1 Possible experimentally estimated values of the parameters a, b, c determining selected linguistic hedges.

Linguistic hedge	a	b	c
Extremely	0.77	0.9	0.99
Significantly	0.71	0.85	0.962
Very	0.66	0.79	0.915
Empty hedge	0.45	0.68	0.851
Typically	0.88	0.95	1
Rather	0.55	0.65	0.82
More or less	0.43	0.6	0.727
Roughly	0.4	0.52	0.619
Quite roughly	0.3	0.42	0.528
Very roughly	0.1	0.2	0.421

pure evaluative expressions given above. Shapes of the corresponding membership functions of these extensions are depicted in Figure 5.7. Recall that the TE-adjective "*medium*" cannot be modified by any of the narrowing hedges *extremely, significantly*, and *very*.

EXAMPLE 5.6 For example, let the context for heights of people be $w_H = \langle 155, 175, 190\rangle$. Then, using the formulas above and Table 5.1, we obtain the following extensions:

$$\mathrm{Ext}_{w_H}(X \text{ is } \textit{very small})(x) = \begin{cases} 1, & x \le 156.7, \\ 1 - \dfrac{(x-156.7)^2}{12.75}, & x \in (156.7, 159.2], \\ \dfrac{(161.8-x)^2}{13.26}, & x \in (159.2, 161.8), \\ 0, & x \ge 161.8, \end{cases}$$

$$\mathrm{Ext}_{w_H}(X \text{ is } \textit{small})(x) = \begin{cases} 1, & x \le 158, \\ 1 - \dfrac{(x-156.7)^2}{27.42}, & x \in (158, 161.4], \\ \dfrac{(161.8-x)^2}{36.89}, & x \in (161.4, 166), \\ 0, & x \ge 166, \end{cases}$$

$$\mathrm{Ext}_{w_H}(X \text{ is } \textit{big})(x) = \begin{cases} 1, & x \ge 187.8, \\ 1 - \dfrac{(187.8-x)^2}{15.43}, & x \in [185.2, 187.8), \\ \dfrac{(x-181.8)^2}{20.75}, & x \in (181.8, 185.2), \\ 0, & x \le 181.8, \end{cases}$$

$$\mathrm{Ext}_{w_H}(X \text{ is } \textit{roughly big})(x) = \begin{cases} 1, & x \ge 184.3, \\ 1 - \dfrac{(184.3-x)^2}{4.88}, & x \in [182.8, 184.3), \\ \dfrac{(x-181)^2}{5.92}, & x \in (181, 182.8), \\ 0, & x \le 181. \end{cases}$$

Similarly, we can compute formulas for extensions of the other evaluative predications. □

Remark 5.3

(a) *Let us emphasize that formulas from Example 5.6 should not be taken absolutely in the sense that they* are *extensions of the given predications but only that they are just mathematical models which we can take as the way how, for example, we can make computer to "understand" the corresponding natural language expressions. Moreover, they are simplifications that can be effectively used in various kinds of technical applications, but if we wanted to model expressions and sentences of genuine natural language, more sophisticated model would be needed (e.g., the parameters in Table 5.1 could be different for different kinds of evaluative expressions in specific contexts).*

(b) *There is a certain subjectivity in the estimation of membership functions. Experimental results confirm, however, that different people estimate these functions similarly, which means that people understand concepts accordingly. Of course, this is not a surprising result, because otherwise, natural language would be useless and it could not serve us as a medium for information transmission. This supports the proclaim that the concept of fuzzy set was introduced meaningfully.*

The theory of evaluative expressions is very important in fuzzy modeling. It is largely applied in approximate reasoning and its applications described below. But it also has applications as such. For example, in [78], an interesting application is described, which mimics the thinking of a geologist when he/she is to determine special sequences of rocks being result of movement of an ancient sea level.

5.3 INTERPRETATION OF FUZZY/LINGUISTIC IF-THEN RULES

5.3.1 Linguistic Description

In Chapter 3, the linguistic description, according to Definition 2.3, was taken as a sort of imprecise description of some function. The surface form has been taken as a coded characterization of the disjunctive (DNF) or conjunctive (CNF) normal forms.

Using the model of the semantics of evaluative linguistic expressions, however, it is possible to extend interpretation of the fuzzy IF-THEN rules in such a way that they can be taken as genuine conditional expressions of natural language. We will call them *fuzzy/linguistic IF-THEN rules*. For the sake of simplicity, we consider only one antecedent variable X here. Generalization to more than one antecedent variable (which is much more common in practice) is not difficult.

Recall from Definition 2.3 that a linguistic description is a finite set of fuzzy/linguistic IF-THEN rules $\mathrm{LD} = \{\mathscr{R}_1, \ldots, \mathscr{R}_m\}$. From a linguistic point of view, a linguistic description LD can be understood as a specific form of structured text. Using it, we describe various situations and processes, control various kinds of processes, make decisions, classify, etc.

Topic and focus. A very important phenomenon studied in linguistics is that of *topic–focus articulation* [46]. It is omnipresent in natural language and cannot be neglected. Due to it, any sentence can be divided into two parts: a *topic*, which is the part of sentence that expresses what we are speaking about, and a *focus*, which is a part of sentence that brings a new information.

For example, consider the sentence "John is a very clever man". This sentence has several readings depending on its division into topic and focus. We can say "JOHN is a very clever man",[13] which means that just John and nobody else (focus) is very clever (topic). Similarly, "John is A VERY CLEVER man" means that the man John (topic) is very clever (focus) and not, for example, very tall or fat. Clearly, there are more readings of this sentence, each of which brings another kind of information.

[13]The text typeset in capitals should be read with accent.

The problem is that this division may not always be clear and heavily depends on the context of use. Luckily, in case of fuzzy/linguistic IF-THEN rules, the situation is much simpler because their topic can be clearly distinguished from their focus. For example, consider the IF-THEN rule

IF *temperature* is *very small* THEN *shift of control lever* is *very big.*

The topic is *temperature is very small*, because it specifies a given situation and precedes the action determined by the consequent—*shift of control lever is very big* (focus).

From it follows that the topic of a fuzzy/linguistic IF-THEN rule is its antecedent, that is, the linguistic predication "X is $\mathscr{A}$", and the focus is its consequent, that is, the linguistic predication "Y is $\mathscr{B}$". At the same time, we can also speak about topic and focus of the whole linguistic description LD. We put

$$Topic_{LD} = \{X \text{ is } \mathscr{A}_i \mid i = 1, \dots, m\}, \tag{5.22}$$

$$Focus_{LD} = \{Y \text{ is } \mathscr{B}_i \mid i = 1, \dots, m\}. \tag{5.23}$$

5.3.2 Intension of Fuzzy/Linguistic IF-THEN Rules

Since we consider a fuzzy/linguistic IF-THEN rule to be a genuine expression of natural language, we must be able to construct its intension. The construction is based on the intensions of evaluative expressions from which the IF-THEN rule is formed.

Let

$$\mathscr{R} := \ \text{IF } X \text{ is } \mathscr{A} \text{ THEN } Y \text{ is } \mathscr{B}$$

be a fuzzy/linguistic IF-THEN rule. Furthermore, let

$$\text{Int}(X \text{ is } \mathscr{A}) : W \longrightarrow \mathscr{F}(\mathbb{R}),$$

$$\text{Int}(Y \text{ is } \mathscr{B}) : W \longrightarrow \mathscr{F}(\mathbb{R})$$

be intensions of the evaluative predications occurring in the rule $\mathscr{R}$. Then the intension of $\mathscr{R}$ is the function

$$\text{Int}(\mathscr{R}) := \ W \times W \longrightarrow \mathscr{F}(\mathbb{R}) \times \mathscr{F}(\mathbb{R}), \tag{5.24}$$

which assigns to any couple of contexts $w, w' \in W$ a fuzzy relation

$$\langle w, w' \rangle \mapsto \text{Ext}_w(X \text{ is } \mathscr{A}) \ominus \text{Ext}_{w'}(Y \text{ is } \mathscr{B}). \tag{5.25}$$

The membership function of this fuzzy relation is

$$(\text{Ext}_w(X \text{ is } \mathscr{A}) \ominus \text{Ext}_{w'}(Y \text{ is } \mathscr{B}))(u, v) = \text{Ext}_w(X \text{ is } \mathscr{A})(u) \rightarrow \text{Ext}_{w'}(Y \text{ is } \mathscr{B})(v) \tag{5.26}$$

for all $u \in w$, $v \in w'$, where $\rightarrow$ is a fuzzy implication (usually, we suppose that it is the Łukasiewicz implication (2.24)). If we realize that $w = \langle v_L, v_S, v_R \rangle$ and $w' = \langle v'_L, v'_S, v'_R \rangle$, then it is clear that

$$\begin{aligned}\mathrm{Ext}_{\langle w,w'\rangle}(\mathscr{R}) = \mathrm{Ext}_w(X \text{ is } \mathscr{A}) \ominus \mathrm{Ext}_{w'}(Y \text{ is } \mathscr{B}) \precsim \\ \precsim ([v_L, v_S] \cup [v_S, v_R]) \times ([v'_L, v'_S] \cup [v'_S, v'_R]).\end{aligned}$$

Schematically, we can write down intension (5.24) in the form

$$\mathrm{Int}(\mathscr{R}) := \mathrm{Int}(X \text{ is } \mathscr{A}) \Rightarrow \mathrm{Int}(Y \text{ is } \mathscr{B}). \tag{5.27}$$

We will sometimes say that the couple $\langle w, w' \rangle$, for $w, w' \in W$, is a *general context* of the rule $\mathscr{R}$.

5.4 APPROXIMATE REASONING WITH LINGUISTIC INFORMATION

5.4.1 Basic Principle of Approximate Reasoning

The theory of approximate reasoning provides general method for a processing of fuzzy/linguistic IF-THEN rules. It consists in finding appropriate conclusion based on a given linguistic description, if a new information is given.

Recall the basic approximate reasoning scheme from (3.54):

$$\begin{array}{ll} \text{Condition:} & \mathscr{R} := \mathsf{IF}\ X \text{ is } \mathscr{A}\ \mathsf{THEN}\ Y \text{ is } \mathscr{B}. \\ \text{Observation:} & X \text{ is } \mathscr{A}'. \\ \hline \text{Conclusion:} & Y \text{ is } \mathscr{B}'. \end{array} \tag{5.28}$$

In the present case, however, the observation is an *evaluative linguistic predication* "X is $\mathscr{A}'$", which can be slightly different from the original predication "X is $\mathscr{A}$". Naturally, it follows that a conclusion is also a linguistic predication "Y is $\mathscr{B}'$", which can be slightly different from "Y is $\mathscr{B}$".

EXAMPLE 5.7 A typical example of an approximate reasoning scheme, which often occurs in fuzzy control, is the following (for two independent variables):

Condition: IF error is *small* AND *change of error is rather big* THEN change of control action is *small* .
Observation: error is *very small* AND *change of error is big.*

Conclusion: change of control action is *rather small.*

In this example, the observation contains expressions *very small* instead of *small* and *big* instead of *rather big*. Because the observation slightly differs from the antecedent of the condition, the conclusion slightly differs from the consequent, namely, the expression *rather small* occurs instead of the expected *small*. □

The reader has certainly noticed that (5.28) is a modification of the rule of modus ponens (2.84). The main difference lies in the fact that $\mathscr{A}$ and $\mathscr{A}'$ in (5.28) are vague expressions of natural language, which can differ (from each other), while in (2.84), the observation A should be identical with the antecedent of the implication $A \Rightarrow B$.

5.4.2 Perception-Based Logical Deduction

The fundamental method for a derivation of a conclusion based on a given linguistic description is called *perception-based logical deduction* (PbLD). Its author is Vilém Novák. We will explain the PbLD method by means of the following example.

EXAMPLE 5.8 Consider the linguistic description:

$$\mathscr{R}_1 := \text{IF } X \text{ is } \textit{small} \text{ THEN } Y \text{ is } \textit{medium},$$

$$\mathscr{R}_2 := \text{IF } X \text{ is } \textit{very small} \text{ THEN } Y \text{ is } \textit{big},$$

$$\mathscr{R}_3 := \text{IF } X \text{ is } \textit{medium} \text{ THEN } Y \text{ is } \textit{small},$$

$$\mathscr{R}_4 := \text{IF } X \text{ is } \textit{big} \text{ THEN } Y \text{ is } \textit{very small}.$$

Such description characterizes, for example, a simple control of a temperature (represented by the variable X) in some furnace. Every fuzzy/linguistic IF-THEN rule provides us with a partial information about an action (represented by the variable Y) that should be taken on the basis of a measured temperature. Moreover, despite the fact that the rules themselves are vague, we can discern among them and, consequently, select the appropriate one according to a given observation.

Let us be given concrete contexts, namely, $w \in W$ for values of X and $w' \in W$ for values of Y. For example, let $w = \langle 150, 330, 600 \rangle$ be the context for the temperature in the furnace (in Celsius degrees) and $w' = \langle 0, 36, 90 \rangle$ be the context for a position of a control cock (in degrees). Then, "small X" are (in this context) values of X around 170–220 (and smaller), "medium X" are values around 300–360, and "big X" are values around 500–550 and bigger. Similarly, "small Y" are values of Y around 5–12 (and smaller), "medium Y" are values around 30–40, and "big Y" around 70–85 (and bigger).

Let the measured value (observation) be, for example, $X = 200$. In order to be able to derive the conclusion based on this observation, we must check whether the linguistic description says something about such value. In other words, we must check whether this value is covered by the topic of this linguistic description. Because the value 200 is apparently *small* in the given context, we know from the rule $\mathscr{R}_1$ that Y (turn of the cock) should be *medium*, that is, it should lie somewhere around 30–40. If $X = 165$, then this value is also small, but, in comparison with 200, significantly smaller. Because the rule $\mathscr{R}_2$ is also present in the linguistic description, we agree

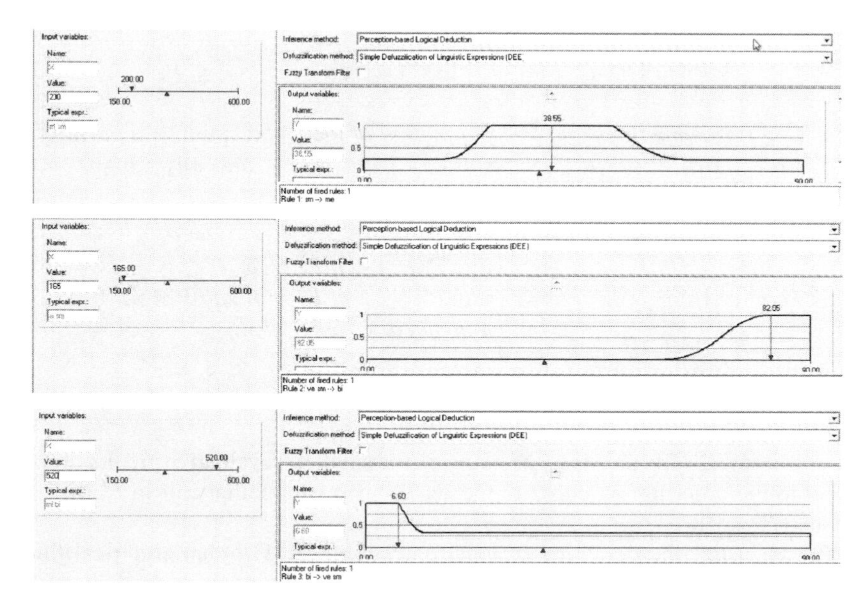

Figure 5.8 Demonstration of the behavior of PbLD on the basis of linguistic description from Example 5.8. **Left:** Three considered input values; **Right:** The resulting fuzzy sets derived from the corresponding rules $\mathscr{R}_1, \mathscr{R}_2, \mathscr{R}_4$ together with marked output values obtained using the defuzzification DEE.

that 165 is a *very small* temperature. Hence, the resulting value of Y should be *big* (lie somewhere around 70–85). If the rule $\mathscr{R}_2$ were not present, then, because 165 is surely also *small* temperature, the resulting value would be *medium* again.

Analogously, $X = 520$ is a *big* value. Then, the rule $\mathscr{R}_4$ says that the value Y should be *very small*, that is, it should be considerably smaller than 12. Demonstration of how PbLD works on this linguistic description is in Figure 5.8. □

Local perception. By *perception*, we understand in this book an *evaluative expression* assigned to a given value (observation) in a given context. The following definition is important for explanation of the PbLD method and has various kinds of applications.

Definition 5.4 *Let W be a set of contexts and $x \in \mathbb{R}$. Let $K \subseteq EvExpr$ be a* finite *set of evaluative predications such that $(K, \lll)$ is* linearly *ordered. Put*

$$P = \{\text{“}X \textit{ is } \mathscr{B}\text{”} \in K \mid \text{Ext}_w(X \textit{ is } \mathscr{B})(x) > 0\}. \tag{5.29}$$

Furthermore, let $P' \subseteq P$ be a set of evaluative predications such that

$$\text{“}X \textit{ is } \mathscr{B}\text{”} \in P' \quad \textit{iff} \quad \text{Ext}_w(X \textit{ is } \mathscr{B})(x) \textit{ is maximal,}$$ [14]

and,

$$P'' = \{\text{“}X \text{ is } \mathscr{B}\text{”} \in K \mid \mathrm{Ext}_w(X \text{ is } \mathscr{B})(x) \geq a_0\} \subseteq P, \tag{5.30}$$

where $a_0 > 0$ is some threshold.[15] *The function of* local perception *is a partial function $LPerc^K : \mathbb{R} \times W \longrightarrow K$, which to any value $x \in \mathbb{R}$ and any context $w \in W$ assigns an evaluative predication as follows:*

$$LPerc^K(x, w) = \text{“}X \text{ is } \mathscr{A}\text{”} = \begin{cases} \min P'', & \text{if } P'' \neq \emptyset, \\ \min P', & \text{if } P'' = \emptyset \text{ and } P \neq \emptyset, \\ \text{undefined}, & \text{otherwise}, \end{cases} \tag{5.31}$$

where minimum is taken with respect to the ordering $\lll$.

Thus, “X is $\mathscr{A}$”$\in P'$ is the *most specific* (i.e., the smallest) evaluative predication with respect to the ordering $\lll$ either in the set P'', if it is nonempty, or in P' (if $P \neq \emptyset$, then also $P' \neq \emptyset$).

We can justify the definition of local perception (5.31) by an empirical finding that any value can be, in a given context, evaluated by some evaluative expression. Because these expressions specifically characterize certain values, we choose the expression that provides the most specific information. If there is no expression in K fulfilling the given conditions, then $\mathrm{LPerc}^K(x, w)$ gives no result.

The assignment of perception is schematically depicted in Figure 5.9. We can see an observer standing at point v_L and watching a person approaching him/her. At each time moment, the observer characterizes the distance of the approaching person by a proper evaluative expression. Extensions of possible evaluative expressions in the given context are indicated in the upper part of the picture. The figure also demonstrates that at the closest point we can characterize the distance of the person by two

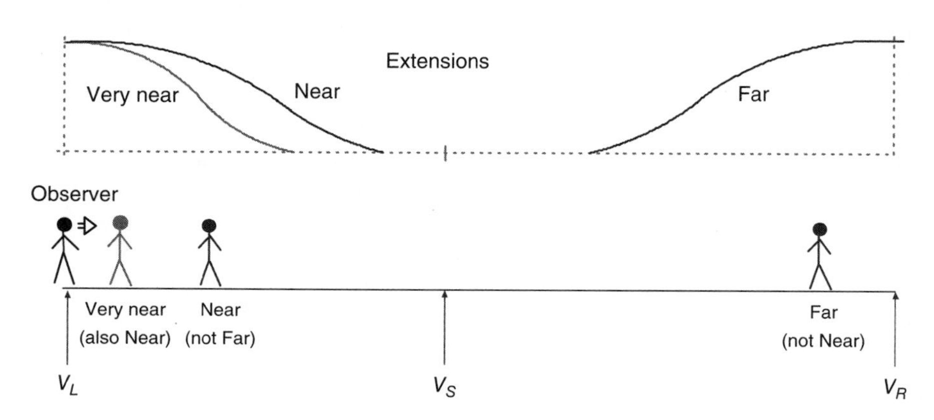

Figure 5.9 Schematic depiction of perceptions.

[14]Precise definition is the following: “X is $\mathscr{B}$” $\in P'$ if for each “X is $\mathscr{C}$” $\in P$, $\mathrm{Ext}_w(X \text{ is } \mathscr{C})(x) \geq \mathrm{Ext}_w(X \text{ is } \mathscr{B})(x)$ implies that $\mathrm{Ext}_w(X \text{ is } \mathscr{C})(x) = \mathrm{Ext}_w(X \text{ is } \mathscr{B})(x)$.

[15]We usually put $a^0 = 0.9$ or $a^0 = 1$.

perceptions: *near* and *very near*. This means that we can distinguish both of them, but the extension of the second, more specific (perception), is included in the first one. Which of them will be used by the observer depends on the *topic* of the given *linguistic description*. If both perceptions are included in it, then he/she uses the second one, because it is more accurate.

Local perception for PbLD. If we want to apply the function of local perception for reasoning on the basis of a given linguistic description LD, we must realize that this is a text that has its topic. Therefore, the choice of the perception is not casual but depends on a topic specified by the given linguistic description.

EXAMPLE 5.9 The topic of the linguistic description from Example 5.8 is

$$Topic_{LD} = \{X \text{ is } small,\ X \text{ is } very\ small,\ X \text{ is } medium,\ X \text{ is } big\}.$$

Then, for example, the evaluative predication "X is *very small*" is the perception of $X = 165$. Analogously, the predication "X is *big*" is the perception of $X = 520$ (of course, with respect to the given context $w = \langle 150, 330, 600\rangle$). □

The function of local perception restricted to the topic of LD is a special case of Definition 5.4, where $K = \text{Topic}_{\text{LD}}$. For simplicity, we will denote it as LPerc^{LD}.

Let us remark that this function is also used for learning of linguistic descriptions on the basis of the given data. The learning method is very powerful and robust. It is used in fuzzy control (see Section 7.4.2) and in forecasting of time series (see Chapter 9).

Evaluation. The concept of local perception is related to the concept of *evaluation*.

Definition 5.5 *Let w be a context, $x \in \mathbb{R}$, and let $\mathscr{A}$ be an evaluative expression. We say that an element $x \in w$ is* evaluated *by $\mathscr{A}$, if there exists a truth value $a > 0$ such that*

$$a \to \text{Ext}_w(X\ is\ \mathscr{A})(x) = 1, \tag{5.32}$$

where $\to$ is the fuzzy implication (2.24). If (5.32) holds true for some $a > 0$, then we formally write

$$Eval(x, w, \mathscr{A}). \tag{5.33}$$

Note that (5.32) is equivalent to the inequality

$$a \le \text{Ext}_w(X \text{ is } \mathscr{A})(x).$$

EXAMPLE 5.10 Consider the linguistic description from Example 5.8 and the contexts $w_X = \langle 150, 330, 600\rangle$ for the variable X and $w_Y = \langle 0, 36, 90\rangle$ for Y.

Then, for example, $Eval(220, w_X, \text{small})$, because the membership degree of 220 in the extension of *small* in the context w is greater than, for example, $a = 0.2$.

On the other hand, it does not hold that $Eval(5, w_Y, \text{big})$, because the membership degree of 5 in the extension of *big* in the context w_Y is equal to 0. At the same time, $Eval(85, w_Y, \text{big})$ holds true. □

5.4.3 Formalization of the Perception-Based Logical Deduction

In this section, we formally describe a procedure of derivation of a conclusion based on a given linguistic description and a perception of the given observation.

Definition of PbLD. Suppose that a linguistic description $LD = \{\mathscr{R}_1, \dots, \mathscr{R}_m\}$ according to Definition 2.3 is given. We can understand it as a set of intensions

$$LD: \quad \begin{array}{l} \text{Int}(\mathscr{R}_1) := \text{Int}(X \text{ is } \mathscr{A}_1) \Rightarrow \text{Int}(Y \text{ is } \mathscr{B}_1), \\ \dots\dots\dots\dots\dots\dots\dots\dots\dots\dots \\ \text{Int}(\mathscr{R}_m) := \text{Int}(X \text{ is } \mathscr{A}_m) \Rightarrow \text{Int}(Y \text{ is } \mathscr{B}_m). \end{array}$$

Further, let us consider a context $w \in W$ for the variable X and a context $w' \in W$ for the variable Y. Let an observation x_0 in the context w be given. Using (5.31), we assign a perception "X is $\mathscr{A}_{i_0}$" for some $i_0 \in \{1, \dots, m\}$ to the observation $x_0 \in w$.[16]

Under the above-mentioned assumptions, we define the PbLD rule as follows (for formal details and justification, see [80, 92, 87]).

Definition 5.6 *The* perception-based logical deduction rule r_{PbLD} *is the scheme*

$$r_{PbLD} : \frac{LPerc^{LD}(x_0, w) = X \text{ is } \mathscr{A}_{i_0}, LD}{Eval(\hat{y}_{i_0}, w', \mathscr{B}_{i_0})}, \tag{5.34}$$

where "X is $\mathscr{A}_{i_0}$" $\in Topic^{LD}$, "Y is $\mathscr{B}_{i_0}$" $\in Focus^{LD}$, and

$$\hat{y}_{i_0} = DEE\left(\left\{ \text{Ext}_w(X \text{ is } \mathscr{A}_{i_0})(x_0) \to \text{Ext}_{w'}(Y \text{ is } \mathscr{B}_{i_0})(y)/y \;\middle|\; y \in w' \right\}\right). \tag{5.35}$$

Explanation of scheme (5.34) is simple. If "X is $\mathscr{A}_{i_0}$" is a perception of the observation $X = x_0$ in the context w and the linguistic description LD is given, then there is an element $\hat{y}_{i_0}$, which is *certainly evaluated by the evaluative expression* $\mathscr{B}_{i_0}$. Furthermore, the local perception "X is $\mathscr{A}_{i_0}$", which is the perception of the element x_0 in the context w, was chosen from the topic of the linguistic description LD. This permits us to fire the rule

$$\mathscr{R}_{i_0} := \mathsf{IF}\ X \text{ is } \mathscr{A}_{i_0}\ \mathsf{THEN}\ Y \text{ is } \mathscr{B}_{i_0},$$

which occurs in *LD*. Its antecedent is "X is $\mathscr{A}_{i_0}$"; therefore, we can execute the modus ponens deduction rule (cf. Section 5.4.1).

[16] In general, there can be more rules with the antecedent "X is $\mathscr{A}_{i_0}$" in LD, see Remark 5.4. In practice, however, usually, only one rule is fired.

Analysis of the conclusion using PbLD. The fired rule $\mathscr{R}_{i_0}$ characterizes the existing local relationship between X and Y. The given value of X, namely, $X = x_0$, induces a fuzzy set $\text{Ext}_{w'}(Y \text{ is } \mathscr{B}'_{i_0}) \subsetneq w'$ given by the membership function

$$\text{Ext}_{w'}(Y \text{ is } \mathscr{B}'_{i_0}) = \text{Ext}_w(X \text{ is } \mathscr{A}_{i_0})(x_0) \rightarrow \text{Ext}_{w'}(Y \text{ is } \mathscr{B}_{i_0})(y), \quad y \in w'. \tag{5.36}$$

If we analyze the predication "Y is $\mathscr{B}'_{i_0}$" in (5.36), we see that its extension was obtained from the extension $\text{Ext}_{w'}(Y \text{ is } \mathscr{B}_{i_0})$ and the truth value $\text{Ext}_w(X \text{ is } \mathscr{A}_{i_0})(x_0)$ using the operation of fuzzy implication $\rightarrow$[17]. But "Y is $\mathscr{B}_{i_0}$" is a consequent of the rule $\mathscr{R}_{i_0}$, which characterizes dependence of Y on X. Hence, the extension $\text{Ext}_{w'}(Y \text{ is } \mathscr{B}'_{i_0})$ must be affected by $\mathscr{R}_{i_0}$. We thus conclude that the extension (5.36) contains all values of Y that are evaluated by $\mathscr{B}_{i_0}$ in the sense of Definition 5.5 where the lower boundary truth value is $\text{Ext}_w(X \text{ is } \mathscr{A}_{i_0})(x_0)$.

EXAMPLE 5.11 We will explain the details of PbLD on a simple example with linguistic description consisting of two rules:

$$\mathscr{R}_1 := \text{IF } X \text{ is } \textit{big} \text{ THEN } Y \text{ is } \textit{small},$$

$$\mathscr{R}_2 := \text{IF } X \text{ is } \textit{small} \text{ THEN } Y \text{ is } \textit{big}.$$

The contexts for antecedent as well as consequent variables are $w = w' = \langle 0, 0.4, 1 \rangle$. Because extensions of "X is small" and "X is big" do not overlap in any context, at most one rule will fire.

For example, let us consider a value $X = 0.85$. The truth degree of this value in the extension of "big" in the context w is 0.82. As there is no other rule in whose antecedent the value $X = 0.85$ would be true in higher or the same degree, only the rule $\mathscr{R}_1$ fires. Using formula (2.24), we obtain the fuzzy set (5.36), which is depicted in Figure 5.10(a). This fuzzy set contains all values of Y evaluated as "small" (in the context w') where the original extension of "small" was modified due to dependence of Y on X characterized by rule $\mathscr{R}_1$.

We now have to choose one value of Y as a consequence of $X = 0.85$ via rule $\mathscr{R}_1$. By PbLD, this value is $\hat{y}_1$ given in (5.35). The defuzzification DEE assures that $\hat{y}_1$ is the worst possible value of Y determined by $X = 0.85$ via $\mathscr{R}_1$, which is evaluated by the evaluative expression "small" in the highest degree 1 (i.e., the membership degree $\text{Ext}_{w'}(Y \text{ is } \textit{small}\,')(\hat{y}_1) = 1$; cf. (5.36)).

Recall that by DEE, $\hat{y}_1$ lies on the right side of the kernel of "small" (see Figure 3.6). As can be seen in Figure 5.10, we obtain $\hat{y}_1 = 0.1$. If we set $X = 0.11$, the output of PbLD due to the fired rule $\mathscr{R}_2$ is $\hat{y}_2 = 0.83$. □

Modifications of PbLD. The DEE defuzzification can give in some cases too pessimistic results. This can be improved by replacing it by the MOM defuzzification in (5.35) (cf. Section 3.2.1). The result for the linguistic description from Example 5.11 is depicted in Figure 5.11. This defuzzification, however, harms the monotone behavior of PbLD (cf. Section 5.4.4).

[17]This interprets implication in the rule $\mathscr{R}_{i_0}$.

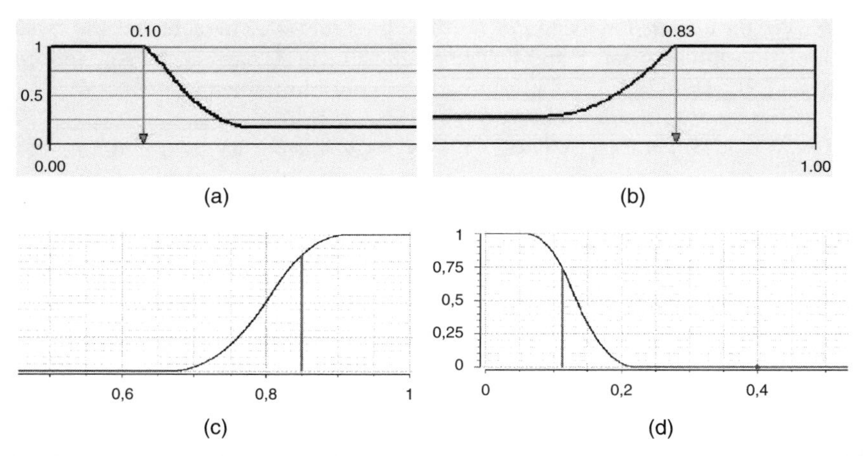

Figure 5.10 (a) The membership function characterizing *small values, provided that they are related with big ones* on the basis of the rule "IF X is big THEN Y is small" and given the antecedent value $X = 0.85$. The marked output value $Y = 0.10$ is result of the defuzzification DEE. (c) The membership function of the extension of *big* with marked value $X = 0.85$. (b) The membership function characterizing *big values, provided that they are related with small ones* on the basis of the rule "IF X is small THEN Y is big" and given $X = 0.11$. The value $Y = 0.83$ is obtained using the DEE defuzzification. (d) The membership function of the extension of *small* with marked value $X = 0.11$.

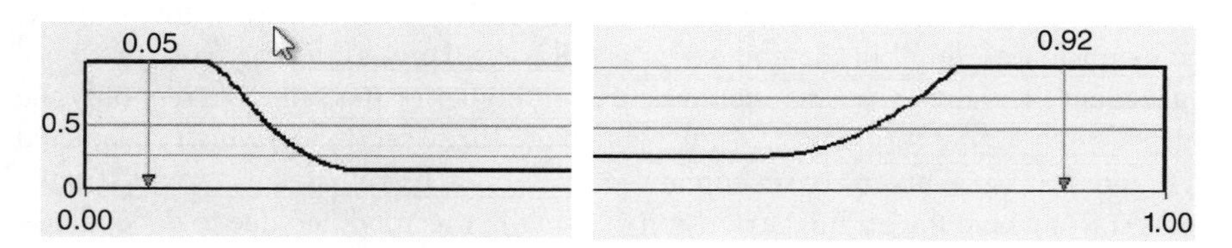

Figure 5.11 Results of the PbLD method on the basis of the same linguistic description as was used in Figure 5.10, where the defuzzification MOM was used instead of the DEE.

Because in PbLD, mostly one rule fires, its output is, in general, discontinuous. This is usually not a problem in practice. However, if we still need continuity, we may use the *smooth DEE defuzzification* described in [94]. This method combines the DEE defuzzification with fuzzy transform (see Chapter 4). The basic idea is first to generate the output using PbLD with DEE defuzzification and then to apply the F-transform on the result. The smooth PbLD method preserves the fundamental behavior of the original PbLD, but it is continuous at the same time. The price, however, is its greater computational complexity.

Other modifications and improvements of the PbLD method were recently published by Antonín Dvořák, Martin Štěpnička, and Lenka Štěpničková in [37] (see also [142]). In the paper, an important problem of finding redundancies in the

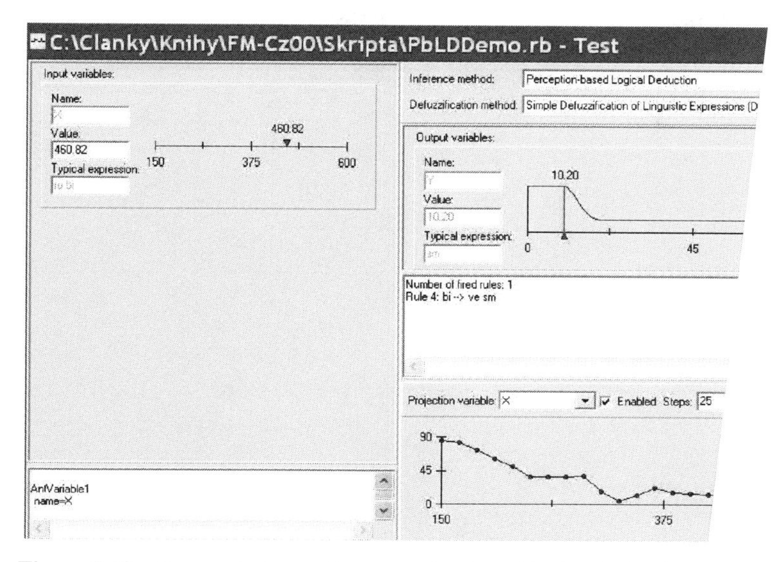

Figure 5.12 Demonstration of the perception-based logical deduction v
defuzzification. In the upper-left corner is the input value (in the given co
in the middle right is the fired rule and the corresponding fuzzy set e
output values, provided that the rule holds. The lower right corner co
output values (after DEE defuzzification) for all the input values in the

given linguistic description is solved. Using these results, it
significantly the number of fuzzy/linguistic IF-THEN rules, w
applications.

Demonstration of the behavior of the PbLD. The behavior of
of the linguistic description considered in Example 5.8 is de
The course of the output can be seen in the bottom-right corn
upper-right corner, the fuzzy set (5.36) and the result of its I
shown. We can see from the course of the output that if X is
(the rule $\mathscr{R}_2$). If X increases, then Y decreases, because it i
very small. At some point, X becomes small but not very s
$\mathscr{R}_1$, which says that Y should be medium, is fired.

If X increases even more, then the rule $\mathscr{R}_3$, saying th
Because the decrease in Y is higher if it is more true tha
extension (fuzzy set) of the evaluative expression "medium
on the output as an arc. But, this can be a problem in son
DEE defuzzification eliminates this behavior (see Figure 5

Remark 5.4 *It can happen that there are two rules,*
antecedent evaluative expressions $\mathscr{A}_i = \mathscr{A}_j = \mathscr{A}$ *and diff*
observation) is such that $LPerc^{LD}(x_0, w) = \mathrm{Int}(X\ is\ \mathscr{A})$, *th*

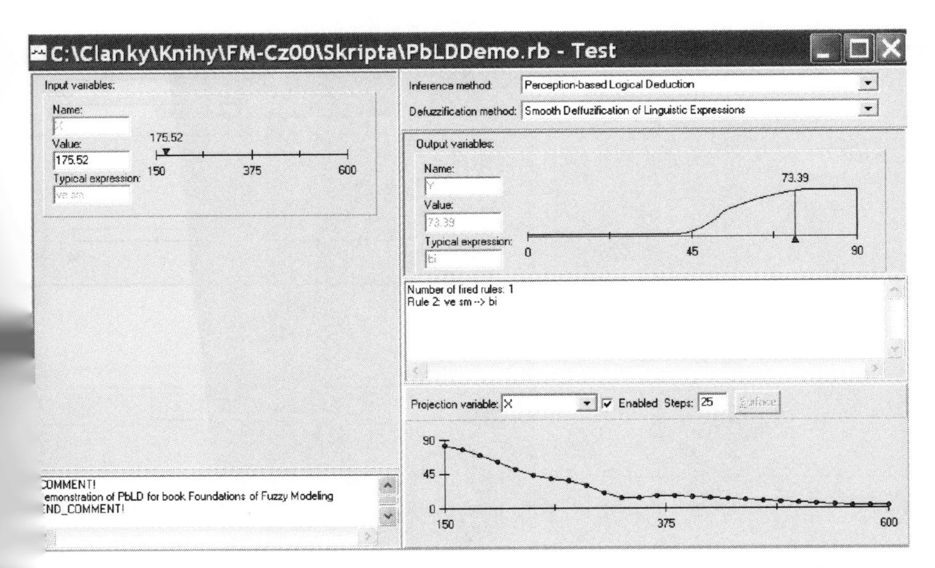

gure 5.13 Demonstration of the same application of PbLD as in Figure 5.12 with the nooth DEE defuzzification.

and $\mathscr{R}_j$ and aggregate the result. Note that this should never happen in well-formed guistic description, otherwise we encounter the presence of inconsistency or redun- ıcy in it. We should also cover the situations when $LPerc^{LD}(x_0, w)$ returns no result. s means that there is no suitable rule in the linguistic description LD for the obser- ion x_0. In other words, the linguistic description is incomplete *in this case.*

4 Comparison of Two Interpretations of Fuzzy IF-THEN Rules

larify the role of both alternatives to the interpretation of fuzzy IF-THEN rules ıtional and linguistic), we must be aware of the fact that the effectiveness of our rt to understand and affect phenomena in the world depends on the quality of our vledge of them. One possibility is that we have an exact mathematical description e studied phenomenon at disposal. The other possibility is to have only rough ription at disposal, using, for example, natural language. The less precise is our ledge, the more we are forced to use natural language. This possibility is always r disposal if anything else fails.

ıe fuzzy approximation methods require more precise knowledge than lin- c/logical methods. If we know, for example, that a modeled process can ccessfully characterized by a functional dependence, we can describe it by fuzzy graph (see Section 3.1.1). Surprisingly, we will see below that the stic/logical description is powerful enough even in cases when we have a ely good mathematical description of the modeled phenomenon at disposal.

he following examples and also in Chapter 7, we used the specialized software ı *LFL Controller*©, which was developed in the University of Ostrava in the

Czech Republic. Extensions of the used evaluative linguistic expressions are in accordance with the theory described in Section 5.2.3. For better readability, we will use the following shorts for components of the evaluative expressions:

Short	Meaning	Short	Meaning	Short	Meaning
Ze	*Zero*	Si	*Significantly*	QR	*Quite roughly*
Sm	*Small*	Ve	*Very*	VR	*Very roughly*
Me	*Medium*	Ty	*Typically*	VV	*Very very roughly*
Bi	*Big*	ML	*More or less*	No	*Not*
Ex	*Extremely*	Ro	*Roughly*		

Monotone linguistic description. The following simple linguistic description characterizes the behavior of PbLD and DNF–COG fuzzy approximation when almost all evaluative expressions available in LFL Controller© are used. The expressions are ordered by the ordering relation $\prec$ defined in (5.5). Therefore, the behavior is monotonous, which means that values of the dependent variable Y monotonically increase in accordance with the increase in values of the independent variable X. The monotone linguistic description is in Table 5.2.

The behavior of this linguistic description for both types of interpretation of fuzzy IF-THEN rules is depicted in Figure 5.14.[18] As can be seen, the DNF–COG fuzzy approximation (i.e., the Mamdani–Assilian method) is not able to distinguish among the rules in case when their antecedent contains various kinds of expressions with the same TE-adjective, namely, “small” and “big”. This method gives constant results on both sides of the given context, even though we intuitively expect that the resulting values of the variable Y should respect the use of linguistic hedges. For example, if the value of the variable X is very small, then, based on rule No. 4, we expect that the value of the variable Y would be very small as well. This is fulfilled by PbLD but

TABLE 5.2 Monotone linguistic description

	X	$\Rightarrow$	Y		X	$\Rightarrow$	Y
1.	Ze		Ze	11.	TyMe		TyMe
2.	ExSm		ExSm	12.	VVBi		VVBi
3.	SiSm		SiSm	13.	VRBi		VRBi
4.	VeSm		VeSm	14.	QRBi		QRBi
5.	Sm		Sm	15.	RoBi		RoBi
6.	MLSm		MLSm	16.	MLBi		MLBi
7.	RoSm		RoSm	17.	Bi		Bi
8.	QRSm		QRSm	18.	VeBi		VeBi
9.	VRSm		VRSm	19.	SiBi		SiBi
10.	VVSm		VVSm	20.	ExBi		ExBi

[18]The short horizontal line in the center of the graph is caused by the rule No. 11 “TyMe $\Rightarrow$ TyMe”.

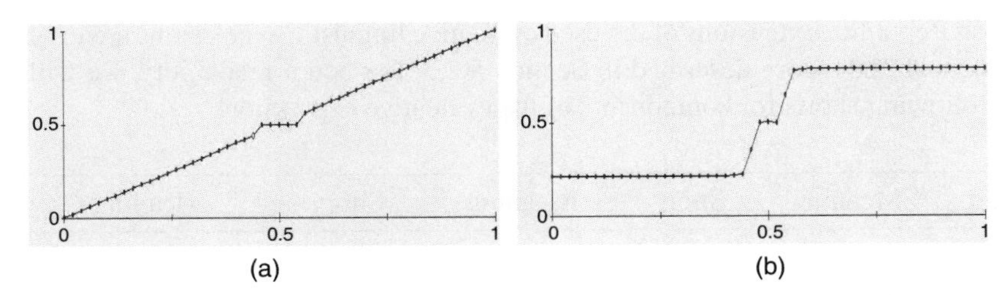

Figure 5.14 Behavior of the monotone linguistic description from Table 5.2. (a) Perception-based logical deduction and (b) DNF–COG (i.e., Mamdani–Assilian) approximation method.

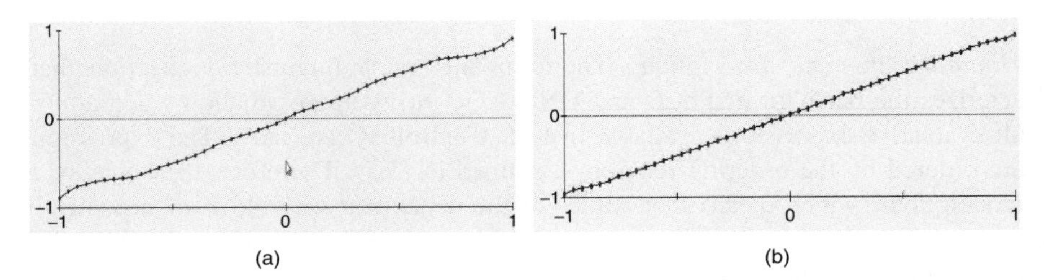

Figure 5.15 The behavior of DNF–COG (Mamdani–Assilian) method for the monotone linguistic description based on uniform triangular fuzzy partition (cf. Figure 5.1) consisting of (a) 7 triangles and (b) 19 triangles.

not by DNF–COG fuzzy approximation, because the latter tries to interpolate among values characterized by the given rules.

It is possible to achieve monotonous behavior using DNF–COG fuzzy approximation, provided that the evaluative expressions in the monotone linguistic description are replaced by predicate symbols $\mathbf{A}_i(x)$, $\mathbf{B}_i(y)$ interpreted by triangular fuzzy sets forming the uniform fuzzy partition depicted in Figure 5.1. Two variants of monotonous behavior of DNF–COG fuzzy approximation are depicted in Figure 5.15. In Figure 5.15(a), only seven predicates from Figure 5.1 are used, that is, the linguistic description consists of seven rules

IF X is NB THEN Y is NB,

IF X is NM THEN Y is NM,

....

IF X is PB THEN Y is PB.

Because the linguistic description from Table 5.2 has 20 rules, in Figure 5.15(b), we present the behavior of DNF–COG (Mamdani–Assilian) on monotone description consisting of 19 predicates, that is, the linguistic description consists of the rules

IF X is p_1 THEN Y is p_1,

$$\mathsf{IF}\ X \text{ is } p_2\ \mathsf{THEN}\ Y \text{ is } p_2,$$

....

$$\mathsf{IF}\ X \text{ is } p_{19}\ \mathsf{THEN}\ Y \text{ is } p_{19}.$$

The behavior of DNF–COG fuzzy approximation on the basis of the latter linguistic description looks fine, though, in the case of seven rules only, it is not too satisfactory. We can improve it by a modification of shapes of the triangles.

The price we pay in comparison with PbLD is the following:

(a) The triangles can be taken as extensions of fuzzy numbers but not of pure evaluative expressions. Moreover, the former will not hold if we modify shapes of the membership functions. In both cases, we are losing human-like interpretability of the linguistic description.
(b) The problem arises when some of the IF-THEN rules are dropped, for example, because of technical faults. In the PbLD method, nothing serious happens, because expressions with narrower meaning are covered by expressions with wider one. For the DNF–COG fuzzy approximation, this can be fatal.

Demonstration of (b) is in Figure 5.16. Figure 5.16(a) depicts the behavior of PbLD on the basis of the linguistic description from Table 5.2, in which Rules 3, 5, 11, 14, 19, and 20 were dropped. Though the monotonicity is slightly harmed, the basic course remains unchanged. This is not the case of DNF–COG fuzzy approximation. Figure 5.16(b) is the result on the basis of linguistic description formed of 7 triangular predicates in which Rules 1, 2, and 5 were dropped. The behavior is fatally destroyed. Figure 5.16(c) is the same when 19 predicates are used but Rules 2, 5, 10, 13, 18, and 19 were dropped. Though better, the result is still unsatisfactory.

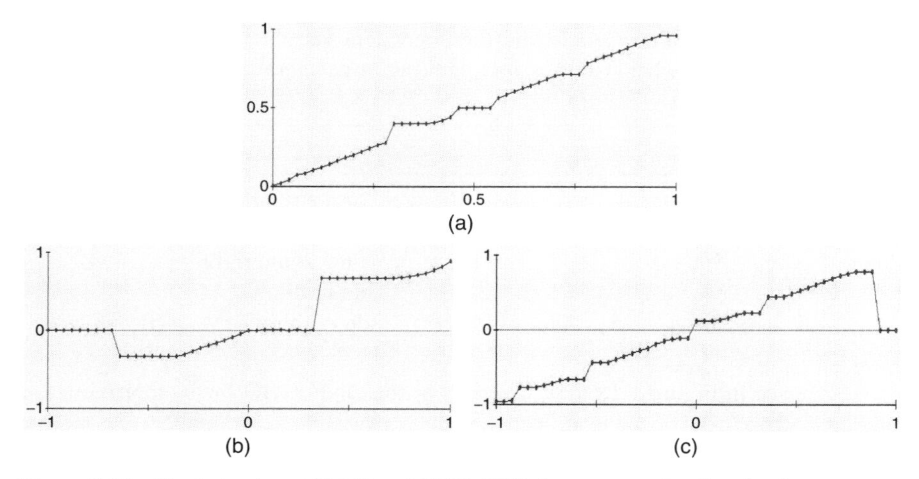

Figure 5.16 The behavior of PbLD and DNF–COG fuzzy approximation for the monotone linguistic description when some rules were dropped. (a) PbLD, linguistic description from Table 5.2. (b) DNF–COG fuzzy approximation, 7 triangles. (c) DNF–COG fuzzy approximation, 19 triangles.

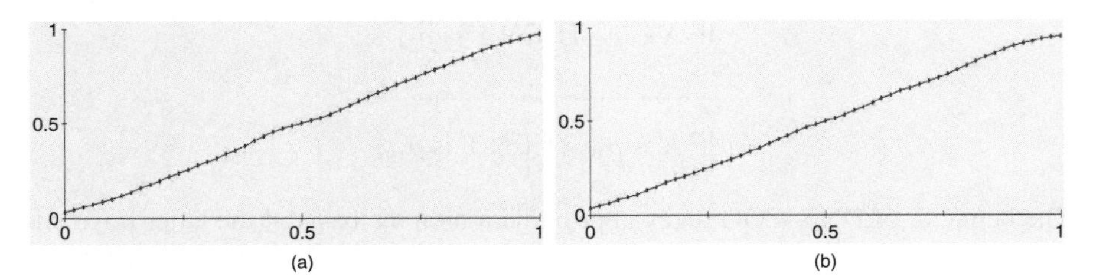

Figure 5.17 The behavior of PbLD with the smooth DEE defuzzification on the basis of linguistic description from Table 5.2. (a) Full description and (b) rules 3, 5, 11, 14, 19, and 20 were dropped.

For comparison, we also present in Figure 5.17 the result of PbLD with the smooth DEE defuzzification. One can see that there is practically no difference between the full description and that one with six rules dropped.

Linguistic description of a simple decision situation. The linguistic description presented below characterizes a simple decision situation when we drive a car and a prompt reaction is necessary. In literature, it is described as follows: our task is to decide where to go depending on the distance of an obstacle in sight. If it is *near*, then we should go to the right, while if it is *very near*, then it can be safer to go to the left.

A similar situation is often faced in practice when, for example, we approach quite fast a crossroad with green signal on. We expect that the light switches to red but we do not know the exact moment when does it happen. Then, if the light switches and we are too near, it can be safer to pass through the crossroad, instead of braking abruptly. Otherwise, if we are far enough, we should stop.

The linguistic description contains four rules. The independent variable X characterizes *distance of the obstacle* and the dependent variable Y corresponds to the rotation of a steering wheel (or, possibly, braking or accelerating).

	X	$\Rightarrow$	Y	*Comment*
1.	ExSm		−VeBi	“turn strongly to the left”
2.	VeSm		−Bi	“turn to the left”
3.	Sm		+Bi	“turn to the right”
4.	Me		Ze	“do nothing”
5.	Bi		Ze	‘do nothing”

We can see in Figure 5.18 that the result of the DNF–COG fuzzy approximation method is striking the obstacle. Indeed, if values of variable X are smaller than approximately 10 m (the distance from the obstacle is very small), then the first rule says that we should turn the steering wheel leftwards. If the distance is slightly larger (it is not very small anymore), then the third rule says that we should turn the steering wheel rightwards. In Figure 5.18, we can clearly see the difference. The PbLD can carry out the fast skip in values of Y, while the DNF–COG fuzzy approximation method tries

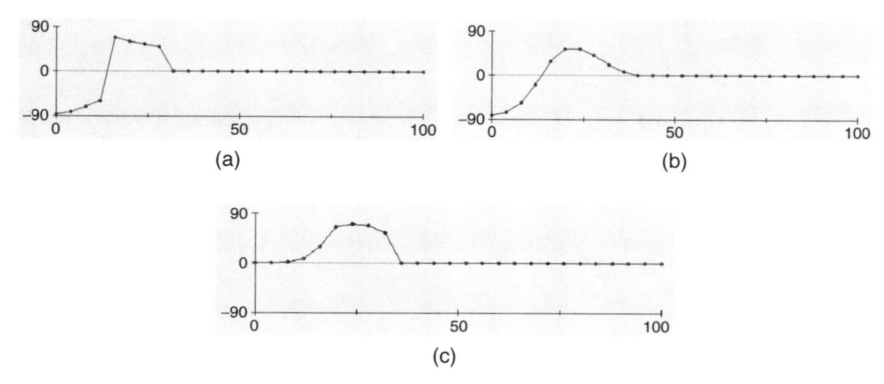

Figure 5.18 Bypassing the obstacle: (a) PbLD with the ordinary DEE defuzzification; (b) PbLD with the smooth DEE defuzzification; (c) Mamdani–Assilian DNF–COG approximation method. On x-axis is the distance of the obstacle, on y-axis is the turn of the steering wheel (negative values denote the leftward turn, positive values denote the rightward turn.)

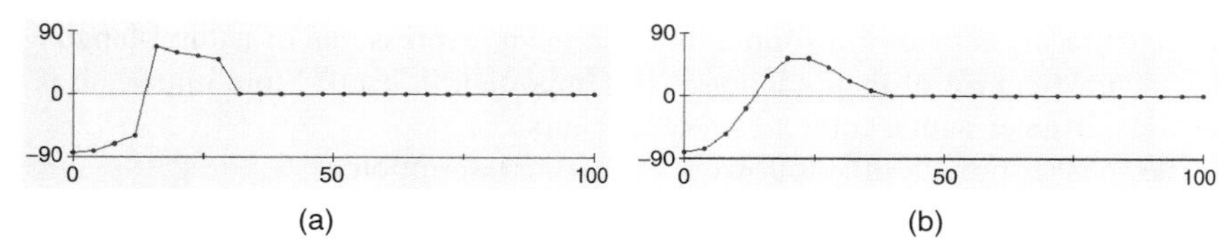

Figure 5.19 Bypassing the obstacle using PbLD based on reduced linguistic description with rule No. 1 omitted. (a) Ordinary defuzzification DEE and (b) smooth defuzzification DEE.

to interpolate between mutually contradictory values +Bi and −Bi with the identical absolute values. The resulting fuzzy set is symmetric. Hence, after defuzzification, we necessarily obtain $Y = 0$.

In Figure 5.19, we also present a situation when Rule 1 is dropped. We see that the behavior of PbLD remains practically unchanged.

The result of fuzzy approximation when we replace evaluative expressions by predicates interpreted by triangular fuzzy sets is shown in Figure 5.20. The triangular fuzzy sets in Figure 5.20(a) replace (from left to right) the evaluative expressions *ExSm, VeSm, Sm, Me*, and *Bi* used in the antecedent of rules, respectively (and similarly for the consequent). We can see that the behavior is now practically the same as for PbLD. However, the problem arises when we drop the first rule. The result can be seen in Figure 5.20(c). The consequence is that in critical situation, we again strike the obstacle. Furthermore, these fuzzy sets do not correspond at all to the meaning of evaluative expressions. For example, value 10 of X is definitely *very small* (VeSm), but not at all *small* (Sm), which contradicts the intuitive understanding of these expressions.

Description of a simple function. The linguistic description discussed below shows that our theory of semantics of evaluative linguistic expressions and its follow-up method PbLD grasp the way how people understand these expressions. Therefore, it is possible to rely on them, for example, in applications in robotics. The goal is to

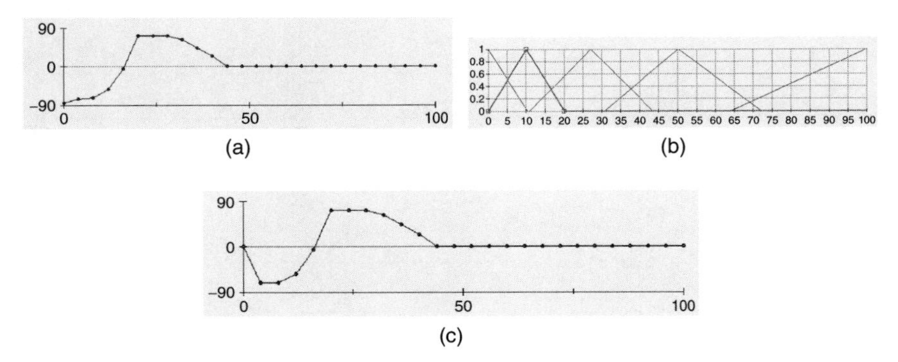

Figure 5.20 Bypassing the obstacle. (a) DNF–COG fuzzy approximation method if a fuzzy partition of a context using triangular membership functions (b) is used. (c) The result of this method based on the reduced linguistic description with Rule 1 being omitted.

characterize the course of a simple function using expressions of natural language. The accurate course of this function is not important in itself. More important is to provide a listener with a correct idea of its course.

The speaker provides the following linguistic description:

	X	$\Rightarrow$	Y
1.	Sm		Sm
2.	Me		Bi
3.	Bi		Sm

According to it, values of the function should be small for small and also for big arguments and big for medium arguments.

The results of PbLD with both the ordinary and smooth DEE defuzzification are in Figure 5.21. Figure 5.22 depicts the result by using the DNF–COG fuzzy approximation applied to the linguistic description above and also the result when triangular fuzzy sets are used instead of extensions of evaluative expressions *small, medium,*

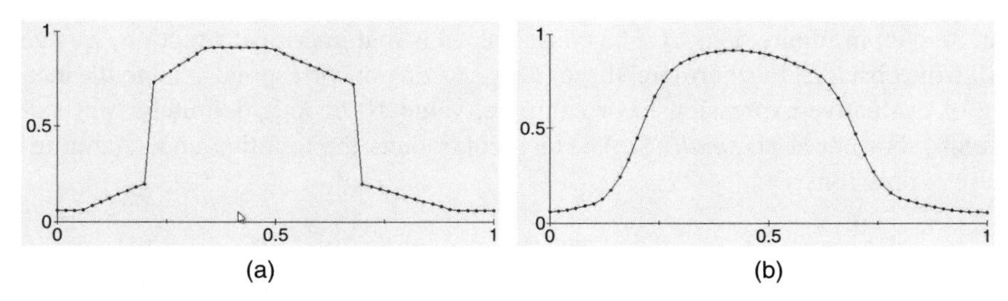

Figure 5.21 The result of the PbLD of the simple linguistic description of a function based on (a) ordinary DEE defuzzification and (b) smooth DEE defuzzification.

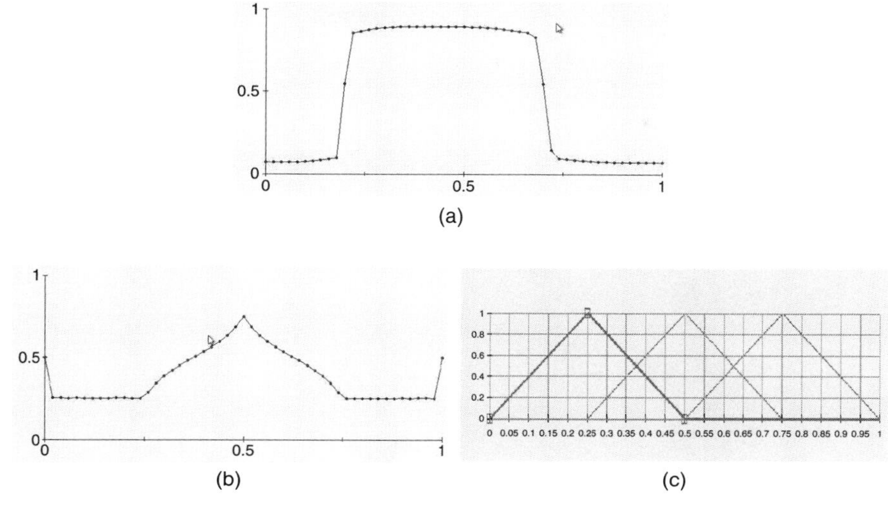

Figure 5.22 The result of the DNF–COG fuzzy approximation on the basis of the linguistic description of a simple function (a) using evaluative expressions; (b) using triangular fuzzy sets depicted in figure (c).

and *big* (these are depicted in Figure 5.22(c)). We can see that the last alternative is quite unsatisfactory.

The discussion above sheds light on differences between both methods. The PbLD method is suitable when an expert knowledge formulated in natural language is at disposal. The DNF–COG fuzzy approximation method is appropriate when we learn that the given description corresponds to some concrete function known only approximately at few points. Furthermore, we should be able to use fuzzy numbers (modeled usually by triangular fuzzy sets) for the description of this function. When we tune the behavior of the method, it is necessary to modify the membership functions of fuzzy sets in accordance with the course of the given function. This is the reason why simple shapes of membership functions (triangular or trapezoidal) are preferred in the DNF–COG fuzzy approximation method.

6

FUZZY CLUSTER ANALYSIS

In this chapter, we describe the so-called *fuzzy cluster analysis*. It is a data analysis method that aims at finding clusters of, in a certain specified sense, mutually similar data. This problem is faced quite often in practice. For example, a specific shoe or clothing size represents a cluster of the corresponding real sizes of people. Similarly, in image recognition, a cluster represents a letter, a certain object on screen, etc.

Cluster analysis has been established in 1950s. In the classical approach, clusters are disjoint subsets of some set of points. Because the origin of clusters can be very diverse, for example, they can be handwritten blurry characters on soiled paper, the requirement that clusters should be disjoint and crisp seems to be too strong. This consideration leads to the fuzzy cluster analysis, where clusters are imprecisely determined classes of objects, namely, partially overlapping fuzzy sets of points.

6.1 BASIC NOTIONS

Let V be a set of objects. The basic notion in cluster analysis is the *distance* between objects, which is a function $d : V \times V \longrightarrow \mathbb{R}$ with the following properties:

$$d(x, x) = 0, \tag{6.1}$$

$$d(x, y) = d(y, x), \tag{6.2}$$

$$d(x, z) \leq d(x, y) + d(y, z) \tag{6.3}$$

for all $x, y, z \in V$. Condition (6.3) is called the *triangular inequality*.

Insight into Fuzzy Modeling, First Edition. Vilém Novák, Irina Perfilieva, and Antonín Dvořák.

Companion Website: www.wiley.com/go/novak/fuzzy/modeling

The typical distance is the Euclidean one. Let V be a set of n-dimensional vectors $\mathbf{v} = (v_1, \ldots, v_n)$. Then the Euclidean distance between $\mathbf{u}, \mathbf{v} \in V$ is

$$d(\mathbf{u}, \mathbf{v}) = \sqrt{\sum_{i=1}^{n} (u_i - v_i)^2} . \tag{6.4}$$

Note that individual methods of cluster analysis differ precisely in the way of the definition of the distance function.

From a mathematical point of view, cluster analysis is a search for a suitable partition $\mathscr{P}_c$ of a set V into c subsets such that

$$\mathscr{P}_c = \{R_j \mid j = 1, \ldots, c, \quad R_j \subset V\}$$

that fulfills the conditions

$$\bigcup_{j=1}^{c} R_j = V,$$

$$R_i \cap R_j = \emptyset, \quad i \neq j, \quad 1 \leq i, j \leq c,$$

and also fulfills the informal condition that elements belonging to the same cluster $R_j, j \in \{1, \ldots, c\}$ should be mutually "closer" to each other than to any other element belonging to the other clusters. This is usually reached by means of minimization of some functional $J(\mathscr{P}_c)$.

In *fuzzy* cluster analysis, $R_j \underset{\sim}{\subseteq} V$ are fuzzy sets. Moreover, they have to fulfill the following conditions (we speak about *fuzzy pseudopartition*): let $V = \{v_1, \ldots, v_r\}$. Then

$$\sum_{j=1}^{c} R_j(v_i) = 1, \quad i = 1, \ldots, r, \tag{6.5}$$

$$0 < \sum_{i=1}^{r} R_j(v_i) < r, \quad j = 1, \ldots, c. \tag{6.6}$$

The first condition says that any element $v_i \in V$ should belong to at least one cluster, and the sum of all its membership degrees to all clusters should be equal to 1. The second condition says that (a) no cluster can be empty and (b) there must not exist a cluster that contains all elements of the set V in the maximal degree.

EXAMPLE 6.1 Let $V = \{v_1, \ldots, v_5\}$. The set $\mathscr{P}_3 = \{R_1, R_2, R_3\}$, where

$$R_1 = \left\{{}^{0.3}\!/v_1, {}^{0.5}\!/v_2, {}^{0.7}\!/v_3, {}^{1}\!/v_4\right\},$$
$$R_2 = \left\{{}^{0.2}\!/v_1, {}^{0.4}\!/v_2, {}^{0.1}\!/v_3, {}^{0.9}\!/v_5\right\},$$
$$R_3 = \left\{{}^{0.5}\!/v_1, {}^{0.1}\!/v_2, {}^{0.2}\!/v_3, {}^{0.1}\!/v_5\right\},$$

is an example of fuzzy pseudopartition. □

Data for fuzzy cluster analysis are real numbers in the following form:

$$\begin{array}{c} \\ o_1 \\ o_2 \\ \vdots \\ o_r \end{array} \begin{array}{c} \begin{array}{cccc} \varphi_1 & \varphi_2 & \cdots & \varphi_n \end{array} \\ \left[\begin{array}{cccc} u_{11} & u_{12} & \cdots & u_{1n} \\ u_{21} & u_{22} & \cdots & u_{2n} \\ \multicolumn{4}{c}{\dots\dots\dots\dots\dots\dots} \\ u_{r1} & u_{r2} & \cdots & u_{rn} \end{array}\right] \end{array}. \tag{6.7}$$

In matrix (6.7), $o_1, \dots, o_r$ are objects on which attributes $\varphi_1, \dots, \varphi_n$ are measured. The objects o_i can be arbitrary things, for example, screen pixels, time instants, people, machines, and others. The measured attributes φ_j can be physical (volume, weight, radiation intensity, etc.) or abstract (income, a grade on a scale, etc.).

The data can be characterized as r points in an n-dimensional space, where elements $u_{i1}, \dots, u_{in}$, $i = 1, \dots, r$, are coordinates of these points. Hence, any point is represented by one row in the matrix (6.7). For the following explication, we denote these rows by symbols $\mathbf{u}_i$ and we will understand them as row vectors

$$\mathbf{u}_i = (u_{i1}, \dots, u_{in}), \quad i = 1, \dots, r. \tag{6.8}$$

If H is a matrix of type (m, n), then H^T denotes its transposition, that is, a matrix of type (n, m). Hence, $\mathbf{u}_i^T$ denotes (6.8) as the column vector

$$\mathbf{u}_i^T = \begin{bmatrix} u_{i1} \\ \vdots \\ u_{in} \end{bmatrix}.$$

6.2 FUZZY CLUSTERING ALGORITHMS

The basic and probably most often used fuzzy clustering method is the *fuzzy c-means*. It originated as a generalization of the classical K-means method that was independently discovered in different scientific fields by several authors [6, 65, 134].

The fuzzy version was introduced by J. C. Dunn [34] and improved by J. Bezdek [10]. Let us remark that majority of other fuzzy clustering methods are derived from fuzzy c-means.

The basic parameter in fuzzy clustering is a given number c of clusters to be found. Let us now suppose for a moment that clusters $\mathscr{P}_c = \{R_1, \dots, R_c\}$ are already known, where $R_j \subsetneq \{\mathbf{u}_1, \dots, \mathbf{u}_r\}$. For any cluster, we compute its *center*. For fuzzy c-means method, the following formula is used:

$$\mathbf{s}_j = \frac{\sum_{i=1}^{r} (R_j(\mathbf{u}_i))^m \mathbf{u}_i}{\sum_{i=1}^{r} (R_j(\mathbf{u}_i))^m}, \quad j = 1, \dots, c, \tag{6.9}$$

where $m > 1$ is a parameter chosen beforehand. In more details, using formula (6.9), we obtain the following vector of centers:

$$\mathbf{s}_j = \left(\frac{\sum_{i=1}^{r} (R_j(\mathbf{u}_i))^m \, u_{i1}}{\sum_{i=1}^{r} (R_j(\mathbf{u}_i))^m}, \ldots, \frac{\sum_{i=1}^{r} (R_j(\mathbf{u}_i))^m \, u_{in}}{\sum_{i=1}^{r} (R_j(\mathbf{u}_i))^m} \right) \tag{6.10}$$

for all clusters $j = 1, \ldots, c$ (note that in (6.10), $R_j(\mathbf{u}_i)$ is the membership degree of the whole vector $\mathbf{u}_i$ in the fuzzy cluster R_j).

The optimal solution of fuzzy cluster analysis is reached if the functional $J_m(\mathscr{P}_c)$ is minimized. For fuzzy c-means, this functional has the form

$$J_m(\mathscr{P}_c) = \sum_{i=1}^{r} \sum_{j=1}^{c} (R_j(\mathbf{u}_i))^m d^2(\mathbf{u}_i, \mathbf{s}_j), \tag{6.11}$$

where d is some distance defined beforehand.

6.3 THE ALGORITHM OF FUZZY c-MEANS

The algorithm of fuzzy c-means computes membership functions of fuzzy clusters iteratively, until the required precision is reached.

1. Choose parameter $m > 1$, precision $\epsilon > 0$, distance measure d, and set an initial fuzzy pseudopartition $\mathscr{P}_c^{(0)}$. Set the iteration number $l := 0$.
2. Compute vectors of the centers $\mathbf{s}_1^{(l)}, \ldots, \mathbf{s}_c^{(l)}$ by (6.10).
3. For any $i = 1, \ldots, r$, modify the fuzzy pseudopartition as follows:
   (a) If $d(\mathbf{u}_i, \mathbf{s}_j) > 0$ for all $j = 1, \ldots, c$, then put

$$R_j^{(l+1)}(\mathbf{u}_i) = \left[\sum_{k=1}^{c} \left(\frac{d(\mathbf{u}_i, \mathbf{s}_j^{(l)})}{d(\mathbf{u}_i, \mathbf{s}_k^{(l)})} \right)^{\frac{2}{m-1}} \right]^{-1}. \tag{6.12}$$

   (b) Let $J = \{j_1, \ldots, j_p\}, p \leq c$, be a set of subscripts such that $d(\mathbf{u}_i, \mathbf{s}_j) = 0$ for all $j \in J$. Then put $R_j^{(l+1)}(\mathbf{u}_i)$ equal to any number from interval $[0, 1]$ in such a way that the condition

$$\sum_{j=j_1}^{j_p} R_j^{(l+1)}(\mathbf{u}_i) = 1 \tag{6.13}$$

   is fulfilled. For the remaining $j \in \{1, \ldots, c\} \setminus J$, put $R_j^{(l+1)}(\mathbf{u}_i) = 0$.
4. Compare fuzzy pseudopartitions $\mathscr{P}_c^{(l+1)}$ and $\mathscr{P}_c^{(l)}$ using the distance

$$\|\mathscr{P}_c^{(l+1)}, \mathscr{P}_c^{(l)}\| = \max_{\substack{j=1,\ldots,c \\ i=1,\ldots,r}} |R_j^{(l+1)}(\mathbf{u}_i) - R_j^{(l)}(\mathbf{u}_i)|. \tag{6.14}$$

5. If $\|\mathscr{P}_c^{(l+1)}, \mathscr{P}_c^{(l)}\| \leq \epsilon$ holds, terminate the process. In the opposite case, set $l := l + 1$ and return to step 2.

The following example illustrates the fuzzy c-means fuzzy clustering algorithm.

EXAMPLE 6.2 Suppose that we have 10 objects and we measure two attributes on them, that is, $r = 10$ and $n = 2$. These data are a result of experimental measurement of some functional dependence $y = f(x)$ in 10 points. Then, data (6.7) have the following form:

$$R = \begin{array}{c} \\ 1 \\ 2 \\ 3 \\ 4 \\ 5 \\ 6 \\ 7 \\ 8 \\ 9 \\ 10 \end{array} \begin{array}{c} \begin{array}{cc} u_i & v_i \end{array} \\ \begin{bmatrix} 9.5 & 2.1 \\ 16.0 & 2.5 \\ 20.0 & 2.3 \\ 24.5 & 2.8 \\ 26.0 & 3.4 \\ 27.0 & 4.3 \\ 32.0 & 4.8 \\ 38.0 & 5.0 \\ 41.0 & 4.7 \\ 47.0 & 5.3 \end{bmatrix} \end{array}. \tag{6.15}$$

These data can be understood as 10 vectors of the form (u_i, v_i), $i = 1, \ldots, 10$. They are graphically depicted in Figure 6.1. From this picture, we can see that the data can be divided to three areas—left-hand flat part, increasing middle part, and again increasing right-hand part.[1] Therefore, we will search for three clusters ($c = 3$).

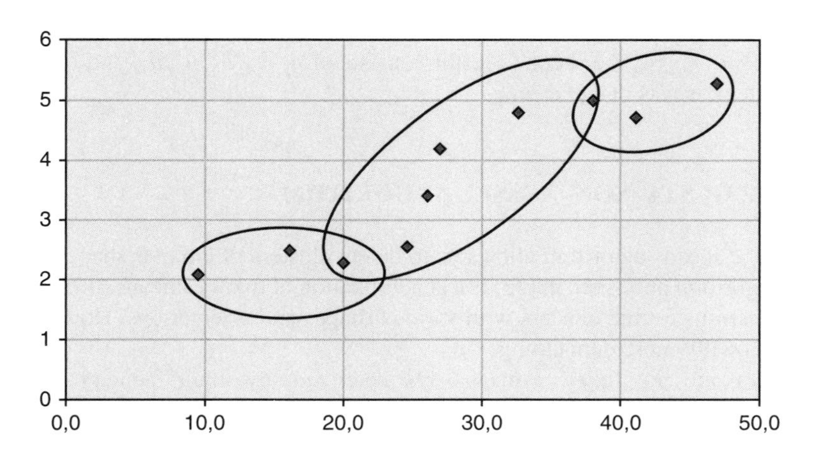

Figure 6.1 Data for fuzzy cluster analysis in Example 6.2.

The distance among objects and centers is Euclidean (6.4):

$$d(\mathbf{u}_i, \mathbf{s}_j) = \sqrt{(u_i - s_{j1})^2 + (v_i - s_{j2})^2},$$

[1]Note that ellipses in Figure 6.1 are only schematic depictions of them. Because c-means computes only centers of clusters, these cluster are in fact circular ones.

where s_{j1} and s_{j2} are coordinates of the center of the jth cluster. Further, we set parameter m to 2 (common value), precision $\epsilon = 0.05$, and initial pseudopartition to three clusters:

	u_1	u_2	u_3	u_4	u_5	u_6	u_7	u_8	u_9	u_{10}
$\mathbf{R}_1$	0.6	0.6	0.6	0.3	0.3	0.3	0.3	0.3	0.3	0.3
$\mathbf{R}_2$	0.3	0.3	0.3	0.6	0.6	0.5	0.5	0.3	0.3	0.3
$\mathbf{R}_3$	0.1	0.1	0.1	0.1	0.1	0.2	0.2	0.4	0.4	0.4

Computed centers of clusters in the zeroth iteration according to (6.9) are

$$\mathbf{s}_1 = (22, 3), \quad \mathbf{s}_2 = (27.5, 3.6), \quad \mathbf{s}_3 = (43.4, 5.1).$$

After six iterations, error decreased under the given precision, and we get the following centers of clusters:

$$\mathbf{s}_1 = (14.2, 2.3), \quad \mathbf{s}_2 = (26.5, 3.5), \quad \mathbf{s}_3 = (42, 5).$$

As a result, we get the following fuzzy clusters:

	u_1	u_2	u_3	u_4	u_5	u_6	u_7	u_8	u_9	u_{10}
$\mathbf{R}_1$	0.91	0.97	0.54	0.04	0	0	0.07	0.02	0	0.02
$\mathbf{R}_2$	0.07	0.03	0.42	0.94	1	1	0.66	0.11	0.01	0.05
$\mathbf{R}_3$	0.02	0	0.04	0.02	0	0	0.27	0.87	0.99	0.93

We can see that we obtained three relatively obvious clusters dividing the given area. If we consider only membership degrees greater than 0.5, then the first cluster consists of u_1, u_2, u_3, the second cluster consists of u_3, u_4, u_5, u_6, u_7, and finally, the third cluster consists of u_8, u_9, u_{10}. □

6.4 THE GUSTAFSON–KESSEL ALGORITHM

The fuzzy c-means algorithm allows us to detect clusters of circular shape. The following algorithm published in [43] is a generalization of fuzzy c-means. Its advantage is that it permits to find clusters with various shapes and orientations. However, it is computationally more demanding.

Clusters are in fuzzy c-means characterized by their centers only. In *Gustafson–Kessel algorithm*, any cluster is additionally characterized by special (positive-definite) matrix H_j, $j = 1, \dots, c$. The meaning of elements of this matrix is in the deformation of coordinates, which also permits to specify the shapes of clusters other than circles. In this algorithm, the so-called *covariance matrix* is defined as

$$S_j = \frac{\sum_{i=1}^{r} (R_j(\mathbf{u}_i))^m (\mathbf{u}_i - \mathbf{s}_j)^T (\mathbf{u}_i - \mathbf{s}_j)}{\sum_{i=1}^{r} (R_j(\mathbf{u}_i))^m}, \quad j = 1, \dots, c. \tag{6.16}$$

Then, we put

$$H_j = (\det(S_j))^{\frac{1}{n}} S^{-1}, \tag{6.17}$$

where $\det(S_j)$ is the determinant of the matrix S_j.

The norm is defined by means of the matrix H_j as $\|\mathbf{u}\|_{H_j} = \sqrt{\mathbf{u}H_j\mathbf{u}^T}$. Then, we define the distance between vectors as

$$d_{H_j}(\mathbf{u}, \mathbf{v}) = \|\mathbf{u} - \mathbf{v}\|_{H_j}.$$

Besides this, the following restriction on the determinant of matrix H_j is set, namely, that $\det(H_j) = \rho$, where ρ is some number specified for all matrices H_j. It means that a constant volume (size) of clusters is specified but not their shape. The disadvantage is that it is necessary to have some a priori information about clusters at disposal.

The algorithm itself is again an iterative one. It is finished when the precision ϵ is reached.

1. Choose parameter m, precision ϵ, and an initial fuzzy pseudopartition $\mathscr{P}_c^{(0)}$. Set the iteration number $l := 0$.
2. Compute centers $\mathbf{s}_1^{(l)}, \dots, \mathbf{s}_c^{(l)}$ using (6.9).
3. Compute all covariance matrices $S_j^{(l)}, j = 1, \dots, c$, using (6.16).
4. For all $i = 1, \dots, r$, compute the distances

$$d_H(\mathbf{u}_i, \mathbf{s}_j^{(l)}) = \sqrt{(\mathbf{u}_i - \mathbf{s}_j^{(l)}) \det{(S_j^{(l)})^{\frac{1}{n}}} (S_j^{(l)})^{-1} (\mathbf{u}_i - \mathbf{s}_j^{(l)})^T}, \tag{6.18}$$

   where $j = 1, \dots, c$.
5. For all $i = 1, \dots, r$, modify the fuzzy pseudopartition as follows:
   (a) Let for all $j = 1, \dots, c$, it holds that $d_H(\mathbf{u}_i, \mathbf{s}_j) > 0$. Then put

$$R_j^{(l+1)}(\mathbf{u}_i) = \left[\sum_{k=1}^{c} \left(\frac{d_H(\mathbf{u}_i, \mathbf{s}_j^{(l)})}{d_H(\mathbf{u}_i, \mathbf{v}_k^{(l)})} \right)^{\frac{2}{m-1}} \right]^{-1}. \tag{6.19}$$

   (b) Let $J = \{j_1, \dots, j_p\}$, $p \leq c$, be a set of subscripts such that $d_H(\mathbf{u}_i, \mathbf{s}_j) = 0$ for all $j \in J$. Then put $R_j^{(l+1)}(\mathbf{u}_i)$ equal to any number from interval $[0, 1]$ in such a way that the condition

$$\sum_{j=j_1}^{j_p} R_j^{(l+1)}(\mathbf{u}_i) = 1 \tag{6.20}$$

   is fulfilled. For the remaining $j \in \{1, \dots, c\} \setminus J$, put $R_j^{(l+1)}(\mathbf{u}_i) = 0$.
6. Compare fuzzy pseudopartitions $\mathscr{P}_c^{(l+1)}$ and $\mathscr{P}_c^{(l)}$ using the distance

$$\|\mathscr{P}_c^{(l+1)}, \mathscr{P}_c^{(l)}\| = \max_{\substack{j=1,\dots,c \\ i=1,\dots,r}} |R_j^{(l+1)}(\mathbf{u}_i) - R_j^{(l)}(\mathbf{u}_i)|. \tag{6.21}$$

7. If $\|\mathscr{P}_c^{(l+1)}, \mathscr{P}_c^{(l)}\| \leq \epsilon$ holds, terminate the process. In the opposite case, put $l := l + 1$ and return to step 2.

6.5 HOW THE NUMBER OF CLUSTERS CAN BE DETERMINED

The disadvantage of methods presented above is that they require to know the number c of clusters in advance. If we are not able to do it, we can apply two methods for its estimation:

1. Construct some global characteristic of the acquired fuzzy pseudopartition $G(\mathscr{P}_c)$. The actual algorithm then runs as follows: First, choose some maximal threshold $c^{\max}$. Then, carry out the clustering algorithm successively for $c = 2, \ldots, c^{\max}$ and form a fuzzy pseudopartitions $\mathscr{P}_c$.
   In each step, compute a characteristic $G(\mathscr{P}_c)$. If it is not improved in comparison with the previous one, take the fuzzy pseudopartition $G(\mathscr{P}_{c-1})$ to be the best.
   There exist a lot of characteristics $G(\mathscr{P}_c)$. Some of them are to be maximized, some minimized. The problem is that there is no optimal characteristics. Therefore, it is appropriate to combine them. The detailed analysis can be found, for example, in book [53].
   An example of characteristics $G(\mathscr{P}_c)$ is the *partition coefficient*:

$$PC(\mathscr{P}_c) = \frac{\sum_{i=1}^{r} \sum_{j=1}^{c} R_j(\mathbf{u}_i)^2}{r}. \tag{6.22}$$

   This characteristic attains values $\frac{1}{c} \le PC(\mathscr{P}_c) \le 1$. Greater values mean that the partition is improved.
2. The second possibility is to evaluate clusters individually according to their "quality" using some local characteristics. If one of two neighboring clusters is significantly worse than the other one, then merge them into one. Because this analysis is more complicated, we are not going into details in our book and refer to specialized literature (e.g., [53]).

6.6 CONSTRUCTION OF FUZZY RULES BASED ON FOUND CLUSTERS

As already mentioned in Section 3.4, it is possible to use fuzzy cluster analysis for finding fuzzy IF-THEN rules on the basis of given data, provided that either the interpretation of these data is functional or the rules are of Takagi–Sugeno type.

Let us assume that sets V_k represent linguistic contexts for individual antecedent variables. They are intervals[2]

$$V_k = [v_{L,k}, v_{R,k}], \quad k = 1, \ldots, n. \tag{6.23}$$

[2]The middle value v_S is not used here; therefore, it is omitted.

A linguistic description whose interpretation based on fuzzy cluster analysis we are going to derive is

$$\begin{aligned}
\mathscr{R}_1 &:= \text{IF } X_1 \text{ is } \mathscr{A}_{11} \text{ AND } \cdots \text{ AND } X_n \text{ is } \mathscr{A}_{n1} \text{ THEN} \\
&\qquad Y = b_{01} + b_{11}X_1 + \cdots + b_{n1}X_n, \\
&\cdots\cdots\cdots\cdots\cdots\cdots\cdots\cdots\cdots\cdots \qquad (6.24) \\
\mathscr{R}_c &:= \text{IF } X_1 \text{ is } \mathscr{A}_{1c} \text{ AND } \cdots \text{ AND } X_n \text{ is } \mathscr{A}_{nc} \text{ THEN} \\
&\qquad Y = b_{0c} + b_{1c}X_1 + \cdots + b_{nc}X_n.
\end{aligned}$$

Note that the number of rules is again denoted by c. From this, it is immediately clear that the number of derived rules will be equal to the number of found clusters.

Further, we suppose that there is a data set at our disposal:

$$\begin{array}{c} \\ \mathbf{u}_1 \\ \vdots \\ \mathbf{u}_r \end{array} \begin{array}{c} \begin{matrix} X_1 & \cdots & X_n & Y \end{matrix} \\ \begin{bmatrix} u_{11} & \cdots & u_{n1} & v_1 \\ \cdots & \cdots & \cdots & \cdots \\ u_{1r} & \cdots & u_{nr} & v_r \end{bmatrix} \end{array}. \tag{6.25}$$

These data correspond to values in the corresponding contexts (6.23), that is, $u_{k1}, \dots, u_{kr} \in V_k$, $k = 1, \dots, n$. Because there is no other information at our disposal, we can set

$$v_{L,k} = \min\{u_{k1}, \dots, u_{kr}\}, \tag{6.26}$$

$$v_{R,k} = \max\{u_{k1}, \dots, u_{kr}\}. \tag{6.27}$$

The clusters $R_1, \dots, R_c$ are now found using fuzzy cluster analysis. Realize that any cluster is, in fact, a fuzzy relation

$$R_j(u_{1i}, \dots, u_{ni}) = R_j(\mathbf{u}_i), \quad i = 1, \dots, r, \quad j = 1, \dots, c.$$

Thus, fuzzy sets $A_{k1}, \dots, A_{kc}$ interpret expressions "X_k is $\mathscr{A}_{k1}$", …, "X_k is $\mathscr{A}_{kc}$", respectively, appearing in (6.24). These fuzzy sets are computed as projections of R_j to the kth variable, that is,

$$A_{kj}(u_{ki}) = \bigvee_{\substack{\mathbf{u}_p, p=1,\dots,r \\ u_{kp}=u_{ki}}} R_j(\mathbf{u}_p), \tag{6.28}$$

where the symbol $\bigvee_{\substack{\mathbf{u}_p, p=1,\dots,r \\ u_{kp}=u_{ki}}}$ means that the maximum is considered over all rows $\mathbf{u}_p$, $p = 1, \dots, r$ in data (6.25), for which kth value u_{kp} is equal to the value u_{ki}.

The fuzzy cluster analysis provides fuzzy sets A_{jk} but it does not provide explicit formulas for their membership functions. We can derive them by approximation, for example, by triangular, trapezoidal, or other membership functions in such a way that they would be in correspondence with generally accepted shapes (see, e.g., [53]). The regression coefficients $b_{0j}, \dots, b_{nj}, j = 1, \dots, c$, are computed using weighted regression analysis method described in Section 3.4.

EXAMPLE 6.3 Let us consider the same problem as in Example 6.2. We will look for three rules of Takagi–Sugeno type that characterize some given functional dependence. Because data (6.15) were obtained as a result of measurement, let us suppose that the context of the independent variable x is $w = [0, 50]$. It was determined as a lower and an upper bound of all values of x.

On the basis of the obtained membership functions of the clusters R_1, R_2, and R_3, we derive (using (6.28)) membership functions of the fuzzy sets $A_1, A_2, A_3 \subsetneq [0, 50]$. Because computation of membership functions of clusters in fuzzy cluster analysis is somewhat complicated, it has no good sense to work with small values scattered among resulting clusters. Therefore, we will consider all membership degrees smaller than 0.1 as being equal to zero. In such a way, we obtain

$$A_1 = \{0.91/9.5, 0.97/16, 0.54/20\}, \tag{6.29}$$

$$A_2 = \{0.42/20, 0.94/24.5, 1/26, 1/27, 0.66/32, 0.11/38\}, \tag{6.30}$$

$$A_3 = \{0.27/32, 0.87/38, 0.99/41, 0.93/47\}. \tag{6.31}$$

These fuzzy sets are approximated by trapezoidal functions depicted in Figure 6.2.

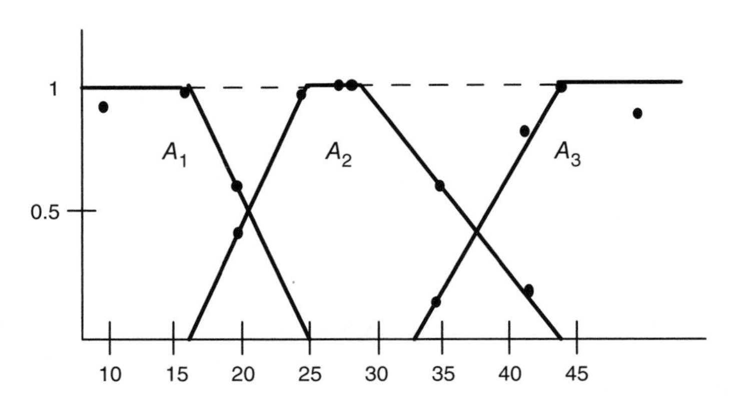

Figure 6.2 Approximation of clusters (depicted by bold dots) obtained using fuzzy c-means algorithm by trapezoidal membership functions of fuzzy sets A_1, A_2, and A_3.

Further, we compute regression coefficients according to (3.74) and (3.75):

$$b_{10} = 1.884, \qquad b_{11} = 0.029, \tag{6.32}$$

$$b_{20} = -1.965, \qquad b_{21} = 0.207, \tag{6.33}$$

$$b_{30} = 3.516, \qquad b_{31} = 0.035. \tag{6.34}$$

We will not look for linguistic expressions for found fuzzy sets, instead we will assign them symbols $\mathscr{A}_1$, $\mathscr{A}_2$, and $\mathscr{A}_3$. Then, we obtain resulting fuzzy rules as follows:

$$\mathscr{R}_1 := \text{IF } X \text{ is } \mathscr{A}_1 \text{ THEN } Y = 1.884 + 0.029X,$$

$$\mathscr{R}_2 := \text{IF } X \text{ is } \mathscr{A}_2 \text{ THEN } Y = -1.965 + 0.207X,$$

$$\mathscr{R}_3 := \text{IF } X \text{ is } \mathscr{A}_3 \text{ THEN } Y = 3.516 + 0.035X.$$

Finally, using formula (3.66) of fuzzy approximation carried out on the basis of found rules, we compute values v_i^A approximating the original values v_i from Example 6.2:

u_i	v_i	v_i^A
9.5	2.1	2.16
16.0	2.5	2.35
20.0	2.3	2.31
24.5	2.8	3.11
26.0	3.4	3.42
27.0	4.3	3.63
32.0	4.8	4.74
38.0	5.0	4.98
41.0	4.7	4.97
47.0	5.3	5.18

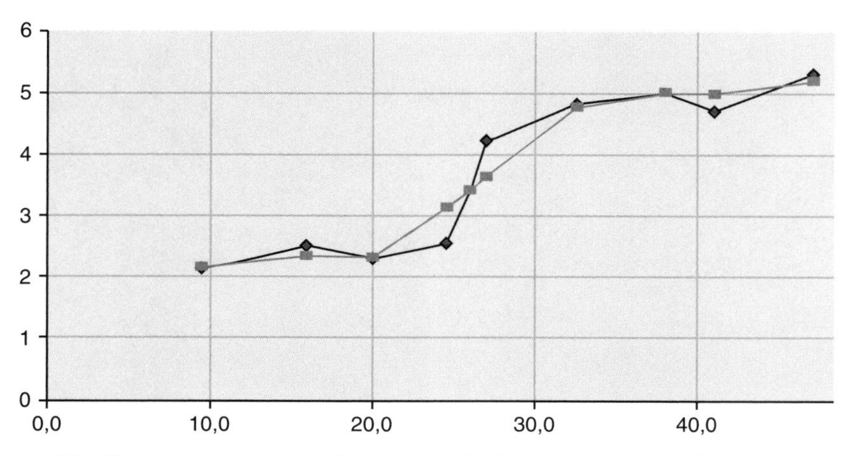

Figure 6.3 Fuzzy approximation using generated Takagi–Sugeno rules. The original data v_i are indicated by diamonds and the results of approximation v_i^A by squares.

The whole situation is depicted in Figure 6.3. For the sake of transparency, the data are connected by line segments. Note that the resulting approximation function can be considered as the fuzzy model of the given data. □

We see that using Takagi–Sugeno rules, we can obtain quality results when searching for the most fitting course of an unknown function. Because classical regression analysis solves a similar problem, the algorithm described above is sometimes understood as one of the methods of *fuzzy regression analysis*. Note that a well fitting model of the data can also be obtained using the F-transform described in Chapter 4.

PART II

SELECTED APPLICATIONS

In this part, we present selected applications of the methods described in Part I. The applications cover control, managerial decision-making, image processing, and time series analysis and forecasting. Our goal is to show the power of the theory presented in Part I and to inspire the reader to search his/her own applications either in the related or in entirely new areas.

Insight into Fuzzy Modeling, First Edition. Vilém Novák, Irina Perfilieva, and Antonín Dvořák.

Companion Website: www.wiley.com/go/novak/fuzzy/modeling

7

FUZZY/LINGUISTIC CONTROL AND DECISION-MAKING

In this chapter, we explain principles of fuzzy control and fuzzy decision-making based on linguistic descriptions. In the description of control, we do not repeat the known and widely described fuzzy control based on Mamdani's ideas. Instead, we return to the original L. A. Zadeh's idea of control realized on the basis of the description using genuine natural language. Therefore, we will concentrate especially on fuzzy control that is based on linguistically interpreted fuzzy IF-THEN rules. Recall that this control method is often called *linguistic control*. In the explanation, we will mostly apply the theory described in Chapter 5.

7.1 THE PRINCIPLE OF FUZZY CONTROL

Fuzzy control can be seen as an application of the theory of approximate reasoning to control of technological processes. To understand the basis of fuzzy control, we must first realize that classical control requires knowledge of a mathematical model of the controlled process. The classical control theory provides reliable solution if the process is described using differential equations. Finding realistic mathematical model, however, can be quite complicated in practice. Moreover, it can happen that the obtained description is too complex. In this situation, it may be almost impossible or too expensive (in the sense of money or time) to design classical controller. Therefore, various simplifications and approximations are used. As a result, the obtained control algorithms need not be satisfactory.

On the other hand, control of complex processes is often manually realized by people who know from their experience how the process should be controlled, for example, crane operators, industrial furnace operators, large mining machine drivers, and others. Their control is satisfactory despite the fact that their knowledge about

Insight into Fuzzy Modeling, First Edition. Vilém Novák, Irina Perfilieva, and Antonín Dvořák.

Companion Website: www.wiley.com/go/novak/fuzzy/modeling

behavior of the controlled process is only approximate. This is just the situation when fuzzy control can be effectively implemented and replace the manual control done by people. Vague instructions given in natural language are transformed into a control algorithm realized by a computer.

A typical school example is stabilization of an *inverted pendulum*, for example, a stick standing on a child's hand. After some practice, the child is able to keep it standing without knowing mathematical model of the inverted pendulum (let us remark that precise model of it consists of four partial differential equations). Design of the classical controller is in this case very complicated. But the child can describe his/her control quite easily using natural language. Hence, a fuzzy controller imitating the behavior of the child can be designed.

Fuzzy controllers can be included in a closed feedback loop or in a complex hierarchical control system whose units represent various algorithms that control subprocesses, which form the whole controlled process. According to concrete problem, these may be fuzzy controllers, classical controllers (e.g., linear PID controller), adaptive controllers, or some other control units. In the root of the hierarchy stays the main control system that controls and switches individual control units depending on the specificity of the given subprocess. We argue that the main controller should be fuzzy, because it is much easier to specify its behavior locally. It can also be seen as a certain expert system that "knows" the best control of each subprocess.

Fuzzy controllers are robust, that is, they are little sensitive to changes in external conditions. This means that a fuzzy controller quite often either needs no modification or requires only slight modification even in cases when either the external conditions are substantially changed or the whole controlled process is changed.

Design of fuzzy controllers is not too difficult. The complexity of their design is, in fact, practically the same independently on the complexity of the process to be controlled. This is in contrast with classical controllers, where complexity of their design dramatically increases with complexity of the controlled process.

The properties mentioned above follow from the fact that the control is described in natural language. Therefore, fuzzy controllers are suitable in situations when the control process is complex, so it is controlled by a skilled operator. Another argument in favor of fuzzy controllers is that in practice, we often either need not too accurate control or the external conditions change so frequently that classical controller would have to be identified or even redesigned again and again.

The advantage not yet mentioned is that fuzzy controllers are, in case of complex control, quite often cheaper than classical ones. The reason is that we do not need identification of the controlled process (i.e., we do not need its mathematical model). This procedure requires testing of the process in various conditions that may not be reached or may even be dangerous. On the basis of all the above mentioned properties, we may conclude that fuzzy controllers are *practically optimal*. Hence, they are very attractive for practice.

To finish this section, we have to note that there are processes that can hardly be controlled on the basis of a rough idea only and so, precise mathematical model of them is necessary. Therefore, though fuzzy control is nice, interesting, and useful control method, we cannot expect it to replace the classical control but to be its convenient complement.

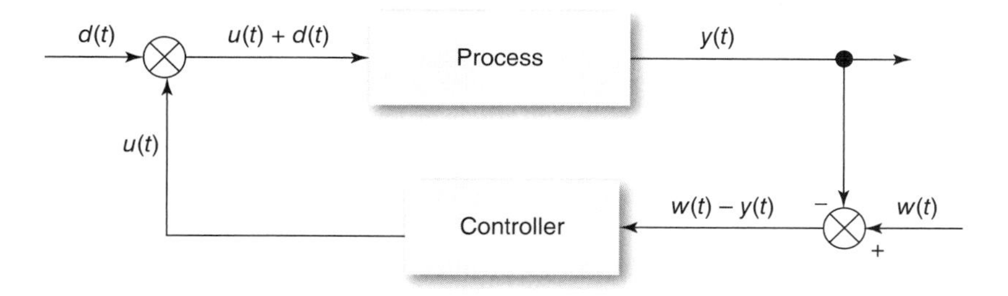

Figure 7.1 The basic scheme of closed-loop feedback control.

7.1.1 Control in a Closed Feedback Loop

The basis of vast majority of control systems is a closed feedback loop control, depicted in Figure 7.1. The $d(t)$ in the figure is a disturbance at time t, $u(t)$ is a control action (manipulated variable) of the controller, $y(t)$ is an output of the process, and $w(t)$ is a set-point value (set-point for short) that the process output should reach and keep. The set-point value is usually constant for some time interval. In general, however, it can vary during time.

The scheme in Figure 7.1 is valid both for classical and for fuzzy control. The difference lies in the implementation of the controller itself. A *classical controller*, from the mathematical point of view, is some function that is derived on the basis of a *mathematical description* of the controlled process. This situation can be mathematically characterized as follows:

$$\dot{\mathbf{y}}(t) = f(\mathbf{y}(t), \mathbf{u}(t)) + \mathbf{d}(t), \tag{7.1}$$

$$\mathbf{u}(t) = g(\mathbf{y}(t), \mathbf{w}(t)). \tag{7.2}$$

Equation (7.1) is a mathematical description of the controlled system, where $\mathbf{y}(t)$ is a vector of system states at time t, $\mathbf{u}(t)$ is a vector of inputs, and $\mathbf{d}(t)$ is a vector of disturbances. Equation (7.2) characterizes a controller subsystem, which on the basis of outputs of a controlled system and a vector of set-points $\mathbf{w}(t)$ gives vector of control actions $\mathbf{u}(t)$ (i.e., inputs to the controlled system). If we introduce an error (with respect to the set-point):

$$\mathbf{e}(t) = \mathbf{w}(t) - \mathbf{y}(t),$$

we can rewrite (7.2) as

$$\mathbf{u}(t) = g^{\star}(\mathbf{e}(t), \mathbf{w}(t)). \tag{7.3}$$

Hence, if we know the function f characterizing the controlled process, then our goal is to find the control function g (or $g^{\star}$) in such a way to make $\mathbf{e}(t)$ (close to) zero in the shortest possible time. This problem has been solved for various types of systems (7.1). The most elaborated solution is available for *linear* systems, that is, for systems with a linear function f. Hence, the fundamental assumption for successful classical control is the knowledge of the function f, because the function g (or $g^\star$) is derived from it.

In practice, the control is not realized continuously, but in time samples. The *sampling period* will be denoted by ΔT. Without a loss of generality, we usually set $\Delta T = 1$.

7.1.2 A General Scheme of Fuzzy Controller

7.1.2.1 The main idea of fuzzy control. As mentioned above, the main idea stems from the assumption that mathematical description of a system (7.1) need not be known. What is known instead is the way how the system can be controlled—a *control strategy*. This strategy is described using fuzzy or fuzzy/linguistic IF-THEN rules.

This assumption is not unrealistic at all, as discussed above. People are able to control successfully a large variety of processes, from furnaces and chemical processes to driving a car, without detailed knowledge of the physical/chemical laws and properties or without being able to describe them mathematically. The control thus works on the basis of more or less rough ideas and experience.

It should be noted that a rough idea about behavior of a controlled process is usually at our disposal, while a mathematical model, even if available, may be so complicated that finding the control function g without considerable simplifications may not be possible. Consequently, even if we know a mathematical model of a controlled system, we need not be able to derive a satisfactory control strategy.

Because the only requirement for the design of a fuzzy controller is at least rough knowledge of a control strategy, we need not look for a mathematical model of the controlled process. This is one of the main advantages of fuzzy controllers, because a difficult and expensive phase of process identification can be omitted. And, as mentioned, this phase need not always lead to satisfactory results.

7.1.2.2 Knowledge base and fuzzy action units. The essential constituent of fuzzy controller is a *knowledge base* schematically depicted in Figure 7.2. It contains information about appropriate control strategy and consists of one or more linguistic descriptions using which we can define actions for special situations or characterize local behavior of some variables. Some variables can be shared, for example, an output variable of one description can be an input variable to another one. In the simplest case, of course, a knowledge base consists of a single linguistic description only.

Each linguistic description is associated with an *inference engine* and, together with it, forms a *fuzzy action unit*. We can also use the term *fuzzy agent*. The inference engine is either fuzzy approximation or perception-based logical deduction. Thus, we can distinguish a relational fuzzy action unit depicted in Figure 7.3 and a linguistic fuzzy action unit—see Figure 7.4. Both fuzzy action units consist of a linguistic description and an inference engine.

Relational fuzzy action unit. This unit realizes the theory explained in Chapter 3. Its inference engine has two components:

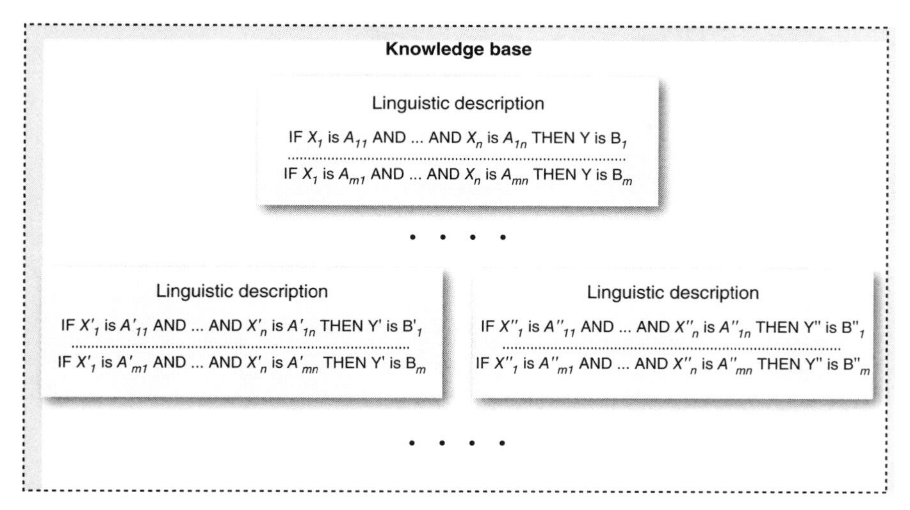

Figure 7.2 Scheme of a knowledge base consisting of one or more linguistic descriptions, each of which characterizes some kind of local knowledge of the control strategy and the use of specific variables. Some of the variables in different linguistic descriptions can be identical.

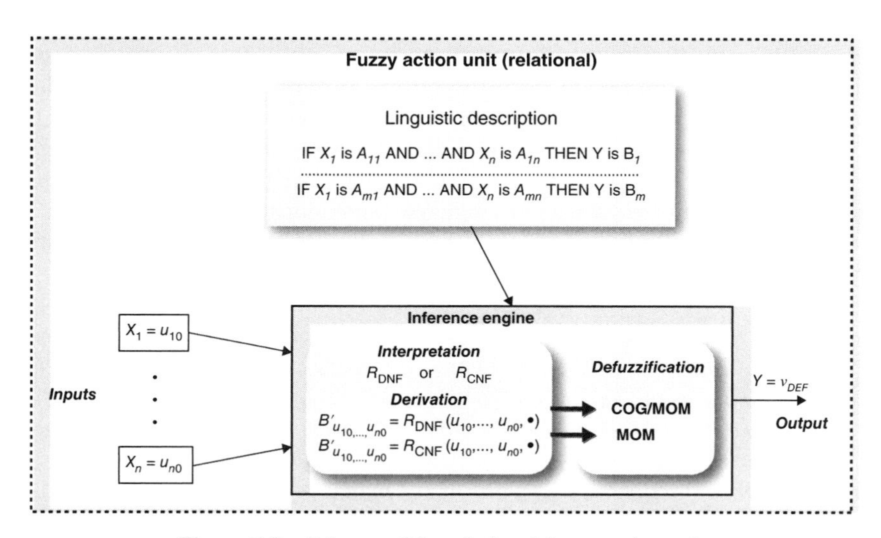

Figure 7.3 Scheme of the relational fuzzy action unit.

(i) Interpretation of the linguistic description either as a DNF or a CNF and construction of the corresponding fuzzy relation R_{DNF} or R_{CNF}. If inputs u_{10}, $\dots, u_{n0}$ are given, the fuzzy set $B'_{u_{10},\dots,u_{n0}}$ is formed using (3.43) or (3.44), respectively.

(ii) The fuzzy set $B'_{u_{10},\dots,u_{n0}}$ is an input to a *defuzzification* operation (see Section 3.2.1). The most convenient methods for the DNF interpretation are COG or MOM, casually also COS. For the CNF interpretation, the most convenient is the MOM defuzzification.

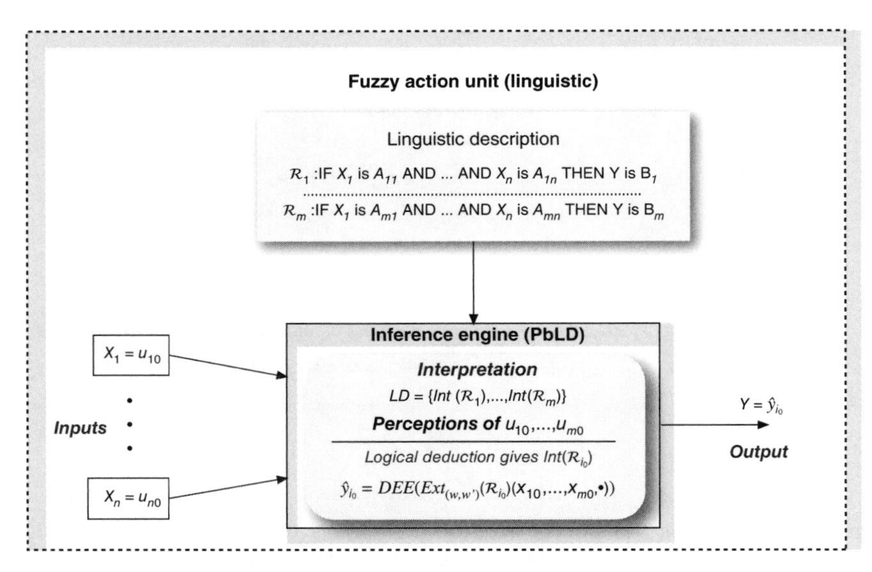

Figure 7.4 Scheme of the linguistic fuzzy action unit.

Note that the fuzzy set $B'_{u_{10},\ldots,u_{n0}}$ represents an action to be performed (e.g., a control action). To be able to perform it, we have to find one individual value. This is analogous, for example, to the situation when somebody tells us "move control lever slightly to the right". Anybody moves it to a position that corresponds to his/her understanding of the expression "slightly to the right". Because people understand natural language expressions in a similar fashion, the difference in control action would not be significant. This intuitive procedure is represented by means of a defuzzification operation.

Remark 7.1 *With respect to Remark 3.4 on page 73, we can also consider formulas (3.56) and (3.57) as the basis for the inference in fuzzy action unit. However, then we should consider a special operation "fuzzifier" which turns the element u_0 into a fuzzy singleton $\{1/u_0\}$. Considering (3.43) and (3.44), such an operation is superfluous.*

Linguistic fuzzy action unit. This unit realizes the theory explained in Chapter 5. A linguistic description is interpreted as a special text and the inference is the perception-based logical deduction (PbLD) method (together with the DEE defuzzification method). The inputs $u_{10} \in w_1, \ldots, u_{n0} \in w_n$ give rise to proper perceptions that are evaluative expressions from the topic (5.22) of the given linguistic description (see (5.31)). By the PbLD rule (5.34), we employ the fuzzy/linguistic IF-THEN rule R_{i_0} and find an element $\hat{y}_{i_0}$ using (5.35).

Structure of a fuzzy controller. We can understand it as a system consisting of one or more interrelated fuzzy action units (Figure 7.5). Its structure is determined by the structure of the knowledge base. Units may be of different kinds. In the simplest case, a fuzzy controller consists of only one fuzzy action unit. The most often applied

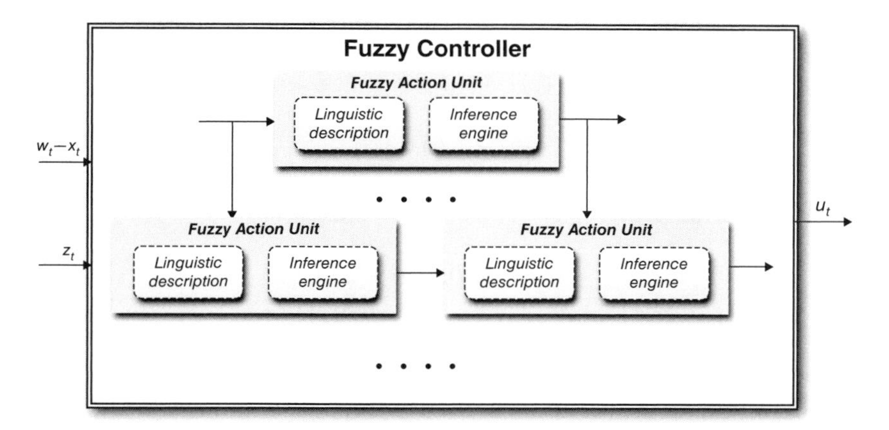

Figure 7.5 Scheme of a fuzzy controller that may consist of several fuzzy action units (fuzzy agents). These units may be of different kinds (relational or linguistic). Of course, units can be joined in various ways; their joining is determined by the structure of the knowledge base. Besides the value $w_t - x_t$, the input may also contain some other kinds of variables (denoted by z_t).

fuzzy controller consists of one relational fuzzy action unit. If a fuzzy action unit is linguistic, then we will call the fuzzy controller linguistic. Because the former (relational) fuzzy controllers are described in many papers and books (see, e.g., [31, 55, 67, 69, 74]), we will focus below on the linguistic ones.

7.2 FUZZY CONTROLLERS

In this and the subsequent sections, we will describe in more detail how fuzzy controllers are designed, how their performance can be improved and how they can be automatically learned. As already mentioned, we will focus mainly on linguistic control, because it is less known. We will also provide examples of various simulations of fuzzy/linguistic control. In these examples, we will model the controlled process by a differential equation, usually a linear one. Of course, we are aware that classical control theory mastered control of this kind of processes, and so we do not need fuzzy control of them. Our goal, however, is twofold: first, it is easy to realize such simulations and the reader can easily see that the fuzzy control indeed works. Second, we want to demonstrate that fuzzy control can be well applied also to processes that can otherwise be controlled classically. Let us emphasize that if these processes were real, the fuzzy control would probably work better because of many nonexpectable random disturbances, to which the fuzzy control is resistant.

7.2.1 Variables

The fundamental method in control theory is control by errors (differences between set-point and process output). More specifically, let $w(t)$ be a set-point, $y(t)$ be an output of the controlled process, and $u(t)$ be a control action at a time t. We will also suppose time sampling with a period ΔT. As stated, we usually set $\Delta T = 1$.

The (control) *error* at time t is given by

$$e(t) = w(t) - y(t). \tag{7.4}$$

Furthermore, the *change of error* (the first difference) is

$$\Delta e(t) = e(t) - e(t - \Delta T). \tag{7.5}$$

This quantity depends on the sampling time. To make it independent, we can use an approximation of the derivative

$$\dot{e}(t) = \frac{\Delta e(t)}{\Delta T}. \tag{7.6}$$

Of course, (7.5) coincides with (7.6) for $\Delta T = 1$. Therefore, we will quite often speak simply about "change of error" instead of "derivative" (alternatively, we will also write "derivative/change").

Finally, we introduce the *change of change of error* (the second difference of error)

$$\Delta^2 e(t) = \Delta e(t) - \Delta e(t - \Delta T). \tag{7.7}$$

Alternatively, we can use an approximation of the second derivative

$$\ddot{e}(t) = \frac{\Delta^2 e(t)}{(\Delta T)^2}. \tag{7.8}$$

As above, both values coincide for $\Delta T = 1$.

7.2.2 Basic Types of Classical Controllers

With the help of the notions defined above, we can introduce the following types of classical controllers.

(i) *P controller* is the simplest one. It is determined by the function

$$u(t) = C_P(e(t)) = K_P \cdot e(t),$$

where K_P is a constant (proportional gain).

(ii) *PD controller* is determined by the function

$$u(t) = C_{PD}(e(t), \dot{e}(t)) = K_P \cdot e(t) + K_D \cdot \dot{e}(t),$$

where K_D is a derivative constant.

(iii) The next possibility is not to set control action directly but just its derivative

$$\dot{u}(t) = \frac{u(t) - u(t - \Delta T)}{\Delta T} \tag{7.9}$$

(for $\Delta T = 1$, we may consider the change of control action $\Delta u(t) = u(t) - u(t-1)$). Then the *PI controller* is the function

$$\dot{u}(t) = C_{PI}(e(t), \dot{e}(t)) = K_I \cdot e(t) + K_P \cdot \dot{e}(t),$$

where K_I is an integration constant.

(iv) The most complicated case is the *PID controller*

$$\dot{u}(t) = C_{PID}(e(t), \dot{e}(t), \ddot{e}(t)) = K_I \cdot e(t) + K_P \cdot \dot{e}(t) + K_D \cdot \ddot{e}(t).$$

7.2.3 Basic Types of Fuzzy Controllers

If we suppose that a fuzzy controller is formed by only one fuzzy action unit (i.e., its knowledge base consists of a single linguistic description), we can, analogously as for classical controllers, distinguish P, PI, PD, and PID fuzzy controllers. Obviously, the type of the controller is determined by independent and dependent variables. From mathematical point of view, the fuzzy controller is a nonlinear function (7.2) whose behavior is determined by the given linguistic description.

(i) *Fuzzy P controller* is given by a linguistic description formed by fuzzy/linguistic IF-THEN rules of the form

$$\text{IF } e \text{ is } \mathscr{A}_e \text{ THEN } u \text{ is } \mathscr{B}_u, \tag{7.10}$$

where "e is $\mathscr{A}_e$" is a linguistic predication that characterizes the extent of error (e.g., "positive big") and "u is $\mathscr{B}_u$" is a linguistic predication that characterizes magnitude of the control action (e.g., "positive very small"). For this type of fuzzy controller, the control action is determined directly by the error.

(ii) *Fuzzy PD controller* is determined by a linguistic description formed by fuzzy/linguistic IF-THEN rules such as

$$\text{IF } e \text{ is } \mathscr{A}_e \text{ AND } \dot{e} \text{ is } \mathscr{A}_{\dot{e}} \text{ THEN } u \text{ is } \mathscr{B}_u, \tag{7.11}$$

where "e is $\mathscr{A}_e$" and "u is $\mathscr{B}_u$" are linguistic predications of the same type as for fuzzy P controller and "$\dot{e}$ is $\mathscr{A}_{\dot{e}}$" is a linguistic predication which characterizes the derivative/change of error. For this type of fuzzy controller, the control action is determined by the error and its derivative.

(iii) *Fuzzy PI controller* is determined by a linguistic description formed by fuzzy/linguistic IF-THEN rules of the form

$$\text{IF } e \text{ is } \mathscr{A}_e \text{ AND } \dot{e} \text{ is } \mathscr{A}_{\dot{e}} \text{ THEN } \dot{u} \text{ is } \mathscr{B}_{\dot{u}}. \tag{7.12}$$

The linguistic predications are similar as in the above case of fuzzy PD controller. The difference is that instead of the control action, the fuzzy PI controller determines a derivative of the control action. This type of fuzzy controller is the most often used, because for people, it is easier to think about two independent variables and to determine the derivative/change of control action. We can imagine the latter as follows: after some time, our control cock is already in some position. Then, for example, "derivative/change of control action is negative small" means that we should turn the cock from this position slightly to the left.

(iv) It is sometimes necessary to use the most complex *fuzzy PID controller*, which is determined by a linguistic description formed by rules such as

$$\text{IF } e \text{ is } \mathscr{A}_e \text{ AND } \dot{e} \text{ is } \mathscr{A}_{\dot{e}} \text{ AND } \ddot{e} \text{ is } \mathscr{A}_{\ddot{e}} \text{ THEN } \dot{u} \text{ is } \mathscr{B}_{\dot{u}}. \tag{7.13}$$

The fuzzy PID controller is usually needed for control of highly nonlinear and unstable processes. Let us note that antecedents of rules for PI, PD, and PID fuzzy controllers are of a conjunctive form. The main problem for fuzzy PID controller is the determination of the fuzzy/linguistic IF-THEN rules, because their number can be quite high. In practice, the PID controller can often be replaced by a combination of fuzzy PI and PD controllers. It means that the fuzzy controller consists of two fuzzy action units, one realizing PI and the second one PD control.

In book [103], a specific approach to the design of fuzzy controller has been analyzed, namely, the approach based on a construction of a *fuzzy model of the controlled process*. It means that it is a sort of a *hybrid approach*, where the requirement of knowing first the model of the controlled process and deriving the way of its control based on it is not completely abandoned. However, both model and control strategy are constructed by means of fuzzy approach described above. It means that we can increase the probability that the resulting control will behave more properly, because we are taking into account the complexity of the process. From this complexity follows a necessity to work with vague or inexact characterization of the situation. In practice, this can look as follows: there are Takagi–Sugeno (TS) rules with the right sides containing more complicated mathematical formulas, for example, (7.1) or (7.2) for fuzzy controller. It means that a given formula characterizes the behavior of the system only locally in an area specified by the left side of TS rules. One of the advantages of this approach is a better chance to mathematically prove the stability of fuzzy control.

Finally, it is necessary to recall that all the introduced types of fuzzy controllers are based on classical control theory, in which the input variables are errors with respect to the set-point, and its derivations. However, in practice, other variables may also be necessary. From our point of view, it means nothing else than an expansion of the antecedent (and, possibly, of the consequent, if there are more than one control actions) of the fuzzy IF-THEN rules by additional variables.

7.3 DESIGN OF FUZZY/LINGUISTIC CONTROLLER

Application of the fuzzy control can be summarized as follows:

1. Determine type of the fuzzy controller and all used variables.
2. Determine linguistic contexts of all the used variables.
3. Decide which kind of the fuzzy action unit(s) will be used.
4. Form a knowledge base that summarizes expert knowledge characterizing the control strategy.

Because relational type of fuzzy controller is in detail described in many books, we will focus here especially on the fuzzy/linguistic one.

7.3.1 Determination of Variables and Linguistic Context

Determination of variables. As the first step, we must decide what kind of the fuzzy controller will be applied, that is, we must choose among fuzzy P, PD, PI, and PID controllers (see Section 7.2.3). This implies what variables will be used:

- error $e(t)$ (7.4),
- first derivative of error $\dot{e}(t)$ (7.6),
- second derivative of error $\ddot{e}(t)$ (7.8),
- control action $u(t)$ or its derivative $\dot{u}(t)$.

In special cases, we may consider also some other variables.

Determination of linguistic context of all variables. The concept of linguistic context was explained in Section 5.2.3. Note that in general control theory the corresponding concept of "scaling" is used. However, because we deal with expressions of natural language, the term "(linguistic) context" is more appropriate and more accurate here.

Recall also that a linguistic context is determined by three numbers $\langle v_L, v_S, v_R \rangle$, where v_L is the smallest, v_S the middlemost, and v_R the greatest value. However, it is specific for the common control variables that they can attain both *positive* and *negative* values. Hence, if, for example, "big positive value" is +10, then "big negative value" is −10.

Consequently, we must distinguish two kinds of contexts:

1. *Simple* $w = \langle v_L, v_S, v_R \rangle$ where $v_L, v_S, v_R \in \mathbb{R}$ and $v_L < v_S < v_R$.
2. *Symmetric*, which has two parts:
   (i) *positive* $w^+ = \langle v_L^+, v_S^+, v_R^+ \rangle$, where $v_L^+, v_S^+, v_R^+ \in [0, +\infty)$ and $v_L^+ < v_S^+ < v_R^+$,
   (ii) *negative* $w^- = \langle v_R^-, v_S^-, v_L^- \rangle$, where $v_L^-, v_S^-, v_R^- \in (-\infty, 0]$ and $v_R^- < v_S^- < v_L^-$.
   We usually set $v_L^- = v_L^+ = 0$, $v_S^- = -v_S^+$ and $v_R^- = -v_R^+$.

In fuzzy control, depending on the type of fuzzy controller, the following contexts must be specified (all of them are symmetric):

(a) Context of *error* $e(t)$:

$$w_e^- = \langle -e_{\max}, -e_S, 0\rangle, \quad w_e^+ = \langle 0, e_S, e_{\max}\rangle. \tag{7.14}$$

(b) Context of *error derivative/change* $\dot{e}(t)$:

$$w_{\dot{e}}^- = \langle -\dot{e}_{\max}, -\dot{e}_S, 0\rangle, \quad w_{\dot{e}}^+ = \langle 0, \dot{e}_S, \dot{e}_{\max}\rangle. \tag{7.15}$$

(c) Context of *error second derivative/change* $\ddot{e}(t)$:

$$w_{\ddot{e}}^- = \langle -\ddot{e}_{\max}, -\ddot{e}_S, 0\rangle, \quad w_{\ddot{e}}^+ = \langle 0, \ddot{e}_S, \ddot{e}_{\max}\rangle. \tag{7.16}$$

(d) Context of *control action* $u(t)$:

$$w_u^- = \langle -u_{\max}, -u_S, 0\rangle, \quad w_u^+ = \langle 0, u_S, u_{\max}\rangle. \tag{7.17}$$

It can happen that negative values of the control action do not have sense. For example, we may heat up or not heat up, but we cannot cool. The context of control action is in this case a simple one.

(e) Context of *derivative/change of control action* $\dot{u}(t)$:

$$w_{\dot{u}}^- = \langle -\dot{u}_{\max}, -\dot{u}_S, 0\rangle, \quad w_{\dot{u}}^+ = \langle 0, \dot{u}_S, \dot{u}_{\max}\rangle. \tag{7.18}$$

For simplicity, we often put $e_S = 0.4\ e_{\max}$ and similarly for the other variables. Note that the context for the control action corresponds to the limit position of a "control cock". In some situations, it can be advantageous to consider "asymmetric" linguistic context in the sense that $|v_R^+| \neq |v_R^-|$. For example, for the context of error, it can be helpful to distinguish contexts for positive errors (we are under the set-point) and negative ones (we are above it).

7.3.2 Choosing Fuzzy Action Unit

The decision about the most appropriate fuzzy action unit in a concrete situation depends on the chosen interpretation of a linguistic description.

The *linguistic fuzzy action unit* can be used in most situations because it provides guaranteed results in accordance with our understanding of the linguistic expressions used in the fuzzy IF-THEN rules and our expectations of what actions should follow. Its disadvantage is the noncontinuous behavior of the control. In practice, this usually does not produce problems. In some cases, however, it could cause more stress to the used actuator. This drawback can be overcome by using smooth DEE defuzzification. However, in this case, we have to take into account higher computation load.

This action unit is very suitable for cases which are of decision-making nature (e.g., bypassing an obstacle described in Section 5.4.4). It is also appropriate when we design supervising control elements in a hierarchical control system. It should also be noted that a linguistic description in this action unit has usually higher number of fuzzy IF-THEN rules than in the relational fuzzy action unit. This is useful for practice because we can influence more accurately behavior of the controller *locally* when a particular situation requires this. The big advantage is that the linguistic description is well comprehensible even after years.

The *relational fuzzy action unit* is effective when we approximate an inexactly given function. Contrary to the linguistic action unit, it is not necessary here to give a special meaning to used fuzzy sets. Shapes of them are chosen to fit the course of an approximated function in the best possible way and to cover the whole domain and range of it. Therefore, membership functions are modified independently on the assigned word labels. For the sake of simplicity, linear membership functions (triangular or trapezoidal) are most often used. More about the difference between both types (linguistic and relational) of interpretation of fuzzy IF-THEN rules can be found in Section 5.4.4.

7.3.3 Formation of Knowledge Base

The knowledge base contains complete information about the control strategy. As already stated, the knowledge base can be formed of one or more linguistic descriptions. For the sake of simplicity, we will in this section consider a simple knowledge base consisting of only one linguistic description.

The form of linguistic description is influenced by the kind of fuzzy action unit. If the latter is linguistic, then the fuzzy/linguistic IF-THEN rules should be formed using evaluative linguistic expressions. If it is relational, then we work with simplified fuzzy sets, usually triangular or trapezoidal, whose underlying linguistic meaning is “categories” or “fuzzy numbers” (these are “word labels”), cf. Remark 5.1. The leading idea is to provide a sufficient fuzzy partition as discussed in Example 3.1.

In the following explanation, we will focus on the case when a fuzzy controller is formed by the linguistic fuzzy action unit.

7.3.3.1 Linguistic description as representation of expert’s knowledge. Recall that the original idea of fuzzy control (cf. [148]) was to use expert’s (human operator’s) knowledge of the control strategy. A linguistic description then represents a sort of schematic record of various details of this knowledge.

The initial formation of a linguistic description can be performed on the basis of interviewing an expert (or an operator) who is sufficiently experienced in the control of a given process. However, we may hardly expect to be successful in acquiring the linguistic description in the first session. Rather, it is an iterative procedure requiring close cooperation of the expert and the (fuzzy) control specialist.[1] Even though we stressed that the knowledge of the mathematical description of a controlled process is not necessary, we must have at least a rough idea about it. This idea usually comes out of the common-sense experience. For example, when controlling a furnace or a boiler, everybody knows that more heating leads to increase of temperature, and less heating stops or slows down its increase. The encountered problems have more or less

[1]The situation is similar to formation of the knowledge base of a classical expert system (its methodology is studied in the so-called *knowledge engineering*).

a local character; namely, it is often necessary to influence the course of the control in a certain region only, for example when approaching a set-point. In fuzzy control, it means that we have to modify just one rule (or a small group of rules). This is not difficult, in particular if we use natural language.

To prepare a linguistic description on the basis of information provided by an expert/operator, we may effectively use linguistic descriptions already formed for control of special classes of processes. From this point of view, the genuine linguistic description (i.e., consisting of fuzzy/linguistic IF-THEN rules) is very convenient, because it is self-explaining and it is easy to understand its message even after a long time. Moreover, this kind of descriptions is robust and works well for a wide class of processes.

The following linguistic description is an example of a sufficiently general fuzzy PI controller for a control of various processes using the linguistic fuzzy action unit. For the meaning of abbreviations of components of evaluative expressions, see Section 5.4.4.

	$e(t)$ &	$\dot{e}(t)$ ⇒	$\dot{u}(t)$		$e(t)$ &	$\dot{e}(t)$ ⇒	$\dot{u}(t)$
1.	Ze	Ze	Ze	19.	−Me	+Sm	−ExSm
2.	+Bi	+NoZe	+Bi	20.	+Sm	+NoSm	+Me
3.	−Bi	−NoZe	−Bi	21.	−Sm	−NoSm	−Me
4.	+Bi	−NoSm	Ze	22.	+Sm	−NoSm	−VeSm
5.	−Bi	+NoSm	Ze	23.	−Sm	+NoSm	+VeSm
6.	+Bi	RoZe	+Me	24.	+Sm	RoZe	+VeSm
7.	−Bi	RoZe	−Me	25.	−Sm	RoZe	−VeSm
8.	+Bi	−Sm	+Me	26.	+Sm	+Sm	+Sm
9.	−Bi	+Sm	−Me	27.	+Sm	−Sm	+ExSm
10.	+Me	+NoSm	+Bi	28.	−Sm	+Sm	−ExSm
11.	−Me	−NoSm	−Bi	29.	−Sm	−Sm	−Sm
12.	+Me	−NoSm	−VeSm	30.	+VeSm	−Sm	−Sm
13.	−Me	+NoSm	+VeSm	31.	−VeSm	+Sm	+Sm
14.	+Me	RoZe	+Me	32.	+VeSm	−VeSm	Ze
15.	−Me	RoZe	−Me	33.	−VeSm	+VeSm	Ze
16.	+Me	+Sm	+Me	34.	RoZe	+VeSm	+ExSm
17.	−Me	−Sm	−Me	35.	RoZe	−VeSm	−ExSm
18.	+Me	−Sm	+ExSm				

When we analyze the above linguistic description, we obtain the following structure:

	$e(t)$ &	$\dot{e}(t)$ ⇒	$\dot{u}(t)$
1. (1.)	Ze	Ze	Ze
2. (2.)	+Bi	+NoZe	+Bi
3. (6.)	+Bi	RoZe	+Me
4. (8.)	+Bi	−Sm	+Me
5. (4.)	+Bi	−NoSm	Ze

6. (10.)	+Me	+NoSm	+Bi
7. (16.)	+Me	+Sm	+Me
8. (14.)	+Me	RoZe	+Me
9. (18.)	+Me	−Sm	+ExSm
10. (12.)	+Me	−NoSm	−VeSm
11. (20.)	+Sm	+NoSm	+Me
12. (26.)	+Sm	+Sm	+Sm
13. (24.)	+Sm	RoZe	+VeSm
14. (27.)	+Sm	−Sm	+ExSm
15. (22.)	+Sm	−NoSm	−VeSm
16. (30.)	+VeSm	−Sm	−Sm
17. (32.)	+VeSm	−VeSm	Ze
18. (34.)	RoZe	+VeSm	+ExSm

The first number is the number of a group of rules and the second one is the number of the first rule belonging to the former.

First, note that there are always pairs of rules in the above table which differ in mutually opposite signs. The reason is that it is necessary to distinguish the behavior of the controller if the controlled system is under the set-point (the error $e(t)$ is positive) or above it (the error $e(t)$ is negative). Similarly, if the difference (derivative) of error $\dot{e}(t)$ is positive, then the output value of the controlled system decreases, and it increases in the opposite case.

The rules are also divided according to the magnitude of error $e(t)$, that is, we have groups of rules for small, medium, or big errors. The last two groups in the table deal with the situation when the output of the controlled system differs only slightly from the set-point. All groups contain rules that react to situation when the process output is moving away from the set-point (i.e., both error and its derivative have the same sign) and when the output is approaching the set-point. The reaction depends on the distance from the set-point.

7.3.3.2 General fuzzy/linguistic PI controller. The above linguistic description can be a core part of a general fuzzy action unit realizing fuzzy PI control using which various kinds of processes can be controlled. Below, we demonstrate the results of fuzzy control of processes having the following transfer functions:

$$\frac{1}{s}, \quad \frac{1}{s+1}, \quad \frac{1}{s(s+1)},$$

$$\frac{1}{(s+1)^2}, \quad \frac{1}{(s+1)^3}, \quad \frac{e^{-5s}}{s+1},$$

$$\frac{1-s}{(s+1)^2}, \quad \frac{1}{s-1}.$$

Recall that these processes are given using the Laplace transform of their transfer functions, that is, in the form

$$\frac{\mathscr{L}(u)}{\mathscr{L}(y)}.$$

Recall also that the power at the s parameter in the denominator corresponds to the order of derivative. From this, we obtain the corresponding differential equation. For example, $\frac{1}{s+1}$ is the Laplace image of the transfer function of the process modeled using the following differential equation:

$$\dot{y} + y = u,$$

and similarly for other processes. Furthermore, it holds that if all solutions of the adjoint equation[2] for s are negative (or have negative real component), then the process is stable. Roughly speaking, it means that a small change in the input corresponds to a small change in the output and the process is able to reach some state and stay in it, if the input remains unchanged. If at least one of the solutions of the adjoint equation is positive, then the process is unstable, that is, a small change of the input causes a very big change of the output. In our case, this holds for the process with the transfer function $\frac{1}{s-1}$.

Notice that the processes given above considerably differ; they include stable processes, processes with delay, as well as unstable processes. For the control of any of these processes, we use *the same linguistic description*! It means that the control strategy is the same for all these processes. The only necessity is to set an individual linguistic contexts of variables for each process.

Let us emphasize that the linguistic description was formed using natural language only, that is, the expert characterized the control strategy in natural language without thinking of membership functions of fuzzy sets. In fact, he/she had no idea about their existence.

Notice also that in the linguistic description, we can well express a certain *caution* of fuzzy controller, that is, its effort not to overshot the set-point: for example, in a situation (Rule 4), when the error is big but the negative derivative is big enough (expressed by linguistic expression *not small*), there is no control action.

Simulation results of the control of processes given above are depicted in Figures 7.6–7.12. Let us repeat that the same general linguistic description of fuzzy PI controller has been used with various linguistic contexts of variables. Because all the variables can attain positive as well as negative values, its (absolute) minimal possible value is 0. Therefore, it is necessary to set only (absolute) maximal value of error $e^{\max}$, its derivative (difference) $\dot{e}^{\max}$, and the derivative of control action $\dot{u}^{\max}$.

When comparing all the figures, one can see that the control of each process can still be improved if we slightly modify the linguistic description separately for each of the process. But this does not deny the basic idea of one essential control strategy common for a wide class of processes.

7.3.4 Tuning Linguistic Description

The way how a given linguistic description should be tuned depends again on its intended interpretation, that is, on the considered fuzzy action unit.

[2]We put the denominator of a transfer function equal to zero and solve the obtained equation with respect to s as an unknown variable.

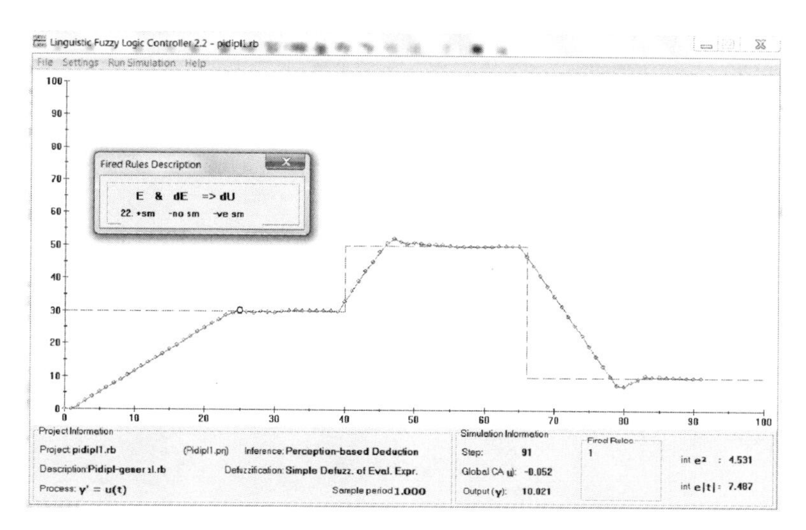

Figure 7.6 Simulation of fuzzy control of the process $\frac{1}{s}$ using general fuzzy PI controller realized by the linguistic fuzzy action unit. The linguistic context is $e^{\max} = 5$, $\dot{e}^{\max} = 2$, and $\dot{u}^{\max} = 3.25$.

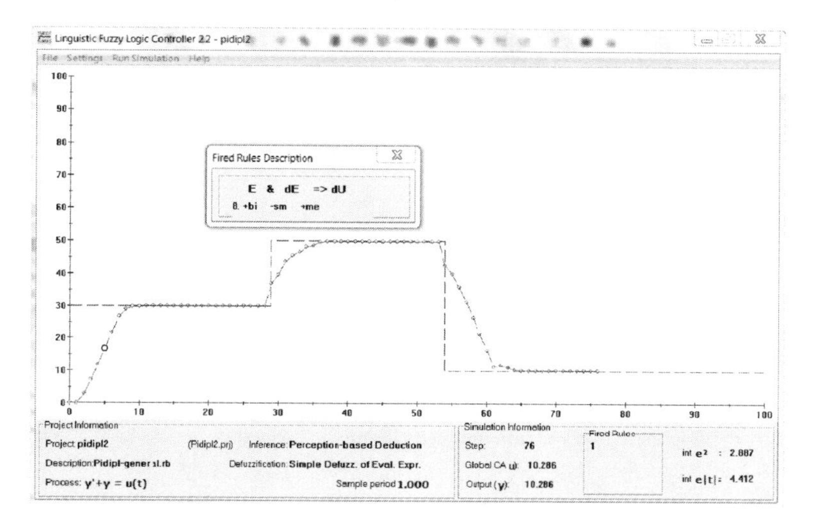

Figure 7.7 Simulation of fuzzy control of the process $\frac{1}{s+1}$ using general fuzzy PI controller realized by the linguistic fuzzy action unit. The linguistic context is $e^{\max} = 13$, $\dot{e}^{\max} = 18$, and $\dot{u}^{\max} = 12$.

7.3.4.1 Linguistic fuzzy action unit.

Extensions (i.e., fuzzy sets) of linguistic expressions are predefined, and only exceptionally, there could be a reason for changing them.[3] This means that tuning of linguistic description is performed by

[3]The authors of this book did not need to change them a single time during many tens of simulations and real use in practice.

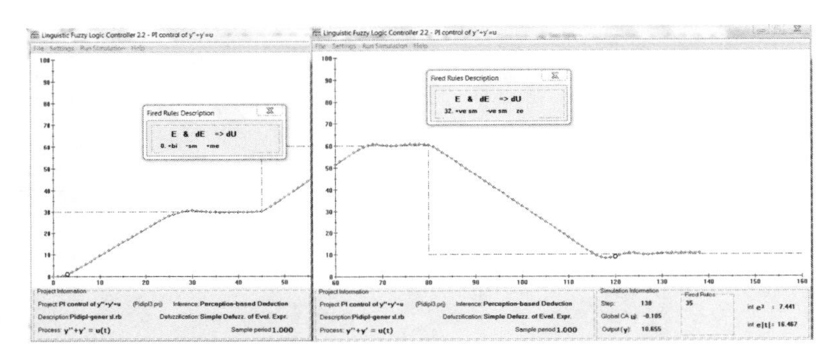

Figure 7.8 Simulation of fuzzy control of the process $\frac{1}{s(s+1)}$ using general fuzzy PI controller realized by the linguistic fuzzy action unit. The linguistic context is $e^{\max} = 9$, $\dot{e}^{\max} = 1.25$, and $\dot{u}^{\max} = 1.55$.

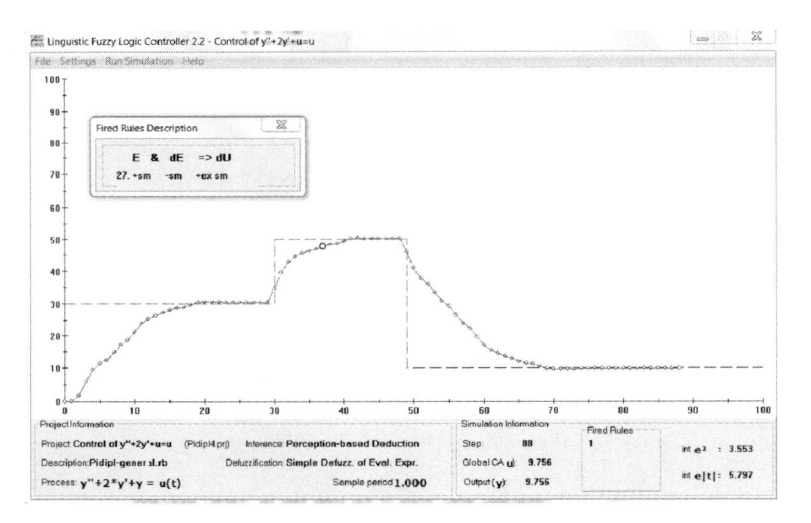

Figure 7.9 Simulation of fuzzy control of the process $\frac{1}{(s+1)^2}$ using general fuzzy PI controller realized by the linguistic fuzzy action unit. The linguistic context is $e^{\max} = 12$, $\dot{e}^{\max} = 6.2$, and $\dot{u}^{\max} = 16.5$.

replacing one linguistic expression by another one which is either more or less precise in the sense of the ordering (5.5). For example, *very small* is more precise than *small* and *roughly small* (the latter is less precise than the former).

The choice of more or less precise expression depends on the reaction of fuzzy controller (either too fine or too coarse). At the same time, it can happen that some rule is superfluous or missing. For example, we encountered in practice the following situation several times: after removal of some of finer rules, the course of the control improves. On the other hand, it was sometimes necessary to add more precise rule, namely, in situations when we got close to the set-point, where the behavior of the given process is crucial. If the process is slow with a long delay, then it is necessary

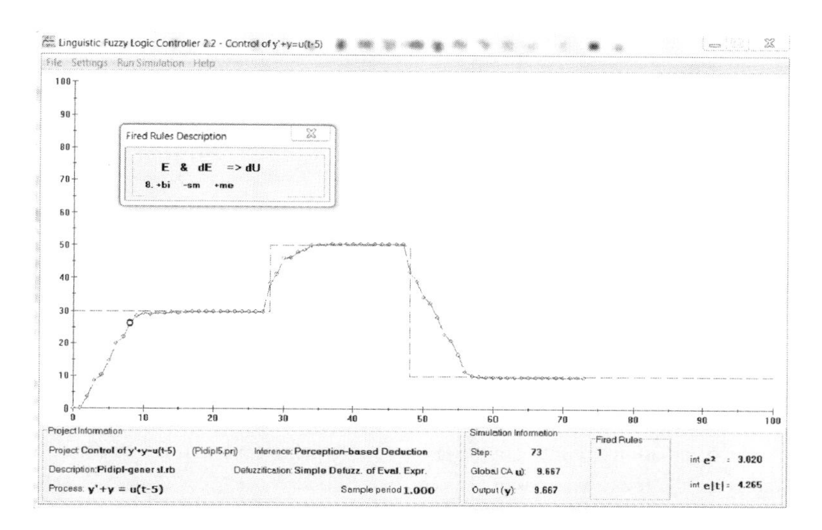

Figure 7.10 Simulation of fuzzy control of the process $\frac{e^{-5s}}{s+1}$ with delay using general fuzzy PI controller realized by the linguistic fuzzy action unit. The linguistic context is $e^{\max} = 12$, $\dot{e}^{\max} = 14$, and $\dot{u}^{\max} = 14$.

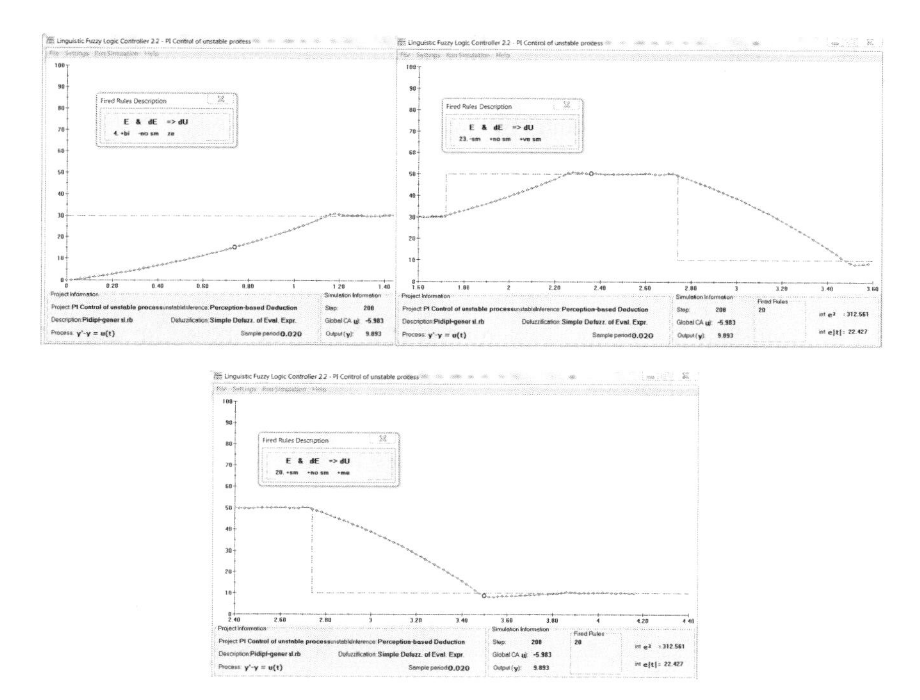

Figure 7.11 Simulation of fuzzy control of the unstable process $\frac{1}{s-1}$ using general fuzzy PI controller realized by the linguistic fuzzy action unit. The linguistic context is $e^{\max} = 1$, $\dot{e}^{\max} = 0.5$, and $\dot{u}^{\max} = 35.9$.

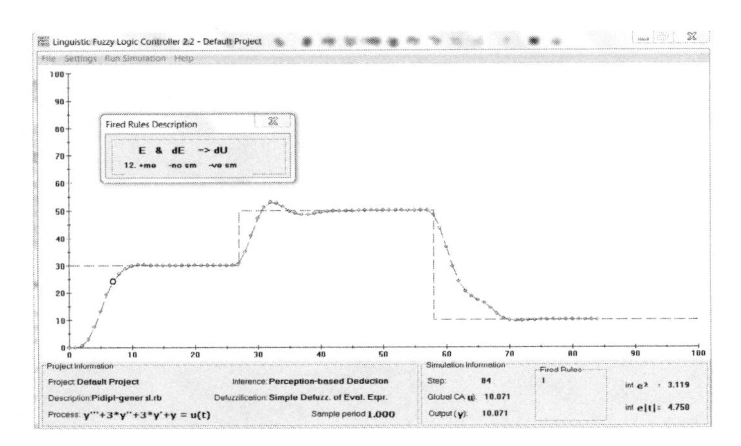

Figure 7.12 Simulation of control of the process $\frac{1}{(s+1)^3}$ using general fuzzy PI controller realized by the linguistic fuzzy action unit. The linguistic context is $e^{\max} = 27.5$, $\dot{e}^{\max} = 16.5$, and $\dot{u}^{\max} = 18.9$.

to be very cautious and to use much finer rules than in other cases. A complicated situation arises for unstable processes, when the tuning is highly dependent on the knowledge of the process behavior and engineering intuition.

Important role is played by the linguistic context. We do not recommend to change both the linguistic context and the linguistic description simultaneously. It is always necessary to set well enough the former and then the latter. In practice, we should first form the linguistic description roughly, then find the context, and only then return to the linguistic description and tune it.

7.3.4.2 Relational fuzzy action unit. In this case, we can choose between DNF and CNF interpretation of the given linguistic description. As discussed in Section 3.2.2, there is no significant difference between their properties.

(a) The linguistic description is interpreted as DNF with a chosen t-norm (minimum, product, or Łukasiewicz t-norm), the defuzzification operation should be center of gravity (COG) or mean of maxima (MOM), possibly also center of sums (COS). Recall that COG gives optimal results and the resulting approximation function f^A is continuous. Fuzzy sets A_i, B_i, $i = 1, \ldots, m$, should be fuzzy numbers preferably having triangular membership functions that form an appropriate fuzzy partition so that condition (3.60) is fulfilled.

(b) The linguistic description is interpreted as CNF with a chosen fuzzy implication. According to our experience, the best results can be obtained with Łukasiewicz implication (2.24). The defuzzification operation should be MOM. Fuzzy sets A_i, B_i, $i = 1, \ldots, m$, should be fuzzy numbers and either A_i or B_i (or both) should have Gaussian membership function. Then, the approximation function f^A is continuous.

Since fuzzy IF-THEN rules should be formed in such a way that the control function is approximated with the best possible accuracy, membership functions should be modified accordingly. Recall that as an auxiliary meaning of these membership functions, we understand them to represent fuzzy numbers.

Remark 7.2 *Let us remark that, according to the theory, we can also use in the DNF interpretation other t-norms than the three ones mentioned above. Practical experiences show, however, that the difference is inessential. Therefore, it may hardly have any sense to use them.*

7.4 LEARNING

In this section, we describe learning abilities of the linguistic fuzzy action unit. The learning task can be realized in two ways:

(a) Modification and learning of linguistic context of all variables occurring in the rules forming a linguistic description.
(b) Learning a linguistic description.

7.4.1 Modification and Learning of Linguistic Context

7.4.1.1 Modification of linguistic context. During fuzzy control, we often meet the situation when the output of the controlled process approaches the set-point but the fuzzy controller typically oscillates around it and only rarely reaches it precisely. The reason is that its behavior is described roughly, so that too small values are simply neglected (evaluated as equal to (fuzzy) zero). In practice, this usually does not cause problems because the limit precision is seldom required. An improvement, however, is always welcome.

An efficient possibility is to modify continuously the linguistic context of variables. The main idea is to modify it in such a way that values originally evaluated as small are now evaluated as big. Consequently, the originally indistinguishable values now become, at least, small; therefore, they can be managed.

The algorithm for modification of the linguistic context described below works nice. Let $w = \langle v_L, v_S, v_R \rangle$ be a context and q be a number. Then we denote $qw = \langle qv_L, qv_S, qv_R \rangle$.

Initialization: Set values of the following parameters:

- The lower limit for a truth value of "small" $k_S \in (0, 1]$ (the usual value is 0.9).
- The lower limit for a truth value of "big" $k_B \in (0, 1]$.
- Ratio for modification of the context $q < 1$ (the usual value is $q = 0.6$).
- The number $m > 1$ of cycles considered for checking whether the set-point is already close. The usual value is $m = 3$.
- Set the initial context w^{IN} and current context $w := w^{IN}$.

The algorithm runs at each sampling time moment t. Its outputs are current linguistic contexts of all the considered variables, that is, in the case of fuzzy control, the contexts (7.14)–(7.18). During its run, we must keep a sequence of recent m time moments $t_m, \dots, t_1$.

Algorithm:

1. Modify the sequence of the recent m time moments $t_m, \ldots, t_1$ as follows: For all $i = 2, \ldots, m$, put $t_{i-1} := t_i$ and $t_m := t$.
2. For the sequence of the recent m time moments $t_m, \ldots, t_1$, compute the truth values

$$r(t_m) = \text{Ext}_w(X \text{ is } small)(|e(t_m)|),$$

$$\ldots\ldots\ldots\ldots\ldots\ldots$$

$$r(t_1) = \text{Ext}_w(X \text{ is } small)(|e(t_1)|).$$

3. If $r = r(t_m) \otimes \cdots \otimes r(t_1) \geq k_S$, then modify the current context w by q, that is, $w := qw$ and exit.
4. For the sequence of the recent m time moments $t_m, \ldots, t_1$, compute the truth values

$$r'(t_m) = \text{Ext}_w(X \text{ is } big)(|e(t_m)|),$$

$$\ldots\ldots\ldots\ldots\ldots\ldots$$

$$r'(t_1) = \text{Ext}_w(X \text{ is } big)(|e(t_1)|).$$

5. If $r'(t_m) = 1$ or $r' = r'(t_m) \otimes \cdots \otimes r'(t_1) \geq k_B$, then return the initial context: $w := w^{IN}$ and exit.

Comments to the algorithm:

(a) We used the property of nilpotency (2.40) of the Łukasiewicz t-norm. Note that this property is quite strict. For example, $0.9 \otimes 0.7 \otimes 0.5 = 0.1$. Therefore, the number m of checked recent time moments should not be too big (practical experiences show that $m = 3$ is sufficient).

(b) Items 2 and 3 check whether the error $e(t)$ (7.4) is already small enough. Similarly, items 4 and 5 check whether it did not suddenly become big. If yes, we must immediately return to the initial context because otherwise the control would become ineffective and nonstable.

(c) This algorithm is very effective especially when controlling stable processes, because it enables to accomplish very precise control. It is implemented in the LFL Controller© software system.[4]

Experimental results of fuzzy control using linguistic fuzzy action unit are depicted in Figures 7.13 and 7.14. They demonstrate the control of an inverted pendulum represented by the following simplified model:

$$-ml^2\ \ddot{y}(t) + mlg \sin y(t) = u(t),$$

where m is the mass of the pendulum, l is its length (both set to 1 in the model), and g is the gravitational constant. The control was realized using fuzzy PD controller.

[4]See the companion website http://www.wiley.com/go/novak/fuzzy/modeling.

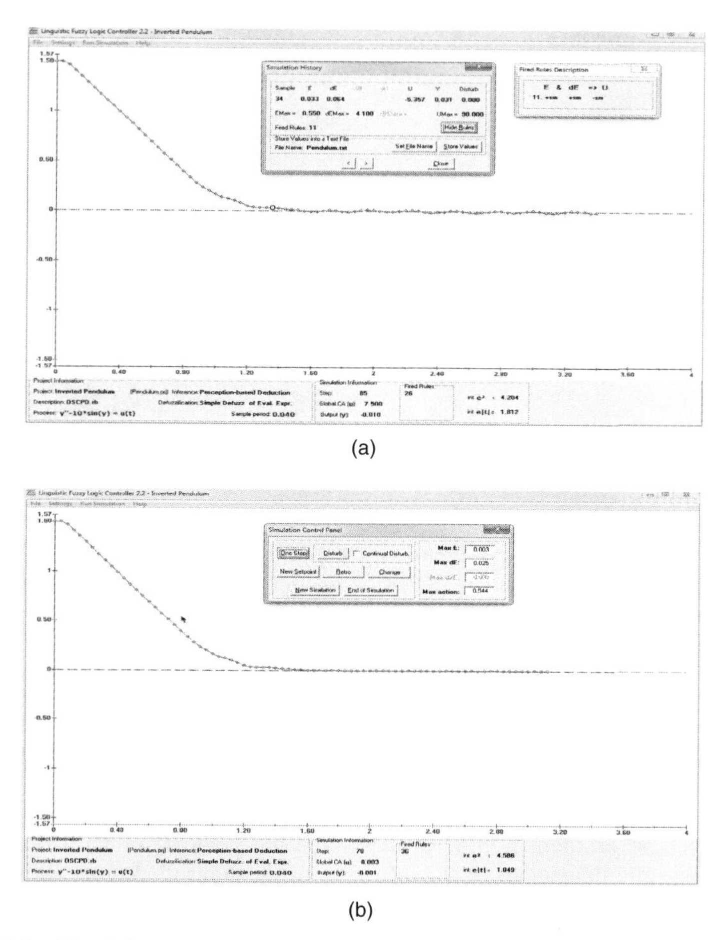

Figure 7.13 Simulation of fuzzy control of the inverted pendulum by fuzzy PD controller using the linguistic fuzzy action unit. The linguistic context is $e^{\max} = 0.55$, $\dot{e}^{\max} = 4.1$, and $u^{\max} = 90$. Figure (a) shows the control without changing the context and (b) the control with continuous change of context.

The error and its derivative are in radians, the control action has physical meaning corresponding to a moment set on the pendulum. The initial position was 1.5 radians. For the sake of comparison, we also demonstrate fuzzy control of inverted pendulum using the PbLD method with smooth DEE defuzzification (Figure 7.14).

We can see in Figure 7.13(b) that the continuous modification of the context using the above described algorithm leads to smoother and more precise course of control (the set-point value is reached almost exactly without oscillation).

7.4.1.2 Learning linguistic context. An exciting possibility is to *learn* the context from the course of control. This is principally possible if a tuned linguistic description is available. A very simple solution suggests itself for independent variables (error and its difference), because linguistic context for them can be set simply from an

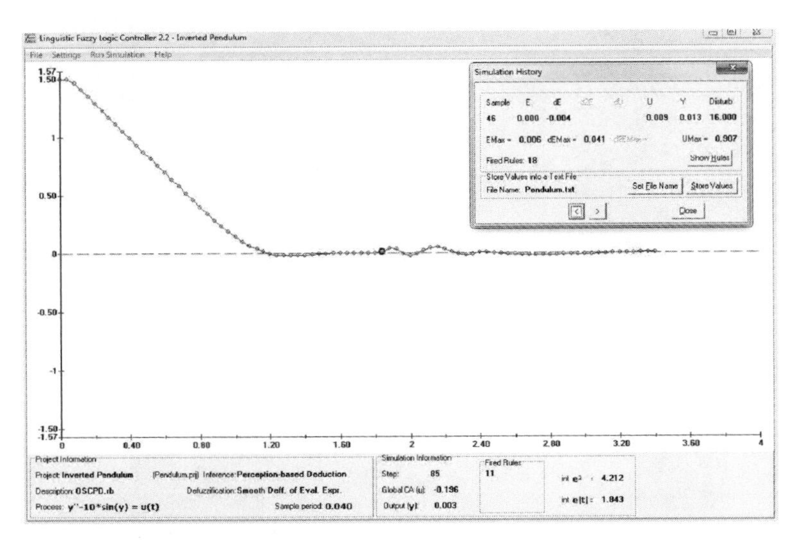

Figure 7.14 Simulation of fuzzy control of the inverted pendulum by fuzzy PD controller using the linguistic fuzzy action unit in which smooth DEE defuzzification has been used in the PbLD method. The jump in the middle is caused by the input disturbance (its magnitude was about 17% of the control action).

assumption that in the beginning of control, *any nonzero difference* from the set-point is *big*. In this case, $e_{\max} = K_e \cdot e(0)$, where $e(0)$ is the initial error with respect to the set-point and K_e is a constant. The suitable value of K_e was experimentally estimated as $K_e = 0.7$. Similarly, we can also learn $\dot{e}_{\max}$ with $K_{\dot{e}} = 2$ and $\ddot{e}_{\max}$ with $K_{\ddot{e}} = 1.2$.

However, the problem is how to determine the context for the control action because it is fully dependent on the character of the controlled process. It could be relatively simple if we know the mathematical description of the process, but this contradicts the basic principle of fuzzy control. Therefore, we cannot assume it. We carried out relatively successful experiments with two given extremal values of control action context $u_{\max_1}$ and $u_{\max_2}$. The desired value $u_{\max}$ of the context is then obtained by interpolation between them (for details, see [18]). A demonstration of fuzzy control on the basis of learned linguistic context can be seen in Figure 7.15.

7.4.2 Learning Linguistic Description

A very attractive possibility is to generate the linguistic description automatically. We can typically do it by monitoring actions of an experienced operator during the control process. Instead of asking him/her, we reconstruct a linguistic description according to his/her individual control actions. We expect that the resulting description would work in a corresponding situation similarly as the operator. For example, if we monitor a technique of a driver during his passage through a curve and generate a linguistic description accordingly, then an automatic driver operating on the basis of this description should pass this curve in a similar way. Moreover, because fuzzy control is robust, it can be expected that this automatic driver can successfully pass also other curves.

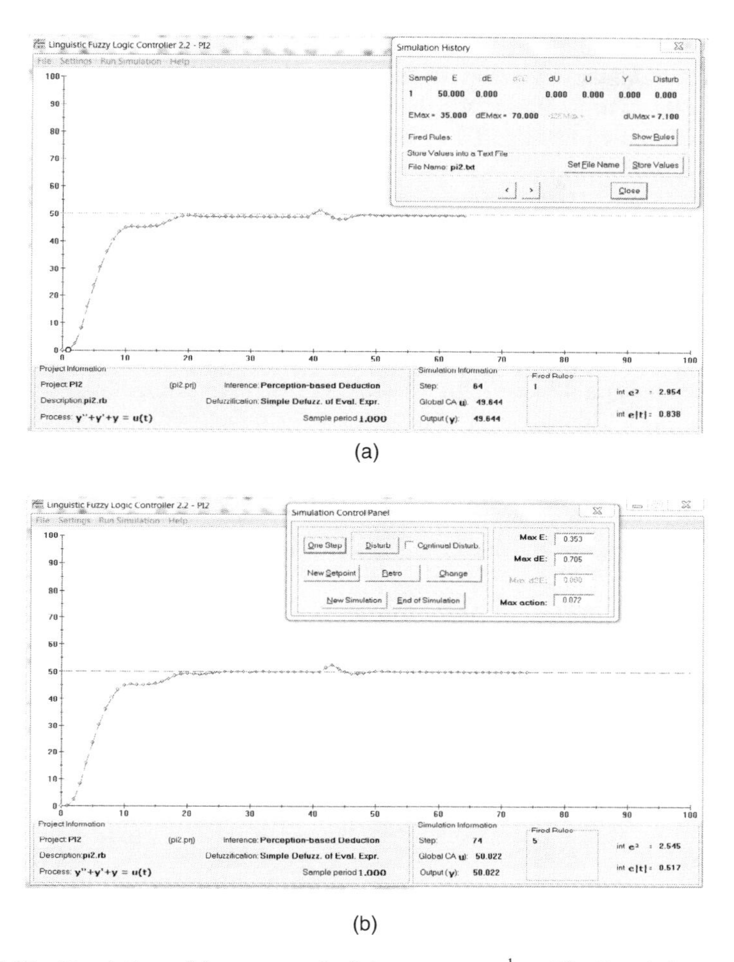

Figure 7.15 Simulation of fuzzy control of the process $\frac{1}{s^2+s+1}$. The linguistic context was automatically learned to $e^{\max} = 35$, $\dot{e}^{\max} = 70$, and $\dot{u}^{\max} = 7.1$. Figure (a) shows the control without changing the context and (b) the control with continuous change of the context. In the middle of both graphs, one can see the reaction of the controller to an input disturbance (its magnitude was 70% of the set-point).

The initial situation can be described as follows. Let P be a dynamical process, which is controlled manually by a human operator. The only known general characteristic of P is that it is a stable process. Our task is to monitor the course of the manual control and to learn a linguistic description on the basis of the obtained data in such a way that the linguistic fuzzy action unit will lead to similar control course as that done by the operator.

Obtained data are supposed to have the following form:

$$\langle u(k), y(k), v(k)\rangle, \qquad k = 1, \ldots, r, \tag{7.19}$$

where $u(k)$ is the input to P in the time moment k, $v(k)$ is the required value, and $y(k)$ is the output $y(k) = P(u(k))$.

Furthermore, we suppose that the control has proceeded on the basis of errors from the set-point and their derivations as in one of the above considered types of fuzzy PI, PD, or PID control. Hence, we should transform data into the proper form. For example, in the case of fuzzy PI control, the transformed data will have the form

$$c'(1) = \langle e(1), \dot{e}(1), \dot{u}(1)\rangle,$$
$$\vdots$$
$$c'(r) = \langle e(r), \dot{e}(r), \dot{u}(r)\rangle. \tag{7.20}$$

Learning algorithm.

1. For each variable in (7.20), specify the corresponding linguistic contexts $w_e, w_{\dot{e}}, w_{\dot{u}}$. The values v_L and v_R can be put equal to minimal and maximal values of the corresponding variable, and v_S is usually (but not necessarily) equal to $v_S = v_L + 0.4(v_R - v_L)$.
2. For each data item $c'(k) = \langle e(k), \dot{e}(k), \dot{u}(k)\rangle$ find, using the function of local perception (5.31), suitable evaluative predications

$$LPerc^{LD}(e(k), w_e) = \text{Int}(e \text{ is } \mathscr{A}),$$
$$LPerc^{LD}(\dot{e}(k), w_{\dot{e}}) = \text{Int}(\dot{e} \text{ is } \mathscr{B}),$$
$$LPerc^{LD}(\dot{u}(k), w_{\dot{u}}) = \text{Int}(\dot{u} \text{ is } \mathscr{C}).$$

   The result is a rule $\mathscr{R}_k$ of the form

$$\mathscr{R}_k := \textsf{IF } e \text{ is } \mathscr{A}_e \textsf{ AND } \dot{e} \text{ is } \mathscr{A}_{\dot{e}} \textsf{ THEN } \dot{u} \text{ is } \mathscr{B}_{\dot{u}}$$

   (see (7.12)).
3. Repeat item 2 for all $k = 1, \ldots, r$ and generate the linguistic description $\mathscr{R}_1, \ldots, \mathscr{R}_r$.
4. Reduce the learned linguistic description obtained due to item 2 as follows:
   (a) If rules $\mathscr{R}_{i_1}, \mathscr{R}_{i_2}, \ldots, \mathscr{R}_{i_k}$, $i_1, i_2, \ldots, i_k \in \{1, \ldots, r\}$, are identical, remove rules $\mathscr{R}_{i_2}, \ldots, \mathscr{R}_{i_k}$.
   (b) Let $\mathscr{R}_i$ and $\mathscr{R}_j$ be two generated rules such that all expressions have the same sign and their consequents be identical. Let the linguistic expressions for both variables in the antecedent of $\mathscr{R}_i$ be wider (in the sense of (5.5)) than those of $\mathscr{R}_j$. Then remove $\mathscr{R}_j$.
   (c) Let $\mathscr{R}_i$ and $\mathscr{R}_j$ be two generated rules such that all expressions have the same sign and their antecedents be identical. Let the consequent of $\mathscr{R}_i$ be narrower (in the sense of (5.5)) than that of $\mathscr{R}_j$. Then remove the latter rule.

   A more sophisticated method for reduction of linguistic descriptions is described in [37].

EXAMPLE 7.1 Let us consider control of the process characterized by the differential equation $3\dot{y}(t) + y(t) - \sqrt{y(t)} = u(t)$. We assume that it is controlled by

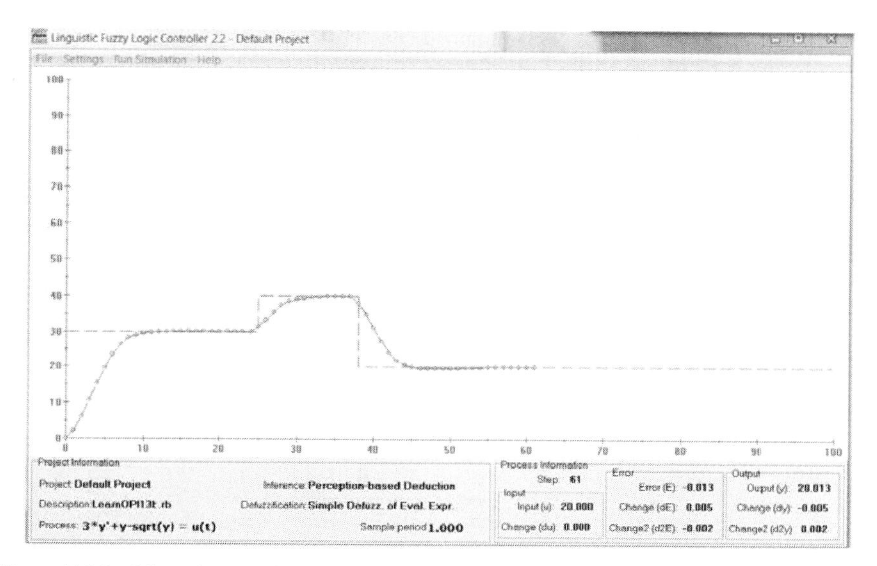

Figure 7.16 Manual control of the process represented by the differential equation $3\dot{y}(t) + y(t) - \sqrt{y(t)} = u(t)$.

(fuzzy) PI controller. It means that we measure error $e(t)$, its derivative $\dot{e}(t)$, and derivative of control action $\dot{u}(t)$.

The process was first controlled manually, namely, the concrete control actions were carried out by a human operator[5]on the basis of the knowledge of the controlled process. The result of the manual control is depicted in Figure 7.16. His control was monitored. Then, the linguistic description presented in Table 7.1 was generated (it uses abbreviations defined in Section 5.4.4).

Using this description, we retrospectively simulated control of the original process. We used PbLD with linguistic expressions and the DEE defuzzification. The linguistic context was set as follows: the corresponding values of v_R are $e_{\max} = 29.9$, $\dot{e}_{\max} = 10$, and $\dot{u}_{\max} = 9.5$. The corresponding values of v_L and v_S were set to $v_L = 0$ and $v_S = 0.4 \cdot v_R$. Simulation results are depicted in Figure 7.17. □

7.4.3 Practical Experiences with Control Using Linguistic Fuzzy Action Unit

There are various kinds of applications of techniques described in this chapter, for example, the control of a plaster kiln, a system for hydraulic transition water–oil, the control of a massive 100-tons steam generator. The most successful application is control of five aluminum smelting furnaces TLP9 in Al Invest company in a small village Břidličná in the Czech Republic. This applications has been in detail described in [91]. Let us recall the main points.

[5]Because this was only simulation, the human operator was a person setting values directly from the computer monitor.

TABLE 7.1 The linguistic description learned on the basis of monitoring of manual control from Figure 7.16.

Rule	e	AND	$\dot{e} \Rightarrow$	$\dot{u}$
1.	ExSm	AND	−ExSm ⇒	About 0.07
2.	−ExSm	AND	−ExSm ⇒	About −2.6
3.	−ExSm	AND	ExSm ⇒	About 2.2
4.	ExSm	AND	ExSm ⇒	About 0.3
5.	ExBi	AND	Ze ⇒	VeBi
6.	SiBi	AND	−RoSm ⇒	VeBi
7.	Bi	AND	−Me ⇒	RoBi
8.	RoBi	AND	−MLMe ⇒	QRBi
9.	VRBi	AND	−MLMe ⇒	VRSm
⋮	⋮	AND	⋮ ⇒	⋮
22.	−Sm	AND	VRSm ⇒	RoSm
23.	−VeSm	AND	MLSm ⇒	VeSm
24.	−ExSm	AND	VeSm ⇒	Sm

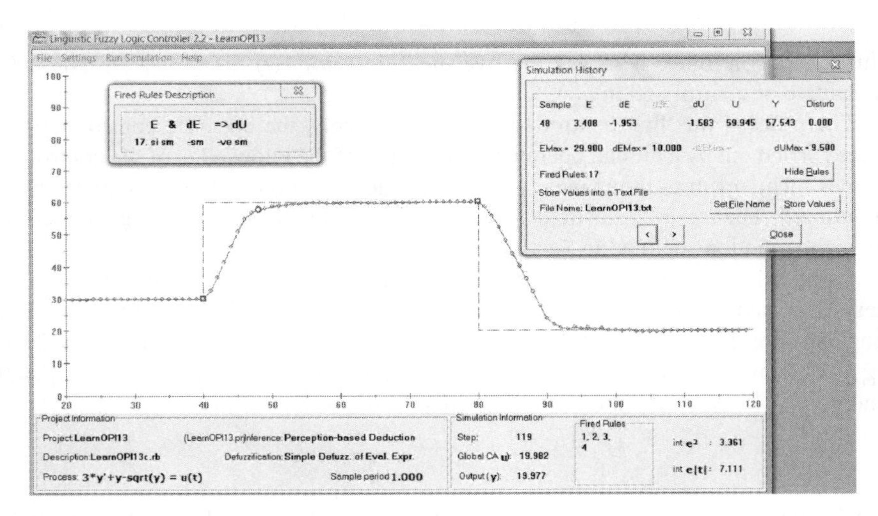

Figure 7.17 Simulation of fuzzy control of the process $3\dot{y}(t) + y(t) - \sqrt{y(t)} = u(t)$ using the linguistic fuzzy action unit with a linguistic description learned on the basis of monitoring of the manual control (see Figure 7.16).

The smelting furnace TLP9 is a highly nonlinear system in which the input is the burning power and the output is the metal temperature. The nonlinearities are caused by physical phenomena occurring inside the furnace as well as by the construction of the furnace. The nonlinearities mainly emerge in the transition of the metal phase from solid to liquid one, change of the sensors' time constants during time, change of the time-delay due to the modification of the distance between the burner flame and the temperature sensor depending on the burning power, the limitation of the range of the power, that is, it is impossible to have arbitrarily small power; temporary changes of the sign of the static amplification of the system, which is caused by the exothermic reaction of the refining salt with the smelted metal; and a few other ones.

The control algorithm had to fulfill the following requirements:

(a) It must be robust in all working states, notwithstanding whether the operation regime is automatic or manual.
(b) It is important to spare the brickwork of the furnace.
(c) The overshoot of the temperature must be minimal.
(d) The total control time and the fuel consumption should be minimized.

The input variables are derived from the measuring of the temperature of the metal on two places, which have been chosen to represent the behavior of the metal temperature in two different stages of its melting—notice that melting begins from top to bottom, so that some parts may be already fully melted while others not. Hence, the first temperature sensor is put into the place that seems to represent the temperature of the whole bulk of metal in the best way. The second sensor is in the place with much higher dynamics, which makes it possible to observe the transition of the system when controlling it.

Fuzzy IF-THEN rules have the following form:

$$\text{IF } e_1 \text{ is } \mathscr{A} \text{ AND } \dot{e}_1 \text{ is } \mathscr{B} \text{ AND } e_2 \text{ is } \mathscr{C} \text{ THEN } u \text{ is } \mathscr{D},$$

where $e_1 = w - T_1$ is the error of the first temperature, $\dot{e}_1$ is its derivative, $e_2 = w - T_2$ is the error of the second temperature, and u is the control action that is the amount of gas brought to the jet.

This form of rules corresponds to the fuzzy PD controller. Their total number is 155. Though fairly high, we decided not to reduce it. First of all, it was not too difficult to prepare all of them. Because of their linguistic nature, tuning the whole linguistic description was quite simple (recall that this means only replacing less suitable expressions inside the rules by more suitable ones according to their linguistic meaning). Furthermore, the system turned out to be able to react correctly also in the emergency situations, such as when only one sensor was working. Of course, the quality of control in such case is worse but still acceptable.

In Figure 7.18, we demonstrate the behavior of linguistic control on one of the five furnaces. Let us emphasize that we have chosen the working days quite randomly. It can be demonstrated that the course of the control is practically the same in all times. At the same time, however, the operation situation in the enterprise may vary, sometimes quite significantly. The areas in which the required value is zero correspond to time periods when the operator has turned off the control system completely (though improper, it is difficult to persuade the operators not to do it).

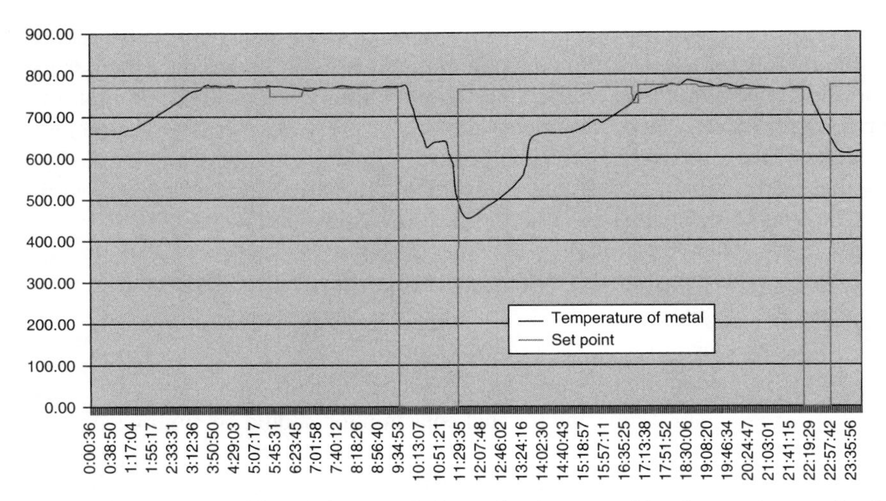

Figure 7.18 Example of linguistic control of one aluminum smelting furnace in a randomly chosen time slot.

7.5 DECISION-MAKING USING LINGUISTIC DESCRIPTIONS

7.5.1 Introduction

Linguistic fuzzy control described above is closely connected with decision-making. Indeed, we can understand control as repeated decision-making on the basis of a linguistic description of the current state of the controlled process.

In this section, we describe an application of linguistic descriptions and perception-based logical deduction in managerial decision-making. This method was published in [97]. Let us remark that managers decide on the basis of complex data consisting of numerical information and information available only in linguistic form. Our technique is thus closer to reality.

There are already thousands of publications devoted to fuzzy decision-making. These works have been initiated by the paper [7]. Among many books on this topic, let us mention, for example, [38, 66, 130].

One of typical problems discussed in these books arises from the fact that criteria, on the basis of which the decision is taken, are not equally important. The usual solution is to express their relative importance using weights. For example, a popular method for weights assignment is the *analytical hierarchy process* (AHP), which itself can be used as a specific fuzzy decision-making method [128]. It involves structuring multiple choice criteria into a hierarchy, assessing the relative importance of these criteria, comparing alternatives for each criterion, and determining an overall ranking of the alternatives.

In this section, we propose not to use numerical weights but to hide various importance of factors into linguistic characterization of the decision situation. As a consequence, the problem of assignment of weights to the criteria disappears. This is one of the great advantages of our approach, because methods for weights assignment, though sophisticated, are rather intricate and still much subjective (though the weights are seemingly objective). Of course, formulation of the problem using fuzzy/linguistic IF-THEN rules is also subjective. However, this subjectivity is here under control, because it is formulated using natural language understandable to everybody. Therefore, people (experts) may agree on the practically optimal formulation reflecting the general knowledge.

Another important problem is how criteria should be aggregated to obtain the final decision. In general fuzzy approach, this is achieved using aggregation operators (cf., e.g., [19, 42]). In our case, we apply the PbLD described in Chapter 5. The gain we obtain is twofold: first, imprecise information contained in natural language is effectively utilized. Second, some problems otherwise necessary to be solved, such as assignment of weights and aggregation of criteria, are replaced, in our opinion, by more natural techniques.

Our decision-making methodology leads to classical preference relation, that is, the alternatives are linearly ordered on the basis of evaluation that behaves as a special utility function.

The assumption that fuzzy/linguistic IF-THEN rules are linguistically described logical implications characterizing a relation among vaguely characterized phenomena makes it possible to distinguish sufficiently subtly and, at the same time, aptly, various degrees of fulfillment of the respective criteria, distinguish their importance, and, moreover, overcome possible discrepancies. If the decision situation is sufficiently well characterized by linguistic descriptions, then the proper decision can be done using the PbLD.

7.5.2 Hierarchy of Linguistic Descriptions in Decision-Making

Multicriterial decision-making problem can be described as follows. Let n criteria $C_1, \ldots, C_n$ be given. On the basis of them, we should decide among alternatives $v_1, \ldots, v_m$. Fulfilling of each criterion is measured by values within a specific scale. This scale determines the linguistic context. There are two possibilities:

(i) The given criterion is measurable by special units. This is the case where a criterion has an objective character such as price and geometrical or physical characteristics. The scale (linguistic context) is then determined by specific application (e.g., economical power of the given company or concrete object).

(ii) The criterion is subjective or abstract. People can still estimate how well it is fulfilled by a given alternative. The scale is then usually set as the abstract interval [0, 1] of spaceless units. Then, according to practical experiences, we should set the corresponding linguistic context to $w = \langle 0, 0.4, 1 \rangle$.

Using fuzzy/linguistic IF-THEN rules, it is possible to characterize clearly how the criteria should be fulfilled so that the global characterization of the given alternative is as good as possible. This practically means that the decision-making situation is described using a linguistic description consisting of rules of the form:

$$\mathscr{R}_1 := \text{IF } C_1 \text{ is } \mathscr{A}_{11} \text{ AND } \cdots \text{ AND } C_n \text{ is } \mathscr{A}_{1n} \text{ THEN } H \text{ is } \mathscr{B}_1,$$

$$\cdots\cdots\cdots\cdots\cdots\cdots\cdots\cdots\cdots\cdots\cdots\cdots$$

$$\mathscr{R}_p := \text{IF } C_1 \text{ is } \mathscr{A}_{p1} \text{ AND } \cdots \text{ AND } C_n \text{ is } \mathscr{A}_{pn} \text{ THEN } H \text{ is } \mathscr{B}_p,$$

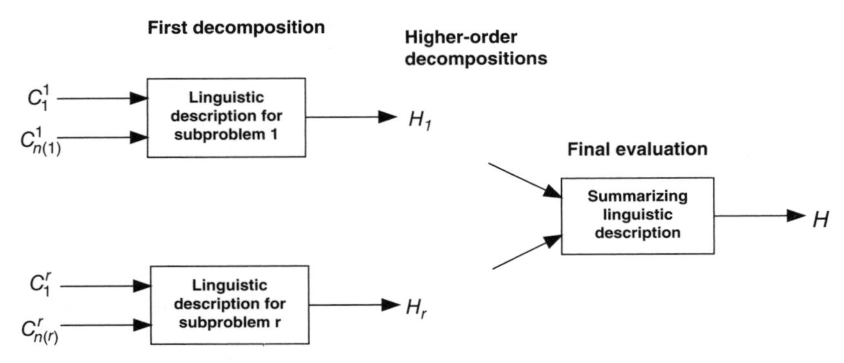

Figure 7.19 General scheme of the hierarchy of decision-making using linguistic descriptions.

where $C_1, \ldots, C_n$ are the criteria, H is the global evaluation, and $\mathscr{A}_{ji}, \mathscr{B}_j, j = 1, \ldots, p$, $i = 1, \ldots, n$, are evaluative linguistic expressions.

However, the number n can be large. Hence, it may not be possible to form such a linguistic description (it could even be hardly understandable). Therefore, we will formally divide the criteria $C_1, \ldots, C_n$ into r groups $H_1, \ldots, H_r$. Let us denote

$$\mathscr{H}_k = \{C_1^k, \ldots, C_{n(k)}^k\}, \quad k = 1, \ldots, r.$$

Then the linguistic description introduced above can be transformed into a hierarchical system of linguistic descriptions:

$$\begin{aligned} \mathscr{R}_1^k &:= \text{IF } C_1^k \text{ is } \mathscr{A}_{11}^k \text{ AND } \cdots \text{ AND } C_{n(k)}^k \text{ is } \mathscr{A}_{1n(k)}^k \text{ THEN } H_k \text{ is } \mathscr{B}_1^k, \\ &\qquad \cdots\cdots\cdots\cdots\cdots\cdots\cdots\cdots\cdots\cdots\cdots\cdots \qquad (7.21) \\ \mathscr{R}_{p(k)}^k &:= \text{IF } C_1^k \text{ is } \mathscr{A}_{p(k)1}^k \text{ AND } \cdots \text{ AND } C_{n(k)}^k \text{ is } \mathscr{A}_{p(k)n(k)}^k \text{ THEN } H_k \text{ is } \mathscr{B}_{p(k)}^k, \end{aligned}$$

$k = 1, \ldots, r$.

The last linguistic description $\{\mathscr{R}_1, \ldots, \mathscr{R}_s\}$ provides a final evaluation (based on evaluations H_k, $k = 1, \ldots, r$ above) using which the proper decision is made:

$$\begin{aligned} \mathscr{R}_1 &:= \text{IF } H_1 \text{ is } \mathscr{A}_{11} \text{ AND } \cdots \text{ AND } H_r \text{ is } \mathscr{A}_{1r} \text{ THEN } H \text{ is } \mathscr{B}_1, \\ &\qquad \cdots\cdots\cdots\cdots\cdots\cdots\cdots\cdots\cdots\cdots \\ \mathscr{R}_s &:= \text{IF } H_1 \text{ is } \mathscr{A}_{s1} \text{ AND } \cdots \text{ AND } H_r \text{ is } \mathscr{A}_{sr} \text{ THEN } H \text{ is } \mathscr{B}_s. \end{aligned}$$

7.5.3 Demonstration of the Decision-Making Methodology Using Linguistic Descriptions

In this section, we demonstrate the above described decision-making technique on a model of a complex decision-making problem. All computations were done using the software system LFL Controller© developed in the University of Ostrava (cf. [35]).

Let us consider the following decision problem: we want to buy a house. After checking the Internet, we found 20 houses offered by a real estate agency in the Czech Republic. Each house has a photo and is described by several characteristics:

1. Economical characteristics, for example, price and reconstruction cost (in Czech crowns).
2. Technical characteristics, for example, number of rooms, bathrooms, and quality of heating.
3. Sizes, for example, size of house and garden and size of additional land (in m^2).
4. Other characteristics, for example, quality of garage and quality of cellar.
5. Infrastructure, for example, distance from city center, accessibility, and neatness of access.
6. Aesthetical characteristics, for example, global appearance, modernness, and elegance.
7. Quality of environment, for example, air pollution and surrounding constructions.

The modeled decision situation is a multicriteria decision-making problem in which alternatives are characterized on the basis of the above-listed characteristics. Note that among them, nonmeasurable characteristics, such as appearance, modernness, and others, are also included. These are evaluated in the standard spaceless context $w = \langle 0, 0.4, 1 \rangle$. Values of these characteristics can be specified either directly or using evaluative expressions, for example, *nice*, *ugly*, and *very modern*.

To give a clear idea about the data, we summarized the data of four selected houses in Table 7.2 (evaluative expressions are written using abbreviations introduced in Section 5.4.4).

As can be seen, some criteria are purely numerical, for example, price, number of rooms, and distance to the center, and the other ones are qualitative. Also criteria such as "garage" or "heating" are evaluated linguistically, because the evaluation also includes quality and state of the garage, or effectiveness of the heating. For simplicity, we used only the basic evaluative expressions. For example, "Heating: *big*" means high effectiveness of the heating.

The hierarchy of linguistic descriptions is depicted in Figure 7.20. One can see that the criteria are gathered into two main groups: *Main characteristics* that are decisive for the choice and *Other characteristics*, whose influence is lower. This distinction can be well expressed in the global evaluation where higher values of the former have more impact on the final evaluation than higher values of the latter. See, for example, Rule 7 in the linguistic description *Global evaluation* below. Though *Other characteristics* are evaluated as *big*, the global evaluation is evaluated as *more or less small* because the *Main characteristics* are evaluated as *small*. Different influence of both groups of characteristics in the other rules is expressed similarly. Consequently, no weight assignment procedure is needed.

TABLE 7.2 Input data of four selected houses (from 20 found on the Internet).

	Economical characteristics		Technical characteristics				Sizes (m^2)	
House	Price 1000CzK	Reconstruction cost	No. Rooms	No. Baths	No. WC	Heating	House size	Land size
1.	2500	Bi	4	1	1	RaBi	72	0
6.	2690	VeSm	5	1	2	Me	114	961
12.	3300	MLBi	7	3	3	Bi	180	1600
19.	3999	ExSm	8	3	2	RaBi	185	505

	Other characteristics		Infrastructure			Aesthetical characteristics			Environment
House	Garage	Cellar	Distance center (km)	Accessibility	Neat access	Appearance	Modernness	Elegance	
1.	Bi	Me	44	Bi	VRSm	Bi	Me	QRSm	SiBi
6.	Bi	bi	0	VeBi	Me	VeBi	VeBi	VeBi	VeBi
12.	Bi	Bi	29	VeSm	Bi	SiSm	Sm	MLSm	Bi
19.	Bi	Bi	4	Bi	Me	Bi	Me	Me	VeBi

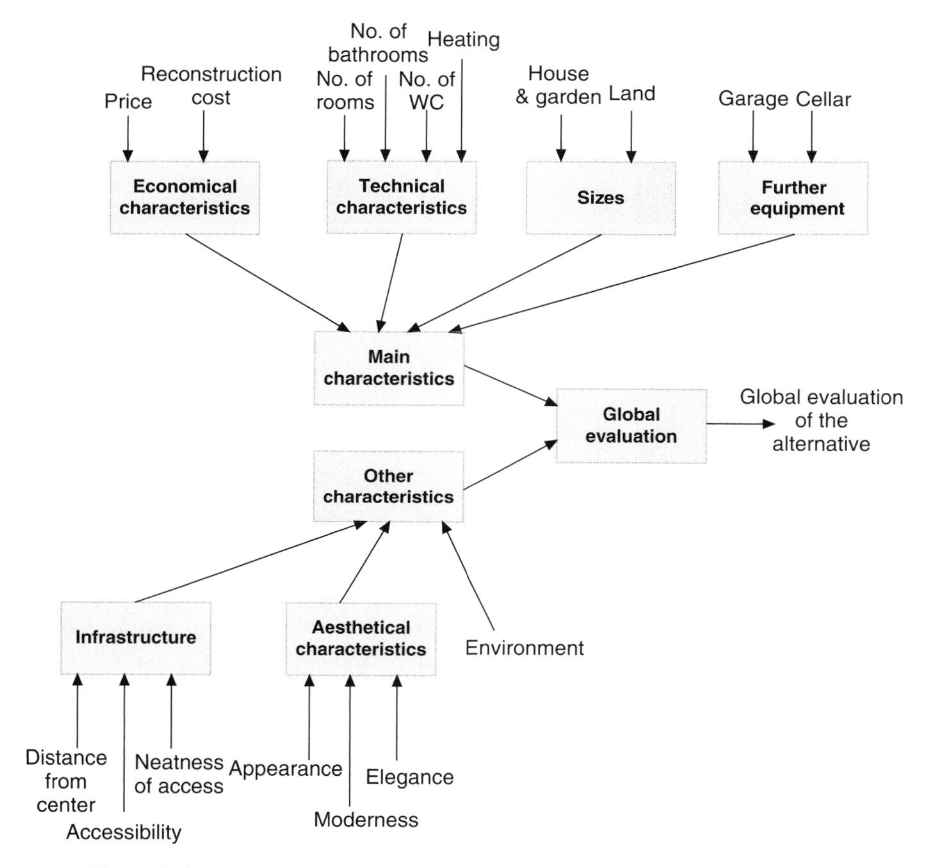

Figure 7.20 Hierarchy of linguistic descriptions for choosing the best house.

Below is an example of the linguistic description *Economical characteristics* (it is an example of linguistic description (7.21)):

Rule No.	Price	Reconstruction cost	$\Rightarrow$	Economical characteristics
1.	VeSm	VeSm	$\Rightarrow$	SiBi
2.	Sm	Sm	$\Rightarrow$	Bi
3.	Sm	Bi	$\Rightarrow$	Me
4.	Me	Me	$\Rightarrow$	MLSm
5.	MLBi	MLBi	$\Rightarrow$	VeSm
6.	Bi	VeSm	$\Rightarrow$	RoBi
7.	VeBi	NoSm	$\Rightarrow$	VeSm
8.	Sm	Me	$\Rightarrow$	MLBi
9.	Me	Sm	$\Rightarrow$	VeBi

(the fuzzy/linguistic IF-THEN rules are formed using evaluative expressions coded by the same abbreviations as in Table 7.2). In a similar way, we also defined the remaining linguistic descriptions with respect to Figure 7.20. The linguistic description for *Global evaluation* is as follows:

Rule No.	Main characteristics	Other characteristics	⇒	Global evaluation
1.	Bi	Sm	⇒	Bi
2.	VeBi	VeBi	⇒	SiBi
3.	SiBi	VeBi	⇒	ExBi
4.	Bi	Sm or Me	⇒	RaBi
5.	Sm	Sm	⇒	VeSm
6.	Bi	Bi	⇒	VeBi
7.	Sm	Bi	⇒	MLSm
8.	Me	Me or Bi	⇒	RoBi
9.	Sm	Me	⇒	Sm
10.	Me	QRSm	⇒	QRSm

Let us remark that rules leading to refinements of the decision can still be added.

The results of the decision-making procedure are summarized in the following table:

House	Main	Other	Global evaluation	
No.	characteristics	characteristics	Numerical	Linguistic
1.	**0.12**	**0.43**	**0.12**	**Sm**
2.	0.41	0.26	0.26	QRSm
3.	0.69	0.22	0.66	RoBi
4.	0.12	0.79	0.19	MLSm
5.	0.83	0.35	0.65	RoBi
6.	**0.71**	**0.76**	**0.81**	**Bi**
7.	0.65	0.88	0.65	RoBi
8.	0.47	0.92	0.54	VRBi
9.	0.46	0.50	0.44	MLMe
10.	0.41	0.88	0.69	RoBi
11.	0.80	0.41	0.76	RaBi
12.	*0.34*	*0.54*	*0.46*	*MLMe*
13.	0.64	0.92	0.64	RoBi
14.	0.52	0.58	0.52	VRBi
15.	0.49	0.73	0.49	VRBi
16.	0.53	0.84	0.53	VRBi
17.	0.51	0.81	0.51	VRBi
18.	0.05	0.88	0.15	RaSm
19.	**0.73**	**0.74**	**0.82**	**Bi**
20.	0.29	0.83	0.29	VRSm

The worst house is No. 1, the best is No. 19. However, the house No. 6 is practically the same. The house No. 12 is the typically medium alternative. The reason why No. 1 is the worst is especially because of its big reconstruction cost, small number of rooms, bathrooms and WC, and also infrastructure.

One can see that different influences of various criteria can be rendered using the evaluative expressions in a way that is well understandable to people. We can make tiny but, at the same time, sufficiently robust variations that lead to differences important for the individual decision-maker.

8

F-TRANSFORM IN IMAGE PROCESSING

In this chapter, we demonstrate how the F-transform introduced in Chapter 4 can be applied to various image processing problems. We will describe the following: image compression, fusion, edge detection, and image reconstruction. Solutions of these problems already exist, of course, but the used methods differ significantly. In this chapter, however, we present successful solution of *all of them using still the same principle*, namely, the fuzzy transform (F-transform). Let us remark that this theory can be used in solution of even more image processing tasks, such as image registration, segmentation, and other ones.

8.1 IMAGE AND ITS BASIC PROCESSING USING F-TRANSFORM

An image is a matrix $N \times M$ of pixels for some natural numbers $M, N > 0$. Because each pixel can be characterized by its intensity, we can understand the image as a function of two variables.

Definition 8.1 *Let us denote*

$$P = \{(i,j) \mid i = 1, \ldots, N,\ j = 1, \ldots, M\}. \tag{8.1}$$

An image of size NM is a function

$$u : P \longrightarrow \mathbb{R}.$$

Insight into Fuzzy Modeling, First Edition. Vilém Novák, Irina Perfilieva, and Antonín Dvořák.

Companion Website: www.wiley.com/go/novak/fuzzy/modeling

The F-transform is applied to the image u in the same way as in the case of the function of two variables described in Section 4.4.

Let u be an image of size NM and $n < N, m < M$ be natural numbers. Furthermore, let $c_0, \dots, c_n \in \{1, \dots, N\}$, where $c_0 = 1, c_n = N$, be h_x-equidistant nodes and $e_0, \dots, e_m \in \{1, \dots, M\}$, where $e_0 = 1, e_n = M$, be h_y-equidistant nodes. Let $A_0, \dots, A_n$ and $B_0, \dots, B_m$ be triangular fuzzy sets over the nodes $c_0, \dots, c_n$ and $e_0, \dots, e_m$, respectively, that establish a uniform triangular fuzzy partition of P. The direct F-transform of u is the matrix

$$\mathbf{U}[u] = \begin{bmatrix} U_{00}[u] & \cdots & U_{0m}[u] \\ \vdots & \vdots & \vdots \\ U_{n0}[u] & \cdots & U_{nm}[u] \end{bmatrix}, \tag{8.2}$$

where

$$U_{kl}[u] = \frac{\sum_{j=1}^{M} \sum_{i=1}^{N} u(i,j)\, A_k(i)\, B_l(j)}{\sum_{j=1}^{M} \sum_{i=1}^{N} A_k(i)\, B_l(j)} \tag{8.3}$$

for all $k = 0, \dots, n$ and $l = 0, \dots, m$. The two-dimensional fuzzy partition and components of the F-transform are schematically depicted in Figure 4.6.

A full-size image $\hat{u}$ is reconstructed from U using the inversion formula:

$$\hat{u}(i,j) = \sum_{k=0}^{n} \sum_{l=0}^{m} U_{kl}[u]\, A_k(i) B_l(j) \tag{8.4}$$

for all $i = 1, \dots, N$ and $j = 1, \dots, M$.

Remark 8.1 *Because the nodes c_k, e_l are taken from the sets $\{1, \dots, N\}$, $\{1, \dots, M\}$, respectively, the condition from Definition 4.6 of sufficient density is automatically fulfilled. Moreover, we assume for simplicity that $N/h_x, M/h_y \in \mathbb{N}$.*

8.2 F-TRANSFORM-BASED IMAGE COMPRESSION AND RECONSTRUCTION

8.2.1 Basic Principles of Image Compression

The problem of image compression is to reduce the image size to save space or transmission time. A desirable size nm (where $n < N$ and $m < M$) of a compressed image can be obtained from the *compression ratio*, $\rho = (nm)/(NM)$. If a compression method is lossy (e.g., JPEG, FEQ, and the F-transform), then the respective reconstruction $\hat{u}$ to a full-size image can be compared with the original image u using two quality indices RMSE (root-mean-square error):

$$\text{RMSE} = \sqrt{\frac{\sum_{i=1}^{N} \sum_{j=1}^{M} (u(i,j) - \hat{u}(i,j))^2}{NM}}$$

and PSNR (peak signal-to-noise ratio) in dB:

$$\text{PSNR} = 20 \log \frac{255}{\text{RMSE}}.$$

In [49], a method of lossy image compression and reconstruction using fuzzy relations was proposed. The dominant idea was a choice of a suitable granulation (represented by a fuzzy relation) of an image domain. We will refer to this method as FEQ (Fuzzy Relation Equations).

The F-transform image compression (referred to as FTR) is also based on the idea of granulation but relates it to the concept of fuzzy partition (see [30, 111]). Two approaches were proposed: a uniform fuzzy partition of the entire domain and a two-step partition in which the entire domain is first partitioned into blocks and then, each block is uniformly partitioned into fuzzy sets.

Both approaches were compared with JPEG and other compression techniques. The conclusion is that the F-transform-based techniques are comparable to JPEG and better than FEQ. Two further improvements of the F-transform-based compression have been proposed in [116, 54]. In a series of experiments, advantage over JPEG was achieved many times. In this section, we explain how a proper choice of a fuzzy partition improves the quality of the reconstructed image. The authors of this method and algorithms are Irina Perfilieva and Petr Hurtík.

8.2.2 Simple F-Transform Compression

A compressed image is represented by the $n \times m$ matrix $\mathbf{U}$ (8.3) of the F-transform components. A full-size image $\hat{u}$ is then reconstructed using the inverse F-transform (8.4) of U. This method does not take possible special properties of the original image into account. Therefore, its quality is not sufficient.

This is illustrated in Figure 8.1, where the original image and its reconstruction using the simple F-transform compression described above and advanced compression described in the next subsection are depicted. The compression ratio is $\rho = 0.25$ and PSNR is equal to 32 dB for the simple F-transform compression and to 37 dB for the advanced F-transform compression introduced below (compare with PSNR = 38.8 dB for JPEG with a similar compression ratio).

A possible improvement of the above simple method is to decompose the entire image into blocks of the same size where the size (chosen experimentally) is such that a certain quality of approximation by the inverse F-transform can be guaranteed. Each block is then uniformly partitioned into cosine-shaped fuzzy sets and compressed by the simple F-transform method according to a compression ratio. In comparison with the simple F-transform compression, this method considers peculiarities of the original images when making the block decomposition. The results are better, but still worse than, for example, results of JPEG.

8.2.3 Advanced Image Compression

The compression method described in this section was published in [54]. It is almost nonlossy and based on a nonuniform generalized fuzzy partition adapted to each particular image.

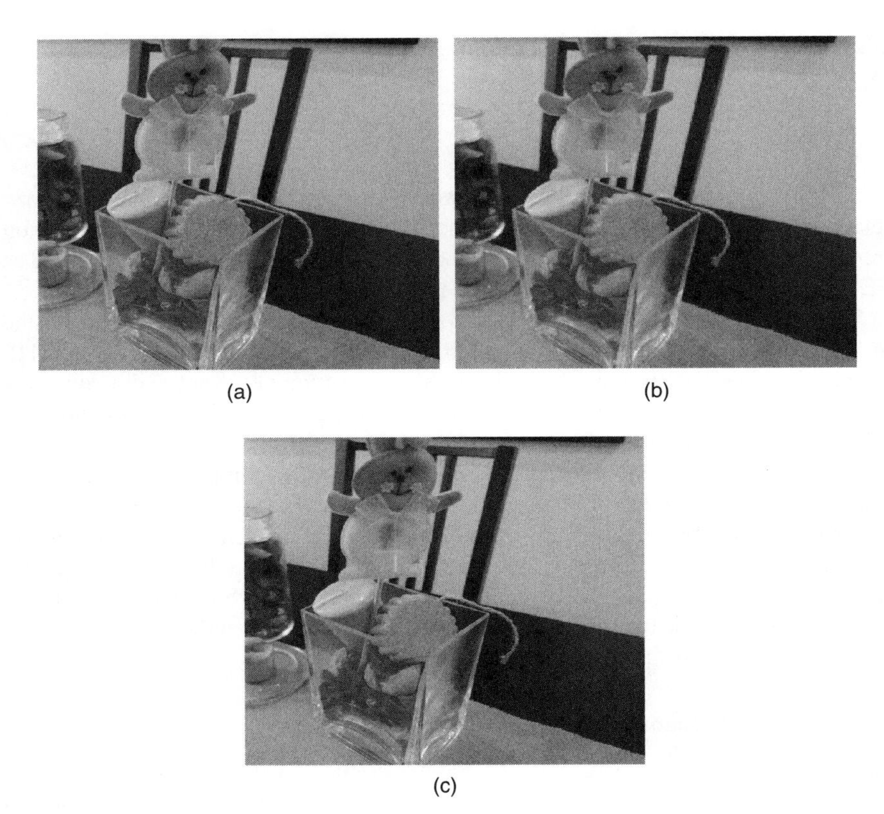

Figure 8.1 (a) The original reference image. (b) Its reconstruction after simple F-transform compression with PSNR = 32 dB. (c) Its reconstruction using the advanced F-transform image compression algorithm, PSNR = 37 dB.

If we analyze the properties of the F-transform (see page 85), then it immediately follows from item (a) that the more the function behaves like a constant, the better is the approximation quality of the inverse F-transform. This suggests the idea to use the generalized fuzzy partition of the domain P, which can guarantee that the difference between extremal values of the image over each $A_k \times B_l$ is not greater than some given limit $\epsilon > 0$ or, if this condition cannot be fulfilled, the area of $A_k \times B_l$ is not greater than a given $\delta > 0$.

Hence, the advanced image compression algorithm uses the following two improvements to increase the quality of the reconstructed image:

- preservation of sharp edges,
- restoration of the histogram of the original image.

In Figure 8.2, we show a special case in which preservation of sharp edges can be very helpful and the circle image can be stored fully nonlossy.

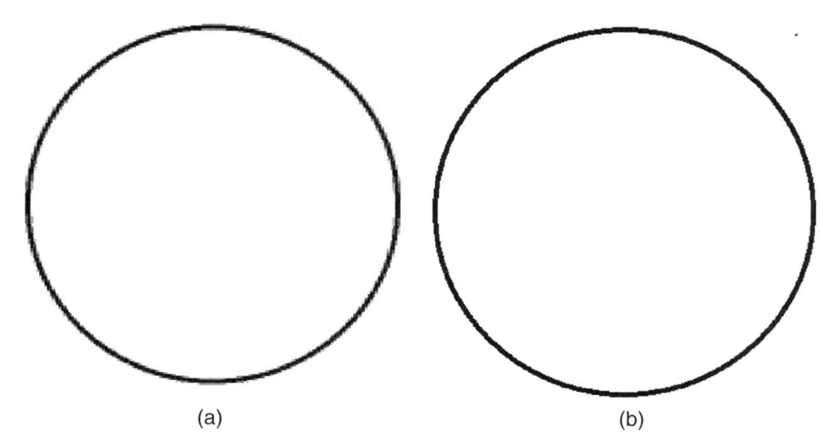

Figure 8.2 Two reconstructions of the circle image after applying the advanced F-transform compression. (a) Without preservation of sharp edges, the compression ratio is 0.013 and PSNR is 27 dB. (b) With preservation of sharp edges, the compression ratio is 0.031 and PSNR cannot be measured.

8.3 F^1-TRANSFORM EDGE DETECTOR

Detection of edges is very important task in image processing. In particular, it is the first step in feature extraction and image segmentation. The most popular method is Canny edge detector [20], which is widely used in computer vision.

We must realize that "edge" is not a precise notion. Hence, it is impossible to say precisely whether the edges are detected well or not. The Canny detector was developed to ensure the following three basic criteria:

- *Good detection:* The algorithm should mark as many real edges in the image as possible.
- *Good localization:* The detected edges should be as close as possible to the edge in the real image.
- *Minimal response:* Each edge in the image should be marked only once, and where possible, the image noise should not create false edges.

Though these criteria are not precisely defined, they enabled to formulate the edge detection task as a search of a function maximizing a given functional using calculus of variations. In this sense, the Canny detector can be considered as an "optimal" edge detector.

The Canny algorithm is a multistep procedure for detecting edges as the local maxima of the gradient magnitude. The first step, performed using a Gaussian filter, is image smoothing and noise filtering. The second step is computation of a gradient of the image function to find the local maxima of the gradient magnitude and the gradient direction at each point. This step is performed using a convolution of the original image with directional masks (edge detection operators, such as those of

Roberts, Prewitt, and Sobel, are some examples of these filters). The next step called nonmaximum suppression [127] selects those points whose gradient magnitudes are maximal in the corresponding gradient direction. The final step is edge tracing and hysteresis thresholding, which leads to the preservation of the continuity of edges.

In [118], an application of the F^1-transform to edge detection was proposed. Its idea consists in replacing the first two steps of the Canny algorithm by computation of approximate gradient values using the F^1-transform. This is justified by Theorem 4.14 according to which the F^1-transform enables to compute approximate values of the first partial derivatives of a given function. Moreover, similarly as the ordinary F-transform, it also filters out noise. The authors of this method and algorithms are Irina Perfilieva, Petr Hurtík, and Petra Hod'áková.

Let u be an image due to Definition 8.1 and the fuzzy sets $A_0, \dots, A_n$ and $B_0, \dots, B_m$ establish a uniform triangular fuzzy partition of $\{1, \dots, N\}$ and $\{1, \dots, M\}$, respectively. By (4.23), the coefficients β_k^1 of the linear polynomials of the F^1-transform components are approximate values of the first partial derivatives of the image function, where by (4.19) and (4.20), the following holds:

$$\beta_{kl}^1[u] = \frac{12}{h_x^3 h_y} \sum_{i=1}^{N} \sum_{j=1}^{M} u(i,j)(i - c_k)A_k(i)B_l(j),$$

$$\beta_{lk}^1[u] = \frac{12}{h_x h_y^3} \sum_{i=1}^{N} \sum_{j=1}^{M} u(i,j)(j - e_l)A_k(i)B_l(j)$$

for all $k = 1, \dots, n-1$ and $l = 1, \dots, m-1$, where c_k and e_l are nodes. Then, we can write approximations of the first partial derivatives as the respective inverse F^1-transforms:

$$G_x(i,j) \approx \sum_{k=1}^{n-1} \sum_{l=1}^{m-1} \beta_{kl}^1[u]A_k(i)B_l(j),$$

and

$$G_y(i,j) \approx \sum_{k=1}^{n-1} \sum_{l=1}^{m-1} \beta_{lk}^1[u]A_k(i)B_l(j),$$

$i = 1, \dots, N, j = 1, \dots, M$. The rest of the Canny algorithm—namely, finding the local maxima of the gradient magnitude and its direction, nonmaximum suppression, tracing edges through the image and hysteresis thresholding—remains unchanged.

In Figure 8.3, results of the F^1-transform edge detector on handmade images are presented. We may observe that many thin edges/lines are detected, as well as their connectedness and smoothness. Moreover, the following properties are retained:

- smoothness of circular lines,
- concentricity of circles,
- smoothness of sharp connections.

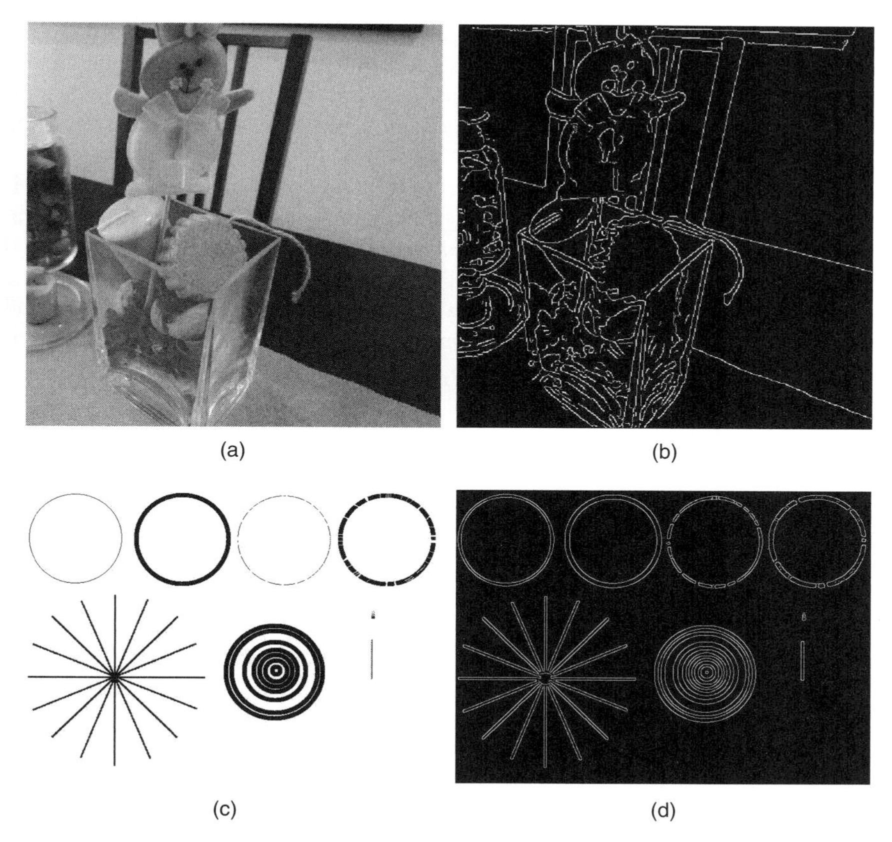

(a) (b) (c) (d)

Figure 8.3 (a,c) Original images. (b,d) Edges found using the F^1-transform.

We may see that the results of edge detection using the F^1-transform are fully comparable with results of the Canny algorithm. The former, however, is algorithmically significantly simpler.

Let us remark that the problem of edge detection is highly subjective because nobody is able to give a definition of edge. Therefore, we can evaluate only subjectively whether the given algorithm detects edges well or not. From this point of view, the F-transform as a general image processing technique provides very convincing results.

8.4 F-TRANSFORM-BASED IMAGE FUSION

8.4.1 Basic Idea of Image Fusion

Image fusion is a procedure that integrates complementary distorted multisensor, multitemporal, and/or multiview scenes into one new image that contains the "best" part of each scene. Thus, the primary problem in image fusion is to find the least distorted scene for every pixel.

There exist various fusion methodologies that can be classified according to the primary technique: aggregation operators [10], fuzzy methods [125], optimization methods (e.g., neural networks and genetic algorithms [70]), and multiscale decomposition methods based on various transforms (e.g., discrete wavelet transforms; see [124]).

The F-transform approach to image fusion was proposed in [115, 114]. The primary idea is to combine (at least) two fusion operators, both based on the F-transform. The first fusion operator is applied to the F-transform components of scenes and based on a rough partition of the scene domain. The second fusion operator is applied to the residuals of scenes with respect to inverse F-transforms with fused components and is based on a finer partition of the same domain. This approach uses a fusion operator that is able to choose an undistorted scene among the available blurred scenes.

The F-transform-based image fusion is based on a certain decomposition of the given image. We will assume the following representation of the image u on P:

$$u(i,j) = \hat{u}_{nm}(i,j) + e(i,j), \tag{8.5}$$

$$e(i,j) = u(i,j) - \hat{u}_{nm}(i,j), \tag{8.6}$$

where $\hat{u}_{nm}$ is the inverse F-transform of u with respect to the fuzzy partition consisting of nm basic functions and e is the first difference with respect to it.

Let us now consider the finest partition of P, which means that it has NM components, so the basic functions A_k, B_l, in fact, become singletons $A_k = \left\{{}^1/k\right\}, B_l = \left\{{}^1/l\right\}$. Let us denote the corresponding inverse F-transform of e by $\hat{e}_{NM}$. If we replace e in (8.5) by the latter, the above representation can be rewritten as follows:

$$u(i,j) = \hat{u}_{nm}(i,j) + \hat{e}_{NM}(i,j). \tag{8.7}$$

We call (8.7) a *one-level decomposition* of u on P. If u is smooth, then the function $\hat{e}_{NM}$ is small (this claim follows from the property (d) on page 85), and we can stop at this level. In the opposite case, we continue with the decomposition of the first difference e in (8.5). We decompose e into its inverse F-transform $\hat{e}_{n'm'}$ with respect to a finer fuzzy partition of P with n' $(n < n' < N)$ and m' $(m < m' < M)$ basic functions and the second difference e'. Thus, we obtain the *second-level decomposition* of u:

$$u(i,j) = \hat{u}_{nm}(i,j) + \hat{e}_{n'm'}(i,j) + e'(i,j),$$

$$e'(i,j) = e(i,j) - \hat{e}_{n'm'}(i,j).$$

In the same manner, we can obtain a *higher-level decomposition* of u on P:

$$u(i,j) = \hat{u}_{n_1m_1}(i,j) + \hat{e}^{(1)}_{n_2m_2}(i,j) + \cdots + \hat{e}^{(k-2)}_{n_{k-1}m_{k-1}}(i,j) + e^{(k-1)}(i,j), \tag{8.8}$$

where

$$0 < n_1 \le n_2 \le \cdots \le n_{k-1} \le N,$$

$$0 < m_1 \le m_2 \le \cdots \le m_{k-1} \le M,$$

$$e^{(1)}(i,j) = u(i,j) - \hat{u}_{n_1 m_1}(i,j),$$
$$e^{(l)}(i,j) = e^{(l-1)}(i,j) - \hat{e}^{(l-1)}_{n_l m_l}(i,j), \tag{8.9}$$

$l = 2, \ldots, k-1$.

In this section, we describe three F-transform-based image fusion algorithms whose authors are Irina Perfilieva, Martina Daňková, and Marek Vajgl:

1. *Simple* F-transform-based fusion algorithm (SA).
2. *Complete* F-transform-based fusion algorithm (CA).
3. *Enhanced simple* fusion algorithm.

The algorithms were published in [115, 142]. The principal role in them is played by a *fusion operator* $\kappa : \mathbb{R}^K \longrightarrow \mathbb{R}$ (for a given K), which is defined as follows:

$$\kappa(x_1, \ldots, x_K) = x_p, \tag{8.10}$$

where $|x_p| = \max(|x_1|, \ldots, |x_K|)$. The choice of a fusion operator is influenced by a type of image degradation encountered. Moreover, this operator is convenient especially for multifocus image fusion.

8.4.2 Simple F-Transform-Based Fusion Algorithm

Assume that we are given $K \geq 2$ input images $u_1, \ldots, u_K$ in the sense of Definition 8.1 with various types of degradation. Our aim is to recognize undistorted parts in the given images and to fuse them into one image. Furthermore, we assume that on each image, a fuzzy partition is defined as described in Section 8.1.

The simple F-transform-based image fusion algorithm (SA) can be summarized as follows:

1. Decompose input images $u_1, \ldots, u_K$ into their inverse F-transforms and error functions using the one-level decomposition (8.5).
2. Apply the fusion operator (8.10) to the respective F-transform components of u_q, $q = 1, \ldots, K$, and obtain the fused F-transform components of a new image.
3. Apply the fusion operator to the respective F-transform components of the error functions e_q, $q = 1, \ldots, K$, and obtain the fused F-transform components of a new error function.
4. Reconstruct the fused image from the inverse F-transforms using the fused components of the new image and the fused components of the new error function.

Description of the SA algorithm:

Setting:

Step 0. Set the number n, $0 < n \leq N$ and m, $0 < m \leq M$ of basic functions in the fuzzy partitions of $\{1, \ldots, N\}$ and $\{1, \ldots, M\}$, respectively.

Initialization:

Step 1. Form fuzzy partitions $A_1^{(1)}, \ldots, A_n^{(1)}$ and $B_1^{(1)}, \ldots, B_m^{(1)}$ of $\{1, \ldots, N\}$ and $\{1, \ldots, M\}$, respectively.

Denote by $A_1^{(2)}, \ldots, A_N^{(2)}$ and $B_1^{(2)}, \ldots, B_M^{(2)}$ the finest partitions of $\{1, \ldots, N\}$ and $\{1, \ldots, M\}$, respectively.

Transformation:

Step 2. For all $q = 1, \ldots, K$, compute the direct and the inverse F-transforms of each input image u_q and obtain the F-transform components $F[u_q]_{11}, \ldots, F[u_q]_{nm}$ and the inverse F-transform $\hat{u}_{q,nm}$.

Step 3. For all $q = 1, \ldots, K$, compute the error functions: $e_q = u_q - \hat{u}_{q,nm}$. Identify values $e_q(i,j)$, $(i,j) \in P$, with the F-transform components $F[e_q]_{ij}$ with respect to the finest partitions of $\{1, \ldots, N\}$ and $\{1, \ldots, M\}$.

Fusion:

Step 4(a). Apply the fusion operator κ to the respective components of the direct F-transforms of the input images u_q, $q = 1, \ldots, K$:

$$\kappa(F[u_1]_{11}, \ldots, F[u_K]_{11}) = \kappa_{11}^{(1)},$$

$$\ldots\ldots\ldots\ldots\ldots\ldots\ldots\ldots\ldots\ldots\ldots\ldots$$

$$\kappa(F[u_1]_{nm}, \ldots, F[u_K]_{nm}) = \kappa_{nm}^{(1)},$$

and obtain the fused F-transform components of a new image

$$(\kappa_{11}^{(1)}, \ldots, \kappa_{nm}^{(1)}). \tag{8.11}$$

Step 4(b). Apply the fusion operator κ to the respective components of the direct F-transforms of the error functions e_q, $q = 1, \ldots, K$, with respect to the finest partitions of $\{1, \ldots, N\}$ and $\{1, \ldots, M\}$:

$$\kappa(F[e_1]_{11}, \ldots, F[e_K]_{11}) = \kappa_{11}^{(2)},$$

$$\ldots\ldots\ldots\ldots\ldots\ldots\ldots\ldots\ldots\ldots\ldots\ldots$$

$$\kappa(F[e_1]_{NM}, \ldots, F[e_K]_{NM}) = \kappa_{NM}^{(2)},$$

and obtain the fused F-transform components of a new error function:

$$(\kappa_{11}^{(2)}, \ldots, \kappa_{NM}^{(2)}). \tag{8.12}$$

Reconstruction:

Step 5. The fused image u is equal to the sum of the two inverse F-transforms with fused components (8.11) and fused components (8.12), that is,

$$u(i,j) = \sum_{k=1}^{n} \sum_{l=1}^{m} \kappa_{kl}^{(1)} A_k^{(1)}(i)\, B_l^{(1)}(j) + \sum_{k=1}^{N} \sum_{l=1}^{M} \kappa_{kl}^{(2)} A_k^{(2)}(i)\, B_l^{(2)}(j)$$

for all $(i,j) \in P$.

8.4.3 Complete F-Transform-Based Fusion Algorithm

The assumptions and summarized description of the *complete F-transform-based* image fusion algorithm (CA) are the same as for the SA algorithm.

Description of the CA algorithm:

Setting:

Step 0. Set:

(a) $k_{\max}$—maximal number of iterations,

(b) *step*—the coefficient for an increment of the number of basic functions in each fuzzy partition,

(c) n_{start}—starting number of basic functions in the fuzzy partition of $\{1, \ldots, N\}$,

(d) m_{start}—starting number of basic functions in the fuzzy partition of $\{1, \ldots, M\}$,

where $step \leq \min\{N - n_{start}, M - m_{start}\}$ and $0 < n_{start} \leq N$, $0 < m_{start} \leq M$. Put $e_q^{(0)} = u_q$ for all $q = 1, \ldots, K$.

For $k = 0$ **to** $k_{\max}$

Initialization:

Step 1.1. Compute $n := n_{start} \cdot k^{step}, m := m_{start} \cdot k^{step}$.

Step 1.2. Form fuzzy partitions $A_1^{(0)}, \ldots, A_n^{(0)}$ and $B_1^{(0)}, \ldots, B_m^{(0)}$ of $\{1, \ldots, N\}$ and $\{1, \ldots, M\}$, respectively.

Transformation:

Step 2. For all $q = 1, \ldots, K$, compute the direct and the inverse F-transforms of each function $e_q^{(k)}$ and obtain the F-transform components $F[e_q^{(k)}]_{11}, \ldots, F[e_q^{(k)}]_{nm}$ and the inverse F-transform $\hat{e}_{q,nm}^{(k)}$

Step 3. For all $q = 1, \ldots, K$, compute the error functions $e_q^{(k+1)} = e_q^{(k)} - \hat{e}_{q,nm}^{(k)}$.

Fusion:

Step 4. Apply fusion operator κ to respective components of the direct F-transforms of the functions $e_i^{(k)}$:

$$\kappa(F[e_1^{(k)}]_{11}, \ldots, F[e_K^{(k)}]_{11}) = \kappa_{11}^{(k)},$$

$$\ldots\ldots\ldots\ldots\ldots\ldots\ldots\ldots\ldots\ldots\ldots\ldots$$

$$\kappa(F[e_1^{(k)}]_{nm}, \ldots, F[e_K^{(k)}]_{nm}) = \kappa_{nm}^{(k)},$$

and obtain the fused F-transform components as follows:

$$(\kappa_{11}^{(k)}, \ldots, \kappa_{nm}^{(k)}). \tag{8.13}$$

Step 5. $k := k + 1$.

End For

Last step of fusion:

Step 6. For all $q = 1, \ldots, K$, identify the values $e_q^{(k_{\max}+1)}(i,j)$, $(i,j) \in P$, with the F-transform components $F[e_q^{(k_{\max}+1)}]_{ij}$ with respect to the finest partitions of $\{1, \ldots, N\}$ and $\{1, \ldots, M\}$. Apply the fusion operator κ to the respective F-transform components of $e_q^{(k_{\max}+1)}$:

$$\kappa(F[e_1^{(k_{\max}+1)}]_{11}, \ldots, F[e_K^{(k_{\max}+1)}]_{11}) = \kappa_{11}^{(k_{\max}+1)},$$

$$\ldots\ldots\ldots\ldots\ldots\ldots\ldots\ldots\ldots\ldots\ldots\ldots\ldots\ldots\ldots\ldots$$

$$\kappa(F[e_1^{(k_{\max}+1)}]_{NM}, \ldots, F[e_K^{(k_{\max}+1)}]_{NM}) = \kappa_{NM}^{(k_{\max}+1)},$$

and obtain the fused F-transform components as follows:

$$(\kappa_{11}^{(k_{\max}+1)}, \ldots, \kappa_{NM}^{(k_{\max}+1)}). \tag{8.14}$$

Reconstruction:

Step 7. The fused image u is equal to the sum of inverse F-transforms with fused components (8.13) and fused components (8.14):

$$u(i,j) = \sum_{k=1}^{n_{start}} \sum_{l=1}^{m_{start}} \kappa_{kl}^{(0)} A_k^{(0)}(i)\, B_l^{(0)}(j) + \cdots$$
$$+ \sum_{k=1}^{N} \sum_{l=1}^{M} \kappa_{kl}^{(k_{\max}+1)} A_k^{(k_{\max}+1)}(i)\, B_l^{(k_{\max}+1)}(j) \tag{8.15}$$

for all $(i,j) \in P$, where $n_0 = n_{start}$ and $m_0 = m_{start}, \ldots, n_{k_{\max}+1} = N, m_{k_{\max}+1} = M$.

TABLE 8.1 Basic characteristics of the three fusion algorithms applied to the image "Castle". The resolution is 1120×840.

Time (s)			Memory (MB)		
CA	SA	ESA	CA	SA	ESA
359	1.9	19.0	160	35	102
MSE			**PSNR (dB)**		
CA	SA	ESA	CA	SA	ESA
9.48	42.48	14.15	40.62	37.51	40.61

The CA algorithm is similar to the SA algorithm except for Step 4 that is repeated in a cycle. Therefore, the quality of fusion is higher, but the CA algorithm is slower and more memory consuming, especially for large images (cf. Table 8.1). Therefore, another and much better algorithm was suggested.

8.4.4 Enhanced Simple Fusion Algorithm

In this section, we describe the *enhanced simple algorithm* (ESA), which in some sense takes the best of both CA and SA algorithms. Namely, it is almost as fast as the SA and its quality is fully comparable with CA. The algorithm was proposed in [142].

The algorithm adds another run of the F-transform over the first difference (8.5). The explanation is as follows: the first run of the F-transform aims at edge detection in each input image, whereas the second run propagates only sharp edges (and their local areas) to the fused image.

Informal description of the ESA:

for all input images u_q, $q = 1 \dots, K$ **do**

Compute the *inverse F-transform* $\hat{u}_q$;

Compute the *first absolute difference* $\Delta u_q = |u_q - \hat{u}_q|$ between the original image and the inverse F-transform of it;

Compute the *second absolute difference* $\Delta^2 u_q = |\Delta u_q - \Delta \hat{u}_q|$ between the first one and its inverse F-transform and set them as the pixel weights $w_q(i,j)$, $i = 1, \dots, N, j = 1, \dots, M$.

end for

for all pixels i,j, $i = 1, \dots, N, j = 1, \dots, M$ in the image **do**

Compute sum $Sw = \sum_{q=1}^{K} w_q(i,j)$ of the weights over all input images

for all input images u_q, $q = 1 \dots, K$ **do**

Compute the value of the ratio $r(i,j) = \frac{w_q(i,j)}{Sw}$ between the weight of the current pixel and Sw

end for

Compute the fused value of a pixel in the resulting image as

$$u(i,j) = \sum_{q=1}^{K} w_q(i,j)u_q(i,j).$$

end for

Justification of the fusion algorithms. By the properties of the F-transform on page 85, it follows from (d) that a smaller modulus of continuity leads to a higher-quality approximation of an input image by its inverse F-transform. If a certain part of the input image is affected by degradation, then by (c), the respective F-transform component captures the weighted arithmetic mean and the error function is close to zero at that part. Thus, by the proposed fusion operator κ, we choose components with maximal absolute values that correspond to those parts of the input image that are least degraded.

Demonstration of the fusion algorithms. The behavior of all three algorithms presented above is in Figures 8.4 and 8.5. In the former, four images, each blurred on some of its part, are given. In Figure 8.5, results of fusion of all four images using all three algorithms SA, CA, and ESA are demonstrated.

Furthermore, in Table 8.1, it is demonstrated that the ESA is more demanding on memory than CA, but it is significantly faster. At the same time, its quality measured in PSNR is practically the same as that of CA.

Figure 8.4 Four inputs for the image "Castle". Each of them is blurred on some part.

(a) (b) (c)

Figure 8.5 Three fusing algorithms applied to the image "Castle". (a) SA, (b) CA, (c) ESA.

8.5 F-TRANSFORM-BASED CORRUPTED IMAGE RECONSTRUCTION

In this section, we show that the F-transform can be efficiently used also in image reconstruction, that is, in reconstruction of an image corrupted by noise, text, scratches, etc. The known solutions of this problem are based on interpolation techniques (cf. [104, 141]). These solutions lead to necessity to solve large systems of linear equations. Therefore, their complexity is very high.

The F-transform, which has a linear complexity, looks for an approximating image that is close to a given one and, at the same time, does not contain what we can recognize as a corruption. This method was published in [123] and its authors are Irina Perfilieva and Pavel Vlašánek.

8.5.1 The Reconstruction Problem

We assume that only one corrupted image is given, together with the information that allows to separate corrupted and noncorrupted pixels. The goal is to replace the former ones. Let us give the technical details below.

In the reconstruction problem, it is assumed that the available set of pixels is D but the given image u is a function $u : P \longrightarrow \mathbb{R}$ where $P \subset D$, and u is not defined (i.e., it is corrupted) on the relative complement $P^c = D \setminus P$. The goal is to extend u to the set D, that is, to propose a method that computes (reconstructs) missing values of u for all $(i,j) \in P^c$. In other words, we want to obtain a new image $u^r : D \longrightarrow \mathbb{R}$ such that $u^r|_P = u$, where $u^r|_P$ is the restriction of u^r on P.

In mathematical literature, the problem of reconstruction is known as *interpolation* or *extrapolation*, depending on whether $(i,j) \in P^c$ is an "internal" point of P or not. In computer science literature, the problem of reconstruction is used to be solved with the help of interpolation methods. For this purpose, a class of interpolating functions is chosen a priori, for example, bilinear, bicubic, and RBF. If u^r is an interpolating function, then it automatically fulfills the restriction $u^r|_P = u$.

Our technique is based on employing the approximating functions. We propose to construct an extension $u^a : D \longrightarrow \mathbb{R}$ such that the restriction $u^a|_P$ *approximates* u. Then the reconstructed image u^r is a combination of two functions u^a and u such that

$$u^r(i,j) = \begin{cases} u(i,j) & \text{if } (i,j) \in P, \\ u^a(i,j) & \text{if } (i,j) \in P^c. \end{cases} \tag{8.16}$$

Note that when working with approximating functions, we are less restricted, which means that we have a wider choice of reconstruction functions at disposal.

8.5.2 F-Transform-Based Reconstruction

Assume that we are given a partially corrupted image, where the corrupted parts are separated from noncorrupted ones. Our purpose is to reconstruct this image, that is, to replace pixels from corrupted areas by new ones, whose values are computed from values of pixels from noncorrupted area. To solve this problem, we propose the technique of the F-transform. We propose two algorithms: one step and multistep. The inputs are the domain D, the noncorrupted image $u : P \longrightarrow \mathbb{R}$, and the mask m of the corrupted part $P^c = D \setminus P$. The output is the reconstructed image u^r.

8.5.2.1 Corrupted and noncorrupted parts. Recall that the image u is not defined on the corrupted part P^c of the whole set D of pixels. This part is identified by its *mask* $m : D \longrightarrow \{0, 1\}$, which is a characteristic function of P^c, that is,

$$m(i,j) = \begin{cases} 1, & \text{if } (i,j) \in P^c, \\ 0, & \text{otherwise.} \end{cases}$$

We will distinguish four types of corruption: "drawing", "noise", "text", and "scratch". Their masks are shown in Figure 8.6. "Noise" is a specific type of corruption that is spread over the whole image. "Scratch" and "drawing" are represented by lines, hand drawings, or scratches themselves. "Text" can be created by some notes or date records or time stamps. It is important to say that from the algorithmic point of view, all types of corrupted parts are processed similarly. However, a quality of reconstruction varies depending on the type of corruption and the applied method of reconstruction.

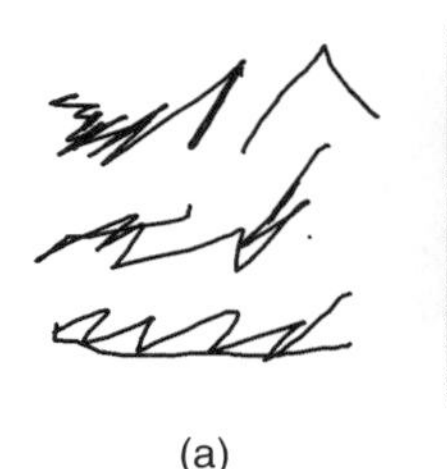

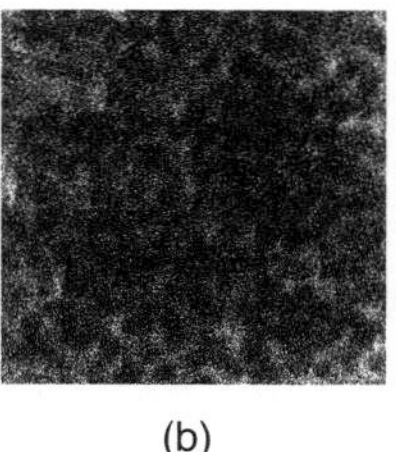

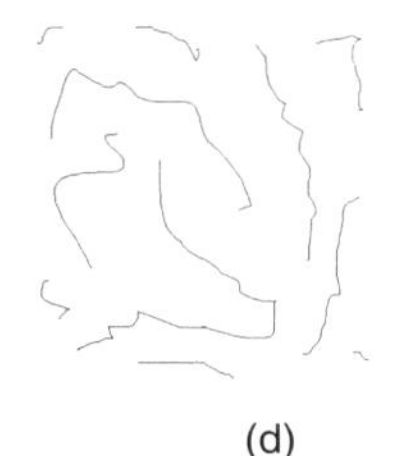

(a) (b) (c) (d)

Figure 8.6 Types of corruption. (a) Drawing, (b) noise, (c) text, and (d) scratch.

8.5.2.2 Reconstruction algorithm. As described in Section 8.1, we assume to be given a uniform fuzzy partition $A_0, \ldots, A_n$ and $B_0, \ldots, B_m$ of P. Moreover, we assume that it fulfills the following property:

Q: To every corrupted pixel $(i,j) \in P^c$, there are basic functions A_k and B_l and there is a pixel $(i',j') \in P$ such that $A_k(i) > 0$, $A_k(i') > 0$, $B_l(j) > 0$, $B_l(j') > 0$.

This means that every corrupted pixel is covered by some combination of basic functions such that this combination also covers at least one noncorrupted pixel. Property **Q** assures that the inverse F-transform $\hat{u}$ of the image u replaces the nondefined value of the intensity function u at every corrupted pixel $(i,j) \in P^c$ by a value of $\hat{u}(i,j)$ that is computed using values of u at noncorrupted pixels $(i',j') \in P$. The reconstructed image u^r is a combination of u and $\hat{u}$:

$$u^r(i,j) = \begin{cases} \hat{u}(i,j), & \text{if } (i,j) \in P^c, \\ u(i,j), & \text{otherwise.} \end{cases} \tag{8.17}$$

The main idea of the reconstruction algorithm is the following: first, we apply the F-transform with a fine partition (the smallest h) and reconstruct those corrupted pixels that fulfill property **Q**, then we recompute the corrupted area P^c by deleting the already reconstructed pixels from it and if P^c is still nonempty, repeat the procedure with a bigger value of h.

Description of the algorithm.

Step 1. Set the distance between nodes to $h = 2$.

Step 2. Establish an h-uniform fuzzy partition $A_0, \ldots, A_n$ and $B_0, \ldots, B_m$ of P.

Step 3. Compute the inverse F-transform $\hat{u}$ of the given image u.

Step 4. Compute the reconstruction u^h corresponding to h.

Step 5. Update the mask m by deleting the reconstructed pixels from it.

Step 6. Update the image u^r in accordance with (8.17). If the mask is nonempty, then increase the distance between nodes $h := h + 1$ and go to Step 2. Otherwise finish.

Figure 8.7 The image "Girl" (the author of this drawing is Renáta Doležalová). (a) The original. (b) Corrupted by text. (c), (d) Reconstructions using inverse F-transform computed on a sequence of triangular h-uniform fuzzy partitions with increasing h.

8.5.3 Demonstration Examples

The F-transform-based reconstruction method is demonstrated on three kinds of corruption of images, namely, hand drawing, noise, and text. The results are in Figures 8.7–8.10. We applied triangular uniform fuzzy partition, because as it is shown in [122], exactly the triangular shape guarantees the best RMSE quality of reconstruction.

To compare the F-transform technique with some classical methods, we present in Table 8.2 some statistics of the RMSE quality measure obtained after processing a set of 55 color images taken from http://decsai.ugr.es/cvg/dbimagenes/c512.php. All these images were artificially corrupted by the three types of corruption and then reconstructed by the F-transform technique, classical techniques "nearest neighbor" and "bilinear interpolation" (cf. [104]), and two inpainting techniques:

Figure 8.8 The image "House". (a) The original, (b) its corruption by hand drawing, and (c) its reconstruction.

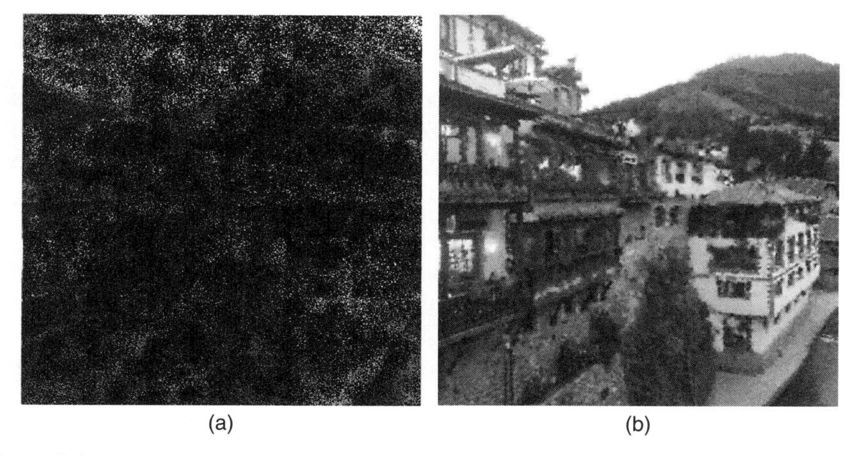

Figure 8.9 The image "House". (a) Its corruption by noise. The corrupted area is above 70%. (b) Its reconstruction.

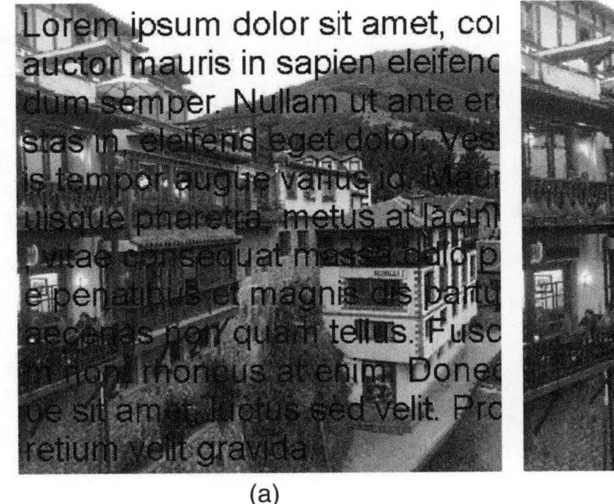

(a)

(b)

Figure 8.10 The image "House". (a) Its corruption by text and (b) its reconstruction.

TABLE 8.2 RMSE values for three kinds of corruption.

	Nearest	Bilinear	Telea	Navier–Stokes	FT
			"Noise"		
Median	21.68	19.046	16.055	15.913	**15.435**
Mean	22.37	19.625	16.928	17.012	**16.615**
			"Drawing"		
Median	6.966	6.276	6.052	5.920	**6.118**
Mean	7.392	6.419	6.086	6.037	**6.163**
			"Text"		
Median	7.096	6.794	6.062	5.808	**6.117**
Mean	7.656	7.097	6.566	6.357	**6.682**

"Navier–Stokes" [9] and fast marching method of Telea [139]. The lowest values of RMSE are put into box. It can be seen from Table 8.2 that the F-transform-based method is fully comparable with the best classical methods. It gives the best results in case of noise corruption, and it is the third best in the two other cases. Its advantages are simplicity and low computational complexity.

9

ANALYSIS AND FORECASTING OF TIME SERIES

Analysis and forecasting of time series is an important problem with numerous applications in economy, industry, meteorology, and other areas (cf. [47]). In this chapter, we address a new methodology for decomposition and forecasting of time series that is based on a combination of two fuzzy modeling techniques: the F-transform presented in Chapter 4 and techniques of fuzzy natural logic (see Chapter 5). The proposed methodology consists of three phases: (1) analysis of time series, (2) prediction of its future course, and (3) evaluation of its current and future course in natural language.

The analysis of time series consists of its decomposition into several components followed by characterization and prediction of each component separately. This approach has one important advantage: its results are well interpretable. This is useful because interpretability is quite often more important than precision of the forecast. Still, as can be demonstrated, precision of our methods is fully comparable with precision of the professional systems such as ForecastPro® that apply classical statistical methods.

At the end of this chapter, we also present a method based on combination of the F-transform and the theory of evaluative linguistic expressions, using which we can recognize trend in various parts of time series and generate linguistic comments that evaluate its course.

Insight into Fuzzy Modeling, First Edition. Vilém Novák, Irina Perfilieva, and Antonín Dvořák.

Companion Website: www.wiley.com/go/novak/fuzzy/modeling

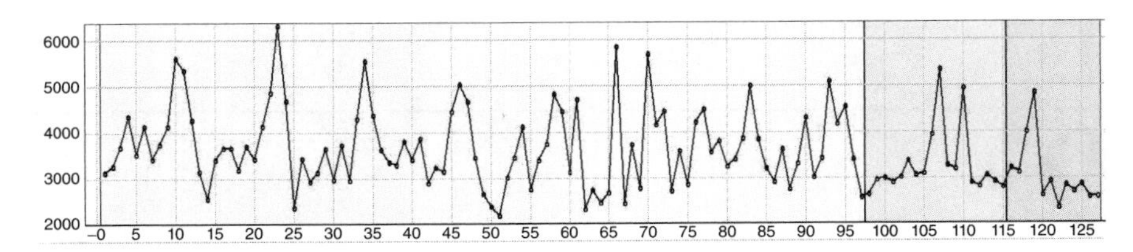

Figure 9.1 Example of a time series.

9.1 CLASSICAL VERSUS FUZZY MODELS OF TIME SERIES

9.1.1 Definition of Time Series

A time series is a function

$$X : \mathbb{T} \times \Omega \longrightarrow \mathbb{R}, \tag{9.1}$$

where $\mathbb{T} = \{0,\dots,N\} \subset \mathbb{N}$ is a finite set of integers interpreted as *time moments* and $\langle \Omega, \mathscr{A}, P \rangle$ is a probability space, where Ω is a set of elementary events, $\mathscr{A}$ is a sigma algebra on Ω, and P is a probability measure defined on $\mathscr{A}$ such that $P(\Omega) = 1$. If we fix some $\omega \in \Omega$, then we obtain a *realization* of the time series (9.1), which is a real discrete function $X(t)$ for $t = 0,\dots,N$. Of course, in reality, we always have only one realization of X at our disposal. A typical example of a time series is depicted in Figure 9.1.

Remark 9.1 *In the theoretical analysis of time series (see [2, 47]), it is advantageous to take a time series as a stochastic process* $X : \mathbb{T} \times \Omega \longrightarrow \mathbb{C}$*, where* $\mathbb{T} = [a, b]$ *is an interval of real numbers,* $\mathbb{C}$ *is a set of complex numbers and each realization of* X *is a* continuous function $X(t)$*,* $t \in [a, b]$.

9.1.2 Classical Models of Time Series

There are two basic classical approaches to the analysis and forecasting of time series. The first one, the so-called *Box–Jenkins methodology* (see [14]), is based on the representation of time series by a combination of autoregressive and moving average models. For instance, the ARMA(p, q) model (see [1]) assumes that each value $X(t)$ of a given time series is determined by

$$X(t) = c + \epsilon(t) + \sum_{i=1}^{p} \varphi_i X(t - i) + \sum_{i=1}^{q} \theta_i \epsilon(t - i), \tag{9.2}$$

where $\varphi_1,\dots,\varphi_p$ are parameters of the *autoregressive model*, $\theta_1,\dots,\theta_q$ are parameters of the *moving average model*, c is a constant, $\epsilon(t)$ is a white noise, and $\epsilon(t-1),\dots,\epsilon(t-q)$ are error terms. These models are powerful and successful in forecasting. Their disadvantage is that they are not transparent, that is, one can hardly understand what happens inside and what is the reason for the resulting forecast.

The second approach, which is called a *decomposition model*, assumes that a given times series can be decomposed into four components, namely, *trend* Tr, *cycle* C, *season* S, and *noise* R:

$$X(t) = Tr(t) + C(t) + S(t) + R(t).$$

As the trend and cycle usually occur together and it is difficult to distinguish one from another, we usually join them and speak about *trend-cycle* TC.

These components have a clear meaning. When constructing the decomposition model using standard statistical methods, the trend-cycle TC (and also the trend Tr) is modeled using some simple function such as linear, quadratic, and exponential. The model is usually obtained using regression analysis (cf. [56]). Note that such an approach is somewhat artificial, because one can hardly expect that the trend-cycle takes a simple course on the whole domain. Therefore, various kinds of improvements have been suggested, such as the STL-method (Season-Trend-Loess regression) [25, 26]. They are, however, computationally demanding. In this respect, the fuzzy modeling methods can bring essential simplification together with preservation of the quality of estimation.

9.1.3 Fuzzy Models of Time Series

There are various approaches to fuzzy modeling of time series. For example, in [3], a study presenting Takagi–Sugeno rules (see Section 3.4) in the view of the autocorrelation Box–Jenkins methodology is presented. This approach can be considered as a special kind of the regression model. Let us also mention a fuzzification of the autoregressive integrated moving average (ARIMA) method [140], where the model parameters φ_i, θ_i are replaced by fuzzy numbers. Other class of approaches is formed by various neuro-fuzzy ones, which lie on the border among neural networks, Takagi–Sugeno models, and evolving fuzzy systems [64, 126].

One of the studied models of time series are the so-called *fuzzy time series*. These were introduced by Song and Chissom in [131, 132]. According to them, a fuzzy time series is a function

$$F : \mathbb{T} \longrightarrow \mathscr{F}(\mathbb{R}),$$

that is, each time $t \in \mathbb{T}$ is assigned a fuzzy set $F(t) \subsetapprox \mathbb{R}$, where the latter is usually taken as a fuzzy number. First, a fuzzy relation characterizing the whole fuzzy time series is constructed as an R_{DNF} in (3.17):

$$R_{\mathrm{DNF}} = \bigcup_{t=0}^{N-1} (F(t) \overset{\mathbf{T}}{\times} F(t+1)).$$

Of course, there are more possibilities how R_{DNF} can be constructed from the time series, for example, we can model a deeper dependence of future values on previous ones such as $F(t-j) \overset{\mathbf{T}}{\times} F(t+1)$ for $j > 1$. The t-norm $\mathbf{T}$ is usually taken as minimum.

The future value is then determined as an image of a fuzzy set in a fuzzy relation (2.64) (or composition of fuzzy relations—cf. Remark 2.4) as follows:

$$F(t+1) = F(t) \circ R_{\mathrm{DNF}}.$$

If we want to obtain one concrete value at the time moment $t + 1$, we must use defuzzification (usually center of gravity, COG, see Section 3.2.1).

One of often discussed contributions of fuzzy models is their good interpretability. However, it happens quite often that after employing an optimization technique, the used, for example, Gaussian fuzzy set has the center, say, at node 5.6989 and the width parameter equal to 2.8893 (see [64]). Such fuzzy sets are undoubtedly far from being interpretable, and so these approaches are closer to the standard regression methods. On the other hand, the decomposition models are more transparent, because one can easily understand the influences inside the separate components and thus understand the structure of the time series and reasons for its forecast. Hence, we find the decomposition models to be more suitable for the application of fuzzy modeling techniques.

In this chapter, we show that the F-transform belongs among suitable fuzzy modeling techniques using which we can plausibly estimate trend or trend-cycle of a time series without the limiting assumption that the shape of the former should be some specific function; still we can obtain their precise analytic expression. Moreover, when combined with techniques of fuzzy natural logic, we can obtain a well-interpretable forecast of time series and mine information in sentences of natural language that gives us a more clear idea about its behavior.

9.2 ANALYSIS OF TIME SERIES USING F-TRANSFORM

9.2.1 Decomposition of Time Series

Our basic assumption is that the time series X in (9.1) can be decomposed into three components, namely,

$$X(t) = TC(t) + S(t) + R(t, \omega), \quad t \in \mathbb{T}, \tag{9.3}$$

where TC is the *trend-cycle*, S is the *seasonal component*, and R is a *random noise*. Both TC and S are usual (i.e., nonrandom) real functions of a real variable. Furthermore, as mentioned, the trend-cycle can be understood as a composition of the trend and the cycle

$$TC(t) = Tr(t) + C(t), \tag{9.4}$$

where the trend Tr may show itself only as a certain kind of a general tendency of the whole time series. We will usually take the trend-cycle TC as a single component.

The component R is a random noise. It is assumed to be a real stationary stochastic process that has a zero mean μ and a finite variance $\sigma^2 > 0$. In this book, we suppose that R can be represented as

$$R(t, \omega) = \xi(\omega)\ e^{i\lambda t + \varphi}, \tag{9.5}$$

where $\xi(\omega)$ is a random variable with zero mean value, λ a real number, and φ is a phase shift (see, e.g., [145]). Below, we will not work with ω in our explanation but only with a single realization $X(t)$. Therefore, we will omit ω in the subsequent formulas.

Remark 9.2 *If convenient for the explanation, we will take the realization $X(t)$ (and consequently, all its constituent components) as a continuous function defined on an interval of reals $\mathbb{T} = [a, b]$ (cf. Remark 9.1).*

The seasonal component $S(t)$ is assumed to be a mixture of real periodic functions

$$S(t) = \sum_{j=1}^{r} P_j \sin(\lambda_j t + \varphi_j), \tag{9.6}$$

where P_j is an amplitude, λ_j is a frequency and φ_j is a phase shift, $j = 1,\dots, r$. Recall that $\lambda_j = \frac{2\pi}{T_j}$, where T_j is the periodicity. Because of the latter equality, we will speak freely either about λ_j or about T_j.

The structure of time series is schematically depicted in Figure 9.2. Recall from Chapter 4 that the F-transform is linear. Therefore, the F-transform components $\mathbf{F}[X]$ are equal to

$$\mathbf{F}[X] = \mathbf{F}[TC] + \sum_{j=1}^{r} P_j \mathbf{F}[\sin(\lambda_j t + \varphi_j)] + \mathbf{F}[R] \tag{9.7}$$

for all fuzzy partitions.

Periodogram. The periodicities considered in (9.6) can be found using the classical method of *periodogram*. This is widely used standard method published in many books and papers—see, for example, [2, 15, 47] and elsewhere. It enables to find distinguished periodicities occurring in time series.

Definition 9.1 *Let X be a realization of the time series (9.1). The periodogram is the function*

$$I_X(\lambda) = \frac{1}{2\pi(N+1)} \left| \sum_{t=0}^{N} X(t) e^{-it\lambda} \right|^2, \quad -\pi \le \lambda \le \pi.$$

This function takes *significantly big values* in selected frequencies $\lambda_1, \dots, \lambda_s$ (i.e., periodicities $T_j = \frac{2\pi}{\lambda_j}$) occurring in the time series X. Note that the periodogram is at disposal in many statistical software packages including R.

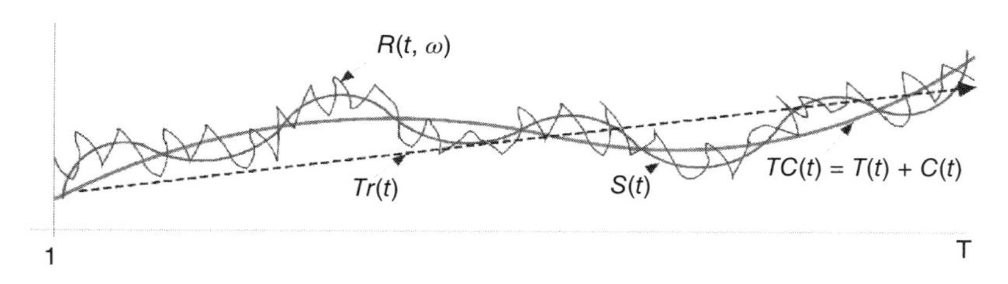

Figure 9.2 Scheme of the structure of time series.

9.2.2 Extraction of Trend-Cycle and Trend Using F-Transform

9.2.2.1 Extraction of trend-cycle. We will apply the F-transform to a realization $X(t)$ of the given time series (9.3) over the time domain $\mathbb{T} = [a, b]$. For simplicity, we will assume that $a = 0$ and construct an h-uniform fuzzy partition consisting of $n + 1$ *triangular fuzzy sets*.

Assumptions:

(i) The time series X can be decomposed as in (9.3), where the seasonal component S consists of periodic functions (9.6) having periodicities T_j, $j \in \{1, \ldots, r\}$. The longest of them is denoted by $\overline{T}$ and the corresponding frequency by $\overline{\lambda}$.

(ii) We will construct an h-uniform fuzzy partition over the domain $[0, b]$, where the distance h between the nodes is related to the periodicity $\overline{T}$ by simple equality

$$h = d \cdot \overline{T}, \tag{9.8}$$

where $d > 0$ is a suitable real number. Thus, we obtain a set of equidistant nodes:

$$\Big\{c_0 = 0,\ c_1 = d\overline{T}, \ldots, c_{k-1} = (k-1)\, d\overline{T}, c_k = k d\overline{T}, \\ c_{k+1} = (k+1)\, d\overline{T}, \ldots, c_n = n d\overline{T} = b\Big\}, \tag{9.9}$$

where $n \geq 2$. The corresponding triangular fuzzy partition $A_0, \ldots, A_n$ is fixed.

(iii) The trend-cycle TC is a function with no clear periodicity or its periodicity is much longer than h so that the modulus of continuity $\omega(h, TC)$ defined in (2.1) is small. This means that TC is smooth with small changes in its course (with respect to h).

Since h is fixed by (9.8), it is easy to see that $h = d_j T_j$, $j = 1, \ldots, r$, for some real numbers d_j. From this, we obtain $h = \frac{2\pi d_j}{\lambda_j}$, that is, $\lambda_j = \frac{2\pi d_j}{h}$. Thus, if h is fixed, then higher values of d_j correspond to higher frequencies λ_j.

Remark 9.3 *The domain $\mathbb{T}$ of real time series can have arbitrary number N of time moments. Then, of course, if we want the nodes c_i to coincide with the time moments $t \in \mathbb{T}$, we may hardly be able to set the distance h between nodes in such a way that the set $\mathbb{T}$ would be covered by the fuzzy partition completely. Because of the forecasting procedure (described below in this chapter), it is recommended to identify the last node c_n with the last available time moment N and then continue from the right to the left.*

Properties of the F-transform of time series. The following results demonstrate that using the F-transform, we can filter out higher frequencies from the time series and obtain a good estimation of the trend-cycle:

(a) Let h be equal to an integer multiple of the corresponding periodicity, that is, $d_j = \frac{h}{T_j} \in \mathbb{N}$, where $j \in \{1,\ldots,r\}$. Then the following holds for all the F-transform components:

$$F_k[P_j \sin(\lambda_j t + \varphi_j)] = 0, \; k = 1,\ldots,n-1.$$

(b) For cases when (a) does not hold, we have the following: Let $I \subset \{1,\ldots,r\}$ be a set of subscripts for which the remainder[1] $d_j' = d_j - [d_j]$ is greater than 0. Then the absolute value of the inverse F-transform of the seasonal component is bounded by

$$|\hat{S}(t)| \leq \sum_{j\in I} \left| \frac{P_j \sin^2(d_j' \pi)}{d_j^2 \pi^2} \right| \tag{9.10}$$

for all $t \in [c_1, c_{n-1}]$.

(c) The noise component R from (9.5) is reduced using the F-transform as follows: Put $\overline{\xi} = \sup\{\xi(t) \mid t \in [a,b]\}$, $\underline{\xi} = \inf\{\xi(t) \mid t \in [a,b]\}$ and

$$\tilde{\xi} = \begin{cases} \overline{\xi} & \text{if } \overline{\xi} \geq 0, \\ \underline{\xi} & \text{if } \overline{\xi} < 0. \end{cases} \tag{9.11}$$

Then for each $k = 1,\ldots,n-1$,

$$|F_k[R]| \leq \frac{|\tilde{\xi}| \sin^2(d\pi)}{d^2\pi^2} \tag{9.12}$$

and

$$|\hat{R}(t)| \leq \frac{|\tilde{\xi}| \sin^2(d\pi)}{d^2\pi^2}, \quad t \in [c_1, c_{n-1}] \tag{9.13}$$

(recall that $\hat{R}$ denotes the inverse F-transform of R). We can see that the F-transform significantly reduces noise of the form (9.5).

For proofs and more details, see [96].

On the basis of (a), it is reasonable to chose d in (9.8) to be equal to some natural number $d \in \mathbb{N}$, provided that $n = \frac{b}{h} \geq 2$.

Trend-cycle extraction method. We will proceed as follows:

(i) Form an h-uniform triangular fuzzy partition over the nodes (9.9) with the distance $h = d\,\overline{T}$ (cf. (9.8)), where $\overline{T}$ is the longest of the periodicities T_j, $j \in \{1,\ldots,r\}$ and $d \in \mathbb{N}$.

(ii) Compute components of the direct F-transform using (4.13) (cf. also (4.8)):

$$\mathbf{F}[X] = (F_0[X],\ldots,F_n[X]). \tag{9.14}$$

[1]Let $\alpha \in \mathbb{R}$ be a real number. Then by $[\alpha]$, we denote the maximal integer that is smaller than or equal to α.

(iii) Compute the inverse F-transform $\hat{X}$ from the direct one (9.14) using (4.11):

$$\hat{X}(t) = \sum_{k=0}^{n} F_k[X]\, A_k(t), \quad t \in [c_1, c_{n-1}].$$

Remark 9.4

(a) In (iii), we consider values of t from the interval $[c_1, c_{n-1}]$ only. The reason is that the fuzzy sets A_0 and A_n are cut in the middle. Hence, the corresponding components $F_0[X]$ and $F_n[X]$ are incomplete.

(b) We can find periodicities of all periodic functions occurring in the decomposition of S (9.6) using periodogram (see Section 9.2.1).

Justification of the trend-cycle extraction method. The inverse F-transform $\hat{X}$ is a good estimation of the trend-cycle TC. Indeed, let us now denote the sum of the bounds in (9.10) and (9.13) by

$$\epsilon = \sum_{j\in I} \left| \frac{P_j \sin^2(d_j' \pi)}{d_j^2 \pi^2} \right| + \frac{|\tilde{\xi}| \sin^2(d\pi)}{d^2 \pi^2}. \tag{9.15}$$

The following theorem gives an estimation of the error if we extract the trend-cycle using the F-transform.

Theorem 9.2 *Let X be the time series (9.3) and the assumptions formulated above are satisfied. Then the inverse F-transform $\hat{X}$ provides estimation of the trend-cycle TC with the error*

$$|\hat{X}(t) - TC(t)| \le 2\omega(h, TC) + \epsilon, \quad t \in [c_1, c_{n-1}]. \tag{9.16}$$

It follows from the analysis above that ϵ is a small number because many F-transform components of the summands in (9.6) are very small or even equal to zero. Moreover, by the assumption, the modulus of continuity $\omega(h, TC)$ is also small. Hence, we may conclude from (9.16) that the *F-transform applied to the time series enables us to extract the trend-cycle TC with a high accuracy.*

EXAMPLE 9.1 Let us artificially form a time series $X(t)$ on the set of integers $\mathbb{T} = \{0, 100\}$ as follows:

$$X(t) = TC(t) + 5\sin(0.63t + 1.5) + 5\sin(1.26t + 0.35) + \\ 15\sin(2.7t + 1.12) + 7\cos(0.41t + 0.79) + R(t), \quad t \in \mathbb{T}. \tag{9.17}$$

The function $TC(t)$ in (9.17) is the trend-cycle given by artificial data without clear periodicity (it is depicted by the dashed line in Figure 9.3). Its modulus of continuity is $\omega(30, TC) = 3.22$. The other four sine members form the seasonal component $S(t)$.

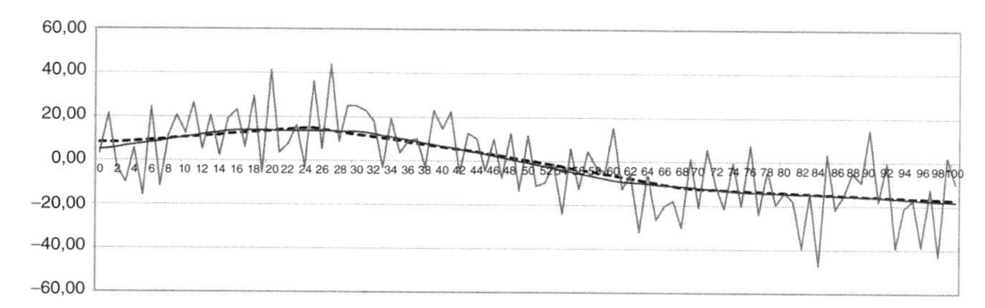

Figure 9.3 F-transform of the artificial time series $X(t)$, $t \in \{0, \ldots, 100\}$ obtained by values of the four sine members and the noise $R(t)$ (formula (9.17)). The dashed line depicts the original trend-cycle $TC(t)$ given by the data. After application of the F-transform to $X(t)$, we obtain approximation $\hat{X}(t)$ of the trend-cycle. This is depicted by the solid line. One can see that it is almost identical with the original $TC(t)$.

Their respective periodicities are $T_1 = 10$, $T_2 = 5$, $T_3 = 2.3$, and $T_4 = 15.4$. Therefore, we set $\overline{T} = [T_4]$ (i.e., the greatest integer part) and $d = 1$, that is, the distance (9.8) is $h = 15$. Consequently, the width of basic functions is $2h = 30$ (the time axis is discrete; hence, fractions are neglected) and $d_1 = 1.5, d_2 = 3, d_3 = 6.5$, and $d_4 = 0.97$. Two of these parameters are close to natural numbers and the parameter d_3 is fairly large. We thus can show that the error ϵ in (9.15) is almost 0. The $R(t)$ is a random noise with the average mean value $\overline{\mu} = -0.24$.

The result of application of the F-transform is depicted in Figure 9.3. One can see that both the seasonal component as well as the noise are almost completely removed. Maximal difference $|TC(t) - \hat{X}(t)| = 3.32$. Hence, we can conclude that the trend-cycle was estimated with the error corresponding to (9.16). □

9.2.2.2 Extraction of trend. In the theory of time series analysis, we quite often extract the trend-cycle TC as a single component. However, it can be interesting to also have the pure trend Tr from (9.4) at disposal. Taking into account that the cyclic component C is also a mixture of periodic functions, we may assume the following decomposition of X:

$$X(t) = Tr(t) + \sum_{j=1}^{s} P_j \sin(\lambda_j t + \varphi_j) + R(t, \omega), \qquad t \in \mathbb{T}, \tag{9.18}$$

where the middle term comprises of all periodic functions forming both the cyclic component C and the seasonal component S. By applying the F-transform to (9.18), we obtain

$$\mathbf{F}[X] = \mathbf{F}[Tr] + \sum_{j=1}^{s} P_j \mathbf{F}[\sin(\lambda_j t + \varphi_j)] + \mathbf{F}[R]. \tag{9.19}$$

We can again apply Theorem 9.2 by considering $\overline{T}$ to be equal to maximal periodicity of the periodic functions in (9.18) and set the distance between the nodes $c_0, \ldots, c_n$ to $\overline{h} = d\overline{T}$ for some d (preferably $d \in \mathbb{N}$). Clearly, $n = b/\overline{h}$. Then, analogously to (9.16), we obtain

$$|\hat{X}(t) - Tr(t)| \leq 2\omega(\overline{h}, Tr) + \epsilon, \qquad t \in [c_1, c_{n-1}], \tag{9.20}$$

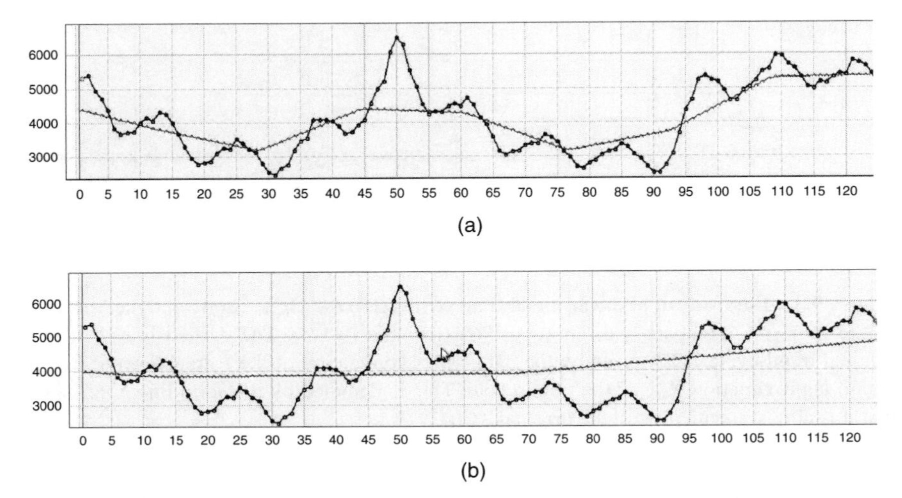

Figure 9.4 Extraction of the (a) trend-cycle TC and (b) trend Tr of a time series using the F-transform.

where the right-hand side of (9.20) is a small number. Hence, the estimation of the trend of X is

$$Tr(t) \approx \sum_{k=0}^{n} A_k(t)F_k[X], \qquad t \in [c_1, c_{n-1}], \tag{9.21}$$

where A_k, $k = 0, \ldots, n$, are fuzzy sets from the fuzzy partition constructed using $\overline{h}$ as described in Section 9.2.2.1 and $F_k[X]$ are the corresponding components. Formula (9.21) is at the same time explicit analytic formula of the trend Tr.

EXAMPLE 9.2 In Figure 9.4, the extraction of the trend-cycle TC and trend Tr of a given time series[2] is depicted. Using periodogram, we found the following periodicities T_j: 3, 6, 11.9, 16.3, 20.9, 25.3, 44.4, 56.5. On the basis of them, we set $h = 16$ to extract the trend-cycle TC and $\overline{h} = 56$ to extract the trend Tr. Using (4.8), we computed the following values of the components:[3]

- *For the trend-cycle TC:* $F_1[X] = 3716, F_2[X] = 3202, F_3[X] = 4561, F_4[X] = 4109, F_5[X] = 3094, F_6[X] = 3947, F_7[X] = 5336, F_8[X] = 5327$.
- *For the trend Tr:* $F_1[X] = 3830, F_2[X] = 3911, F_3[X] = 4836, F_4[X] = 4810$.

Then, using formulas (4.2)–(4.4), we can obtain from (9.21) the analytic expression both for the trend-cycle TC as well as for the trend Tr. □

[2]The time series was taken from the *M3-Competition* provided by the *International Institute of Forecasters*.
[3]The components are computed over uniform fuzzy partition starting from right to left—cf. Remark 9.3.

9.3 TIME SERIES FORECASTING

9.3.1 Decomposition of Time Domain

Time series is defined in (9.1) and (9.3) on the domain $\mathbb{T} = \{0,\ldots,N\} \subset \mathbb{N}$. Let us now consider a set $\mathbb{T}^F = \{N+1,\ldots,N+M\}$. Then, forecasting of the time series X means to find its *future values* $\{X(t) \mid t \in \mathbb{T}^F\}$ on the basis of the *known values* $\{X(t) \mid t \in \mathbb{T}\}$. The number M is also called *forecasting horizon.*

To be able to find such a forecast, we first divide the time series $\{X(t) \mid t \in \mathbb{T}\}$ into two subsets:

(i) *learning set* $\{X(t) \mid t \in \mathbb{T}^L\}$, where $\mathbb{T}^L = \{0,\ldots,N^L\}$ for some $N^L < N$,
(ii) *validation set* $\{X(t) \mid t \in \mathbb{T}^V\}$, where $\mathbb{T}^V = \{N^L+1,\ldots,N\}$.

Furthermore, on the basis of the learning set, we find the best model $\tilde{X}$ of the time series, where we measure the quality of the model on the validation set. Finally, we compute the estimation $\{\tilde{X}(t) \mid t \in \mathbb{T}^F\}$ of the future values of the time series using the best model above.

To see how good is our estimation, we should now wait M time moments (e.g., M months) until we learn the real values $\{X(t) \mid t \in \mathbb{T}^F\}$ to be able to compare them with the results of our model. This inconvenience can be overcome if the set of time moments $\mathbb{T}$ in which the time series is known is long enough. Then its last part can be cut off and we can pretend that it is unknown. Thus, we have the following situation:

(i) We suppose that we are given the time series $\{X(t) \mid t \in \mathbb{T} \cup \mathbb{T}^F\}$. The set $\{X(t) \mid t \in \mathbb{T}\}$ is called *in-samples* and the set $\{X(t) \mid t \in \mathbb{T}^F\}$ is called *out-samples*.
(ii) The in-samples are divided into the learning set $\{X(t) \mid t \in \mathbb{T}^L\}$ and the validation set $\{X(t) \mid t \in \mathbb{T}^V\}$.
(iii) The out-samples $\{X(t) \mid t \in \mathbb{T}^F\}$ become the *testing set* using which we can measure how good is our forecast. Let us emphasize that this data are never used in computation of the model but only to check how good is the model using which we obtained the estimation $\{\tilde{X}(t) \mid t \in \mathbb{T}^F\}$ of the future values.

The situation is schematically depicted in Figure 9.5.

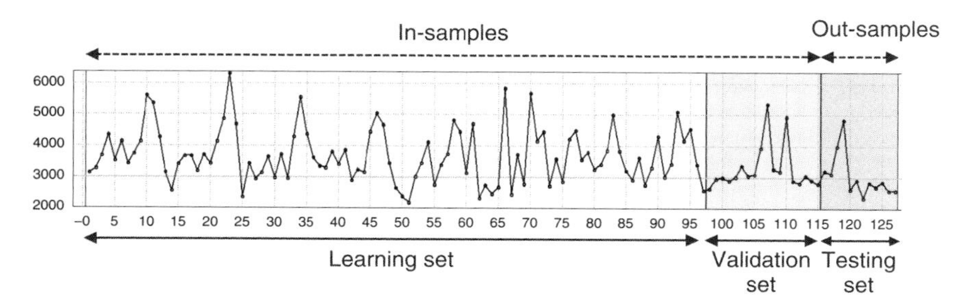

Figure 9.5 Scheme of the division of time series for analysis and forecast.

The quality of forecast is measured using various kinds of measures. The most often used one is the *symmetric mean absolute percent error* (SMAPE) defined by

$$\text{SMAPE} = \frac{1}{M}\sum_{t=\text{N}+1}^{N+M} \frac{|X(t) - \tilde{X}(t)|}{\frac{|X(t)|+|\tilde{X}(t)|}{2}}. \tag{9.22}$$

This measure gives results from the interval $[0, 200]$. It is used both in the validation set to find the best model $\tilde{X}(t)$ as well as in the testing set to check the quality of the forecast (provided that the testing set is defined).

Below, we will explain how the time series can be forecasted. Because of the decomposition (9.3), we can forecast each component separately.

9.3.2 Forecast of Trend-Cycle

The trend-cycle TC also contains the cyclic component; therefore, we cannot forecast TC by simple prolongation of its previous values. The fact that the F-transform components contain information about values of the time series in areas determined by the basic functions A_k suggests the following: first, we *forecast the F-transform components* and then compute the future values $TC(t)$ of the trend-cycle using the inverse F-transform.

Forecasting procedure. Let us suppose that we have components of the direct F-transform of X

$$\mathbf{F}[X] = (F_1[X], \ldots, F_{n-1}[X]) \tag{9.23}$$

at disposal. These components will be used to predict the future ones

$$F_n[X], \ldots, F_{n+\ell}[X]. \tag{9.24}$$

Note that in (9.23), we did not include the first component $F_0[X]$ and the last one $F_n[X]$ because they are computed using only halves of the basic functions and, consequently, they are incomplete. Instead, $F_n[X]$ should be predicted too.

Note that the components (9.24) correspond to future nodes $c_n, \ldots, c_{n+\ell+1}$ over the time interval $\mathbb{T}^F$. This means that $c_n = N$ and $c_{n+\ell+1} = N + M$. We must also extend the fuzzy partition to cover all the nodes $\{c_0, \ldots, c_n, \ldots, c_{n+\ell+1}\}$, that is, we deal with the new fuzzy partition $A_0, \ldots, A_n, \ldots, A_{n+\ell}$.

On the basis of the latter and the components (9.24), we compute future values of the trend-cycle using the inverse F-transform:

$$TC(t) = \hat{X}(t) = \sum_{k=1}^{n+\ell} F_k[X]\, A_k(t), \quad t = N, \ldots, N + M. \tag{9.25}$$

Scheme of the trend-cycle forecast is in Figure 9.6.

Prediction of future F-transform components. The future F-transform components (9.24) are predicted using the perception-based logical deduction (PbLD, see

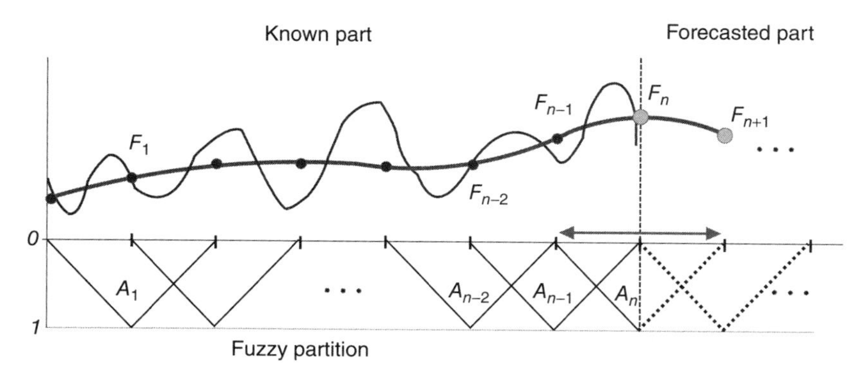

Figure 9.6 Scheme of the trend-cycle forecast. The predicted components in this figure are F_n and F_{n+1}. The fuzzy partition is depicted upside-down below the x-axis. The dotted lines denote its prolongation to cover also the testing part.

Section 5.4.2) applied to a *linguistic description learned from the data*. The learning procedure is the same as that described in Section 7.4.2.

The variables occurring in the learned linguistic description are the given F-transform components $F_1[X],\ldots,F_{n-1}[X]$ as well as their first and second differences:

$$\Delta F_i[X] = F_i[X] - F_{i-1}[X], \qquad i = 2,\ldots,n-1,$$
$$\Delta^2 F_i[X] = \Delta F_i[X] - \Delta F_{i-1}[X], \qquad i = 3,\ldots,n-1.$$

There are two principal ways to forecast the F-transform components:

(i) Forecast the next F-transform component on the basis of all the previous ones and their corresponding first and second differences. The generated fuzzy/linguistic IF-THEN rules take, for example, the form

$$\text{IF } \Delta^2 F_{i-1}[X] \text{ is } \mathscr{A}_{i-1} \text{ AND } \Delta F_{i-1}[X] \text{ is } \mathscr{B}_{i-1} \text{ AND } F_{i-1}[X] \text{ is } \mathscr{C}_{i-1} \text{ THEN } F_i[X] \text{ is } \mathscr{D}_i, \qquad (9.26)$$

$i = n,\ldots,n+\ell$.

(ii) Forecast some of the next F-transform components (not necessarily the immediate next one) from some of the components (9.23) and their first and second differences. The linguistic descriptions then consist of rules of the form

$$\text{IF } \Delta^2 F_i[X] \text{ is } \mathscr{A}_i \text{ AND } \Delta F_i[X] \text{ is } \mathscr{B}_i \text{ AND } F_i[X] \text{ is } \mathscr{C}_i \text{ THEN } F_{n+j}[X] \text{ is } \mathscr{D}_{n+j} \qquad (9.27)$$

for a suitable $i < n$ and $j = 0, 1, 2,\ldots,\ell$.

The rules (9.26) or (9.27) may describe both dynamic behavior as well as logical dependencies inside the trend-cycle (hidden cycle influences).

In case (i), we take into account the components $F_1[X], \ldots, F_{n-1}[X]$ and their differences and, using several last ones (their number depends on the length of the antecedents of the rules), forecast the component $F_n[X]$. Then, using the same linguistic description, we forecast $F_{n+1}[X]$ from $F_2[X], \ldots, F_n[X]$, and their differences, etc. Obviously, there is a danger of propagation of the forecast errors, because we forecast from already forecasted values. The longer is the forecasting horizon, the more dubious is the forecast.

Case (ii) overcomes this problem, because we build a finite number of independent trend-cycle forecasting linguistic descriptions—one for each forecasted component from (9.24) (because each of them can be the result of different cause).

EXAMPLE 9.3 We performed the analysis and forecast of the time series *Monthly gasoline demand in Ontario in 1960–1975* taken from `www.datamarket.com`. The forecasting horizon is 18 months. The result is in Figure 9.7 and was obtained using the software LFL Forecaster.[4]

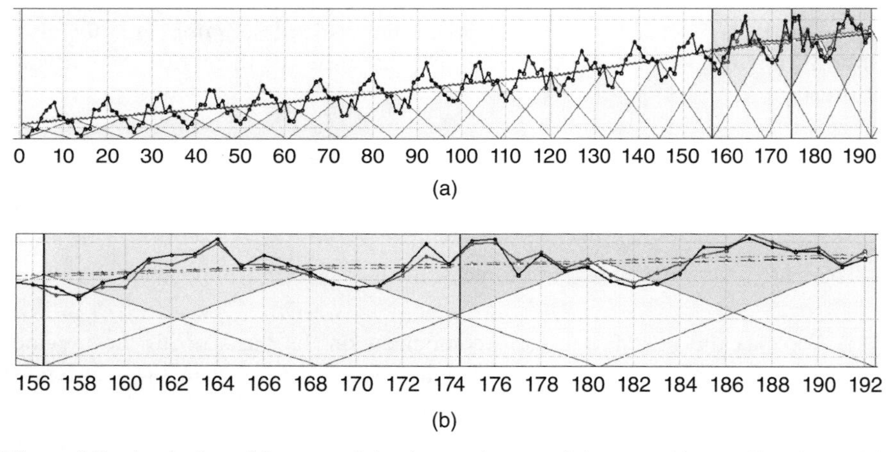

Figure 9.7 Analysis and forecast of the time series containing monthly gasoline demand in Ontario in 1960–1975. (a) The whole time series. (b) Its detail over validation and testing sets. The lower dot-and-dash line is estimation of the trend-cycle using the F-transform, the upper one is its forecast. The real data are black, the forecast is gray.

The linguistic context is $w = \langle 86800, 154000, 256000 \rangle$ and the partition period, that is, the width of the basic functions, is 24 months. The forecasting error in the validation set is 2.9% and that in the testing set is 3.3%. The forecast is based on two linguistic descriptions learned independently from the known part, that is, we considered the case (ii) above. The learned linguistic descriptions characterizing the future F-transform components are the following:

[4]The software can be downloaded for testing in the website page http://irafm.osu.cz, see also the companion website http://www.wiley.com/go/novak/fuzzy/modeling.

	First Description				Second Description			
Nr.	$F_i[X]$	$\Delta^2 F_i[X]$	$\Rightarrow$	$\Delta F_{i+1}[X]$	$F_i[X]$	$\Delta^2 F_i[X]$	$\Rightarrow$	$\Delta^2 F_{i+2}[X]$
1	Ze	-MLMe	$\Rightarrow$	SiSm	Ze	-MLMe	$\Rightarrow$	SiSm
2	**VeSm**	**Sm**	$\Rightarrow$	**Sm**	**VeSm**	**Sm**	$\Rightarrow$	**VeSm**
3	Sm	Sm	$\Rightarrow$	VeSm	MLSm	Sm	$\Rightarrow$	RoSm
4	MLSm	-SiSm	$\Rightarrow$	VRSm	RoSm	-SiSm	$\Rightarrow$	RoSm
5	QRSm	VRBi	$\Rightarrow$	Sm	VRSm	VRBi	$\Rightarrow$	VRSm
6	MLMe	-Me	$\Rightarrow$	Me	Me	-Me	$\Rightarrow$	MLMe
7	MLMe	QRBi	$\Rightarrow$	RoSm	VRBi	QRBi	$\Rightarrow$	MLSm
8	VRBi	-Me	$\Rightarrow$	Sm	RoBi	-Me	$\Rightarrow$	Me
9	QRBi	-MLSm	$\Rightarrow$	RoBi	MLBi	-MLSm	$\Rightarrow$	MLBi
10	MLBi	ExBi	$\Rightarrow$	QRBi	Bi	ExBi	$\Rightarrow$	SiBi
11	Bi	-ExSm	$\Rightarrow$	SiBi	ExBi	-ExSm	$\Rightarrow$	Bi
12	ExBi	QRBi	$\Rightarrow$	VRBi				

Since the chosen width of the basic functions is 24 months, the linguistic description characterizes periods of this length. Let us emphasize that each rule can be taken as a *sentence of natural language*. Hence, for example, the rule No. 2 says the following:

First description: If average gasoline demand in the **previous two years** is *VERY SMALL* and its average acceleration is *SMALL*, then its average change in the **upcoming two years** will be *SMALL*.

Second description: If average gasoline demand in the **previous two years** is *VERY SMALL* and its average acceleration is *SMALL*, then its average acceleration **three years later** will be *VERY SMALL*. □

The time series in Figure 9.7 has a simple trend cycle that is not so difficult to predict. Another time series with more complicated trend-cycle is depicted in Figure 9.8.

More details about the forecasting method described in this section can be found in [98, 100, 101, 102, 121]. Let us also remark that the trend *Tr* can be forecast using the same method as that used for forecast of the trend-cycle (cf. also Section 9.2.2.2).

9.3.3 Forecast of Seasonal Component

Once we obtain the estimation of the trend-cycle $TC = \hat{X}$, we can compute the seasonal component S on the basis of (9.3) by

$$S(t) = X(t) - \hat{X}(t), \qquad t \in \mathbb{T}. \tag{9.28}$$

There are several methods that can be used for forecasting of the seasonal component. Among most successful belong artificial neural networks. Their disadvantage, however, is their high computational demand.

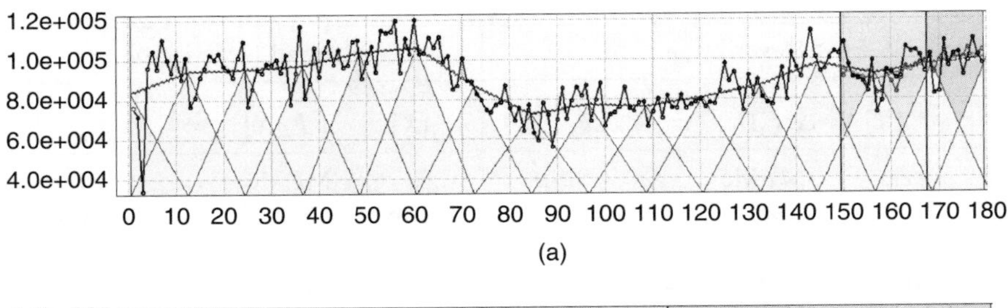

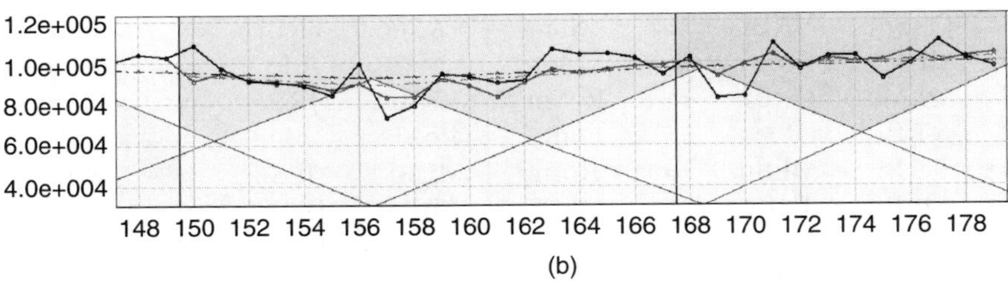

Figure 9.8 Analysis and forecast of the monthly time series. (a) The whole time series and (b) its detail over validation and testing sets. The figure contains clearly depicted trend-cycle estimated using F-transform. The real data are black, the forecast is gray.

The following method is very simple and surprisingly quite successful. Notice that the seasonal component is related to some fixed period p (e.g., $p = 12$ months). Therefore, we will consider $K > 0$ vectors of values (9.28)

$$\mathbf{S}_j = (S(t) \mid t = 1 + (j-1) \cdot p, \ldots, p + (j-1) \cdot p), \quad j = 1, \ldots, K. \tag{9.29}$$

If we assume that the seasonal component is stationary (standard assumption that can easily be checked), we can compute the future vector $\mathbf{S}_{K+1}$ as a linear combination of K previous vectors $\mathbf{S}_1, \ldots, \mathbf{S}_K$. It determines the following system of linear equations:

$$\mathbf{S}_{K+1} = \sum_{j=1}^{K} d_j \cdot \mathbf{S}_j. \tag{9.30}$$

Of course, the coefficients $d_1, \ldots, d_K$ are overdetermined. However, we can estimate them by minimizing the SMAPE measure over the validation set.

It follows from (9.30) that p future values of the seasonal component are computed simply by

$$\tilde{S}(t) = \sum_{j=1}^{K} d_j \cdot S(t - jp), \quad t = N + 1, \ldots, N + p. \tag{9.31}$$

If it holds that the forecasting horizon M is greater than p, then we compute the values $\tilde{S}(t)$ for $t = N + p + 1, \ldots, N + M$ again from the known values of $S(t)$ in accordance with (9.30).

9.3.4 Forecast of the Whole Time Series

Forecast of the whole time series is obtained as a composition of the forecasted trend-cycle as well as the seasonal component. If we put together (9.3), (9.25), and (9.31), we obtain the following forecast of the whole time series:

$$\tilde{X}(t) = \hat{X}(t) + \tilde{S}(t), \quad t \in \mathbb{T}^F. \tag{9.32}$$

EXAMPLE 9.4 Example of the forecasting using the above described methods is in Figures 9.7 and 9.8 (cf. Example 9.3). In these figures, both validation and testing parts are clearly distinguished. The detail of the forecast is depicted in the lower part of the figures, where one can see the following:

- The trend-cycle $\{TC(t) = \hat{X}(t) \mid t \in \mathbb{T}^V \cup \mathbb{T}^F\}$ computed using the F-transform from the real data (9.25) (black line) and its prediction (gray line).
- The real data $\{X(t) \mid t \in \mathbb{T}^V \cup \mathbb{T}^F\}$ (black) and their prediction $\{\tilde{X}(t) \mid t \in \mathbb{T}^V \cup \mathbb{T}^F\}$ (9.32) (gray). □

As mentioned above, the described forecasting method is in its precision fully comparable with the professional systems. Extensive tests by comparing our results with results obtained by the system ForecastPro® were made. The details can be found in [101].

If one needs to have as precise forecast as possible, we can suggest the method of ensemble forecasting. The idea is that there is no single forecasting method suitable for forecasting all kinds of time series. Namely, we always face a danger of choosing a method that is inappropriate for a given time series. Therefore, a reasonable solution is to forecast the given time series using several different methods (*ensemble* of methods) and then put all their forecasts together using some kind of aggregation. This can be either simple or weighted arithmetic mean. In the latter case, the weights are determined using the PbLD method applied to a linguistic description (see Chapter 5) generated on the basis of time series *features* such as strength of trend, seasonality, or stationarity. The ensemble-based forecasting method including an extensive testing example was published in [18, 137].

9.4 CHARACTERIZATION OF TIME SERIES IN NATURAL LANGUAGE

In this section, we demonstrate how the theory described in Chapters 4 and 5, namely, the F-transform and fuzzy natural logic, can be employed in formation of natural language sentences characterizing time series.

Recall that in Section 9.2.1, we decomposed the time series into several components, among which we considered also its trend *Tr*. It can be understood either as a function having a specific shape (cf. Section 9.2.2.2) or as a general tendency telling us whether the time series (the whole or some of its parts) *increases, decreases*, or is *stagnating* and, possibly, in what degree. The real time series, however, can be largely volatile, so recognizing its trend can be a difficult and ambiguous task. We propose to solve it using the F^1-transform, because it enables us to estimate the average tangent using formula (4.20) (cf. also (4.23)). The methods of fuzzy natural logic then make it possible to automatically generate sentences of natural language characterizing the trend.

In this section, we use the following notation. Let X be a time series (9.1) and $\overline{\mathbb{T}} \subseteq \mathbb{T}$ be a subinterval of $\mathbb{T}$. By $X|\overline{\mathbb{T}}$, we denote the time series X with the domain restricted to $\overline{\mathbb{T}}$. For consistency of this notation, we will sometimes also write $X|\mathbb{T}$, which is, in fact, the whole time series (9.1).

9.4.1 Sentences Characterizing Trend

Trend is in natural language characterized in two possible ways. First, we must realize that the basic characteristic of trend is its *slope*. Its direction is characterized by a special word, namely, *increasing* (or *increase*) and *decreasing* (or *decrease*). This expression can be further supplemented by special words characterizing its degree. Moreover, the obtained expressions can be apparently ordered similarly as the "standard" evaluative expressions considered in Chapter 5. We conclude that the general syntactic form of expressions characterizing direction of trend is either direct, that is, (i) *characterization of trend*, or indirect (ii) by characterizing the feature *sign of trend*. We thus have the following syntactic form of both possibilities:

(i) *Characterization of trend:*[5]

$$\textit{Trend} \text{ is } \langle\text{direction}\rangle, \tag{9.33}$$

where

$$\langle\text{direction}\rangle := \text{stagnating}|\langle\text{hedge}\rangle\langle\text{sign}\rangle, \tag{9.34}$$

$$\langle\text{sign}\rangle := \text{increasing}|\text{decreasing} \tag{9.35}$$

and

$$\langle\text{hedge}\rangle := \langle\text{empty hedge}\rangle|\text{negligibly}|\text{slightly}|\text{somewhat}|\text{clearly}|$$
$$\text{roughly}|\text{sharply}|\text{significantly}.$$

The ⟨empty hedge⟩ is used if more specific characterization of the direction of the trend is not required.

(ii) *Characterization of increase (decrease) of trend*:

$$\langle\text{sign of trend}\rangle \text{ is } \langle\text{special hedge}\rangle, \tag{9.36}$$

where

$$\langle\text{sign of trend}\rangle := \text{increase}|\text{decrease}, \tag{9.37}$$

[5]The pedantic characterization should be stated as "direction of trend", that is, "direction of trend is increasing". The real language, however, is not so strict, and so we commonly say "the trend is increasing".

$$\langle\text{special hedge}\rangle := \text{negligible}|\text{slight}|\text{small}|\text{clear}|\text{rough}|\\ \text{large}|\text{fairlylarge}|\text{quitelarge}|\\ \text{sharp}|\text{significant}|\text{huge}. \tag{9.38}$$

Expression (9.33) is more general because it also includes the possibility that the trend is stagnating. Otherwise, we can choose between (9.33) and (9.36) because the meaning of both corresponding expressions is the same.

The model in this book is based on the assumption that both special evaluative predications (9.33) and (9.36) are semantically tantamount to the standard form

$$\textit{Trend of } X|\overline{\mathbb{T}} \text{ is } \pm Ev[X|\overline{\mathbb{T}}], \tag{9.39}$$

where $Ev[X|\overline{\mathbb{T}}] \in EvExpr$ is an evaluative expression in which the TE-adjective is canonical (i.e., one of "small", "medium", "big", or "zero").

To obtain the model of the meaning of (9.39), we must first specify the context in which (the direction of) the trend is evaluated (cf. the discussion in Chapter 5). For example, increase of temperature in an Arctic area by 3 °C in, say, 2 hours, can be taken as a "sharp increase", while the same in Sahara is a "very small increase". Therefore, we start with a specification of what is extreme increase (decrease). This should be determined as the largest acceptable (or meaningful) difference in time series values with respect to a given (basic) time interval. The usual basic time interval is, for example, 12 months, 31 days, and 1 hour, depending on the kind of time series. In correspondence with Section 5.2.3, the context is determined by the three distinguished values v_L, v_S, v_R of the tangent. The largest tangent v_R is the mentioned extreme increase (decrease), while the smallest one is typically (but not always) $v_L = 0$. The typical medium value v_S is determined analogously as v_R. The result is the linguistic context for the (course of) trend that will be denoted by $w_{tg} = \langle v_L, v_S, v_R\rangle$. Moreover, because we distinguish between an increase and a decrease in time series, we must also distinguish a positive context $w_{tg}^+ = \langle v_L^+, v_S^+, v_R^+\rangle$ and a negative one $w_{tg}^- = \langle v_R^-, v_S^-, v_L^-\rangle$. Similarly as in Section 7.3.1, we usually put $v_L^- = v_L^+ = 0$, $v_S^- = -v_S^+$, and $v_R^- = -v_R^+$.

Definition 9.3 *Let X be a time series (9.1) and $\overline{\mathbb{T}} \subseteq \mathbb{T}$ be a time interval. Let $\beta^1[X|\overline{\mathbb{T}}]$ be the coefficient (4.20) that provides an estimation of the slope of trend of X over the period $\overline{\mathbb{T}}$. Finally, let w_{tg}^-, w_{tg}^+ be the corresponding negative and positive parts of the context, respectively. Then, the evaluative expression $\pm Ev[X|\overline{\mathbb{T}}]$ in (9.39) is obtained as follows:*

$$+Ev[X|\overline{\mathbb{T}}] := LPerc(\beta^1[X|\overline{\mathbb{T}}], w_{tg}^+) \quad \text{if } \beta^1[X|\overline{\mathbb{T}}] \geq 0, \tag{9.40}$$

$$-Ev[X|\overline{\mathbb{T}}] := LPerc(\beta^1[X|\overline{\mathbb{T}}], w_{tg}^-) \quad \text{if } \beta^1[X|\overline{\mathbb{T}}] < 0. \tag{9.41}$$

From (9.40) or (9.41), we can construct intension of (9.39) and then its extension in the context w_{tg}^+ or w_{tg}^-, depending on the sign of $\beta^1[X|\overline{\mathbb{T}}]$:

$$\text{Ext}_{w_{tg}}(\textit{Trend of } X|\overline{\mathbb{T}} \text{ is } \langle\text{direction}\rangle) \subsetapprox \mathbb{R}. \tag{9.42}$$

Recall that (9.42) is a fuzzy set of values of the slopes of $X|\overline{\mathbb{T}}$ estimated by (4.20). A few more details can be found in [87].

9.4.2 Automatic Generation of Sentences Characterizing Trend

In this section, we briefly describe how linguistic evaluation of a direction of trend of time series can be automatically generated.

Algorithm:

1. Set the interval $\overline{\mathbb{T}} \subseteq \mathbb{T}$ and the contexts w_{tg}^{+} and w_{tg}^{-}.
2. Compute the coefficient $\beta^1[X|\overline{\mathbb{T}}]$ defined in (4.20). Depending on its sign, find the evaluative expression $Ev[X|\overline{\mathbb{T}}]$ using (9.40) or (9.41).
3. If $Ev[X|\overline{\mathbb{T}}]$ ="Zero" or $\pm Ev[X|\overline{\mathbb{T}}]$ ="±extremely small", then generate the predication (9.33) in the form

   Trend is stagnating

   and *finish.* Otherwise, generate either of the expressions (9.33) or (9.36) according to items 4 or 5.
4. The predication of the form

   $$Trend\ of\ X|\overline{\mathbb{T}} \text{ is } \langle\text{direction}\rangle \tag{9.43}$$

   (see (9.33)) can be generated as follows:

   (a) Set ⟨sign⟩ of the direction of the trend from (9.35) according to the table

Coefficient (4.20)	⟨sign⟩
$\beta^1[X\|\overline{\mathbb{T}}] > 0$	*increasing*
$\beta^1[X\|\overline{\mathbb{T}}] < 0$	*decreasing*

   (b) Set the rest of (9.43) according to the following table:

$\pm Ev[X\|\overline{\mathbb{T}}]$	⟨direction⟩
± *significantly small*	*negligibly* ⟨sign⟩
± *very small*	*slightly* ⟨sign⟩
± *rather small*	*somewhat* ⟨sign⟩
± *medium* or *very roughly small*	*clearly* ⟨sign⟩
± *very roughly big*	*roughly* ⟨sign⟩
± *very big*	*sharply* ⟨sign⟩
± *significantly big*	*significantly* ⟨sign⟩

5. The predication

   $$\langle\text{sign of trend}\rangle \text{ is } \langle\text{special hedge}\rangle \tag{9.44}$$

   (see (9.36)) can be generated as follows:

(a) Set ⟨sign of trend⟩ according to the table

Coefficient (4.20)	⟨sign of trend⟩
$\beta^1[X\vert\overline{\mathbb{T}}] > 0$	*increase*
$\beta^1[X\vert\overline{\mathbb{T}}] < 0$	*decrease*

(b) Set the rest of (9.44) according to the table:

$Ev[X\vert\overline{\mathbb{T}}]$	⟨special hedge⟩
significantly small	*negligible*
very small	*slight*
small	*small*
very roughly small or *medium*	*clear*
very roughly big	*rough*
roughly big	*fairly large*
rather big	*quite large*
big	*large*
very big	*sharp*
significantly big	*significant*
extremely big	*huge*

EXAMPLE 9.5 In Figure 9.9, demonstration of linguistic evaluation of a trend of a monthly inflation measure is presented. The curve is smooth, therefore, one can clearly see the slopes.

The contexts for the tangent are $w^+_{tg} = \langle v_L = 0, v_S = 2/12, v_R = 5/12\rangle$ and $w^-_{tg} = \langle v_L = 0, v_S = -2/12, v_R = -5/12\rangle$. This means that typically extremely small tangent is 0, typically medium is $2/12$ (i.e., increase 2% per 12 months), and typically extremely big is $5/12$ (i.e., increase 5% per 12 months). Note that Slot 2 is wider than Slot 5. Hence, the program generated the comment *fairly large decrease* for Slot 2 and *huge decrease* for Slot 5. The evaluations are generated on the basis of objectively

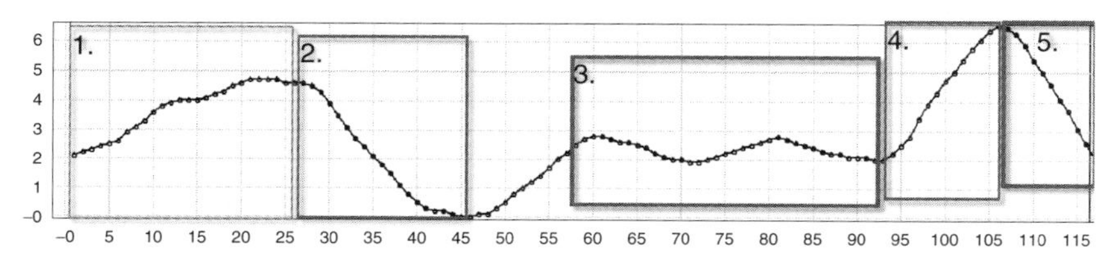

Figure 9.9 Automatically generated linguistic evaluations of monthly trend of inflation measure over 10 years. Evaluation of trend in the marked areas is the following: Slot 1: *clear increase*, Slot 2: *fairly large decrease*, Slot 3: *stagnating*, Slot 4: *significant increase* and Slot 5: *huge decrease*.

computed average tangent. We argue that they comply well with the course of the time series. □

EXAMPLE 9.6 Another example of evaluation of a trend is in Figure 9.10. One can see that the time series is very volatile and it is difficult to distinguish the (direction of) trend in various time slots. The coefficient $\beta^1[X|\overline{\mathbb{T}}]$ provides objective estimation of it. The linguistic contexts for the trend evaluation were set to

$$w_{tg}^{+} = \langle v_L = 0, v_S = 1200/12, v_R = 3000/12\rangle,$$
$$w_{tg}^{-} = \langle v_L = 0, v_S = -1200/12, v_R = -3000/12\rangle,$$

because the time series has clear periodicity $T = 12$ (this was obtained using the periodogram—see Definition 9.1).

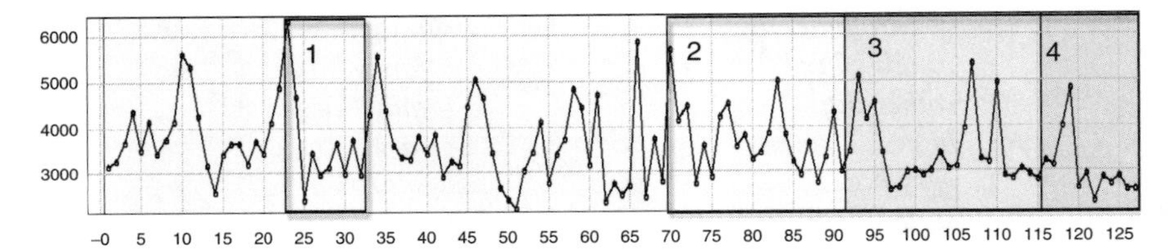

Figure 9.10 Demonstration of the evaluation of trend of various parts of a complicated time series. Trend of the whole series is *stagnating*. Slot 1 (time 23-32): *clear decrease*, Slot 2 (time 70-127): *negligible decrease*, Slot 3 (time 92-115): *small increase*, Slot 4 (time 116-127): *fairly large decrease*.

In Figure 9.10, special slots of the time series are also marked, namely, Slot 3 (time 92–115) is the validation part on which the quality of the forecast is tested and the best one is chosen and Slot 4 (time 116–127) is the testing part that is not used for computation of the forecast but only for comparison of the forecast with real data. Then, the generated evaluation of trend of the predicted values in the testing part is *rough decrease*, while the evaluation of trend of the real data is *fairly large decrease*. Both evaluations are in good agreement, which is another argument supporting the quality of our forecasting method. Thus, instead of presenting concrete predicted numbers, the user (manager) might be satisfied with the information that a "*rough decrease is expected*". □

9.4.3 Mining Information from Time Series

An interesting area of research is that of *mining information* from time series. This topic comprises various methods using which we can get a detailed information about behavior of time series. We may thus better understand natural principles that led to its formation. A nice overview of various mining methods is provided in paper [39]. In this section, we briefly mention how methods of fuzzy modeling described in this book can contribute to solution of this task. More details can be found in [87, 95].

Reduction of dimensionality of long time series. This can be effectively realized on the basis of results presented in Section 9.2.2.1 for the following reasons: (a) reduction of the dimension is accompanied by reduction of noise and by removing some of the high frequencies contained in the time series; (b) the reduced time series can be anytime recovered to a time series of the original dimension but still with reduced noise and free of selected high frequencies.

We can proceed as follows. Let a time series X from (9.1) be given, where the number of time points N is very high (hundreds to thousands). First, we choose some *periodicity* T_q[6] *taken as boundary* and set the distance between nodes to $h = d\ T_q$ $(d \in \mathbb{N})$. Then we form a fuzzy partition of the domain $\mathbb{T}$ and compute the F^1-transform of X

$$\mathbf{F}^1[X] = (F_1^1[X], \ldots, F_{n-1}^1[X]). \tag{9.45}$$

Finally, we replace the original time series X by the reduced one

$$X^{[h]} = \{\beta_1^0[X], \ldots, \beta_{n-1}^0[X]\}, \tag{9.46}$$

where the coefficients β_i^0 are computed using formula (4.19).

The new time series (9.46) has the following properties:

1. Its dimension $n - 1$ is significantly lower than the original dimension N.
2. By (9.16), it fits well the shape of the original time series X. Moreover, it keeps all its salient points, that is, points in which the time series exhibits some significant change of its direction.
3. It is free of all frequencies higher than $\lambda_q = \frac{2\pi}{T_q}$ and variance of its noise is lower than the variance of the original noise $R(t, \omega)$.

Example of the size reduction of time series is in Figure 9.11.

Perceptionally important points. Interesting problem also discussed in [39] is the recognition of perceptionally important (salient) points. As the F-transform provides estimation of the first and second derivative in a specified area, it can also be used for finding these points. A possible algorithm is the following:

(i) Determine a fuzzy partition with the distance between nodes set to $h = dT_k$ for some $d \in \mathbb{N}$ and $T_k > T_q$. Compute the F^2-transform[7] of X.

(ii) Find all nodes c_k with the highest $|\beta_k^2[X]|$. Mark such c_k-s as potentially important points.
However, note that not all potentially important points may be indeed important. Their importance also depends on the width of the corresponding basic functions A_k, which depends on the value of h.

[6]Found using periodogram.
[7]The F^2-transform was not presented in this book. In general, its components are formed of special polynomials of degree 2. The details can be found in [113].

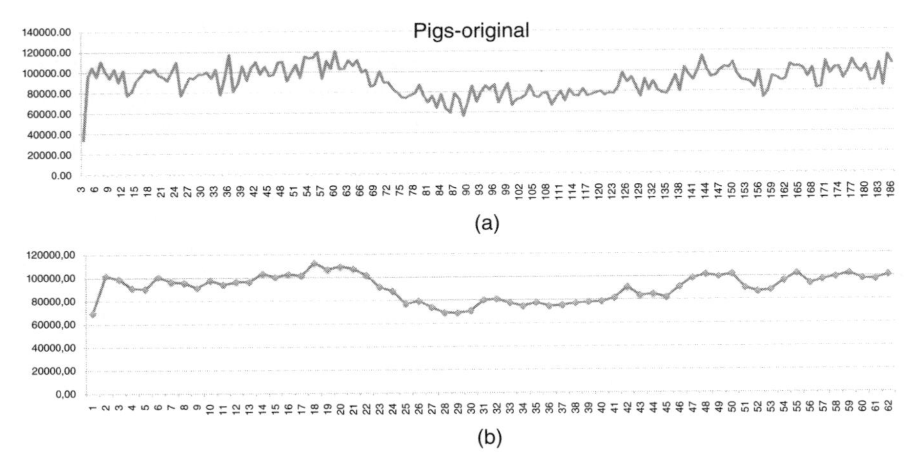

Figure 9.11 Demonstration of dimensionality reduction of a time series using F-transform with $h = 3$. (a) is the original time series (consisting of 186 points). (b) is the reduced time series consisting only of the components $\beta_1^0, \ldots, \beta_{62}^0$.

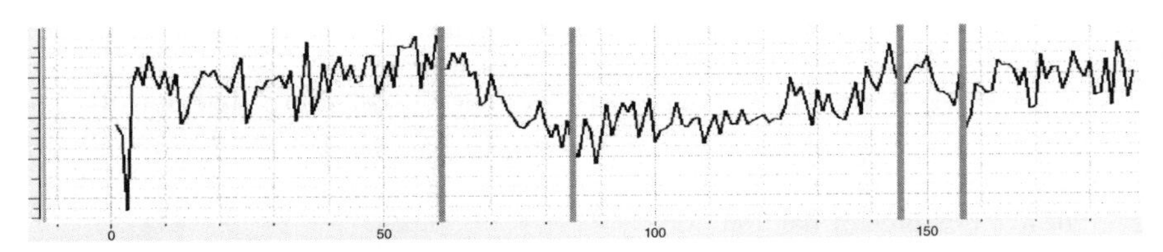

Figure 9.12 Example of perceptionally important points found in a time series from Figure 9.11 using the outlined method. The points are marked by thick vertical lines. They clearly correspond to areas of more radical change.

(iii) Check areas around the potentially important points around which several (at least two) components $\beta_k^1[X], \beta_{k+1}^1[X], \ldots$ have the same sign followed by one or more components $\beta_j^1[X]$ that have the opposite sign and/or are sufficiently large.

Example of the result of this algorithm is in Figure 9.12.

Summarization. To summarize information about time series, we suggest to apply the sophisticated formal theory of *intermediate quantifiers* developed in fuzzy natural logic. Because this theory is not included in this book, we will only outline what kind of information can be obtained and refer the reader to specialized papers [71, 73, 84].

Example of possible information mined from a single time series:

- In *most (many, few)* cases, the time series was *stagnating (slightly increasing, decreasing)*.
- In *almost all (most, many, few)* cases, *if the time series is increasing (decreasing), then the increase (decrease) is very slight (sharp, clear)*.

Let a set $\{X_i \mid i = 1, \ldots, s\}$ of time series (9.1) be given. Using the theory of intermediate quantifiers, we can model the meaning of sentences such as

- *Most (many, few)* analyzed time series stagnated recently but their future trend is *slightly increasing*.
- There is an evidence of *huge (slight, clear)* decrease of trend of almost all time series in the recent quarter of the year.

Another possibility is to mine interesting information from the given set of time series, summarize their properties, and also summarize their possible future development. Namely, we start with analysis and forecasting of all these time series. Then we generate comments to interesting time slots, or we can also determine time slots in which behavior of the time series is interesting for us, for example, "in which period was the time series sharply increasing" and "how long was the time series stagnating or decreasing before sharp increase". Finally, we can summarize the results using intermediate quantifiers and derive further properties on the basis of valid syllogisms.

Moreover, we can also apply syllogistic reasoning with such expressions, for example,

Major premise: *In few cases the increase of time series is not small.*

Minor premise: *In many cases the increase of time series is clear.*

Conclusion: *In few cases the clear increase of time series is not small.*

The used intermediate quantifiers are "few" and "many".

It is important to note that the latter is example of the valid generalized Aristotle's syllogism. Such syllogism is true in all situations (models). There are over 120 valid syllogisms with few selected basic quantifiers (all, almost all, most, many, few, some); the details can be found in [71, 73].

There are more possibilities how methods of fuzzy modeling presented in this book can be employed in mining information from time series. For example, in [87], an algorithm whose task is opposite to the evaluation of the trend presented in the previous section is presented. Namely, it enables us to find intervals in which trend of the time series X exhibits monotonous behavior (increasing, decreasing, stagnating).

One can see that the repertoire of possible applications of the methods of fuzzy modeling presented in this book is quite wide. We hope that the reader is motivated to find his own solutions to problems that he/she meets in the future.

REFERENCES

1. A. Al-Smadi. Automatic identification of ARMA systems. *International Journal of General Systems*, 38:29–41, 2009.
2. J. Anděl. Statistical Analysis of Time Series. SNTL, Praha, 1976 (in Czech).
3. J. Aznarte, J. Benítez, and J. Castro. Smooth transition autoregressive models and fuzzy rule-based systems: functional equivalence and consequences. *Fuzzy Sets and Systems*, 158:2734–2745, 2007.
4. R. Babuška. Fuzzy Modeling for Control. Kluwer Academic Publishers, Boston, MA, 1998.
5. M. Baczyński and B. Jayaram. Fuzzy Implications, volume 231 of *Studies in Fuzziness and Soft Computing*. Springer-Verlag, Berlin, 2008.
6. G. Ball and D. Hall. ISODATA, a novel method of data analysis and pattern classification. Technical Report AD 699616. Stanford Research Institute, Stanford, CA, 1965.
7. R. Bellman and L. A. Zadeh. Decision making in a fuzzy environment. *Management Science*, 17:140–164, 1970.
8. R. Bělohlávek and V. Novák. Learning rule base in linguistic expert systems. *Soft Computing*, 7:79–88, 2002.
9. M. Bertalmio, A. L. Bertozzi, and G. Sapiro. Navier-stokes, fluid dynamics, and image and video inpainting. In Proceedings of IEEE Computer Vision and Pattern Recognition (CVPR), pages 355–362, 2001.
10. J. Bezdek. Pattern Recognition with Fuzzy Objective Function Algorithms. Plenum Press, New York, 1981.
11. R. S. Blum. Robust image fusion using a statistical signal processing approach. *Information Fusion*, 6:119–128, 2005.

Insight into Fuzzy Modeling, First Edition. Vilém Novák, Irina Perfilieva, and Antonín Dvořák.

Companion Website: www.wiley.com/go/novak/fuzzy/modeling

12. U. Bodenhofer, M. Daňková, M. Štěpnička, and V. Novák. A plea for the usefulness of the deductive interpretation of fuzzy rules in engineering applications. In Proceedings of the FUZZ-IEEE'07 Conference, pages 1567–1572, London, 2007.
13. B. Bouchon-Meunier and Y. Jia. Linguistic modifiers and imprecise categories. *International Journal of Intelligent Systems*, 7:25–36, 1992.
14. G. Box and G. Jenkins. Time Series Analysis: Forecasting and Control. Holden-Day, San Francisco, CA, 1976.
15. P. J. Brockwell and R. A. Davis. Time Series: Theory and Methods. Springer-Verlag, New York, 2006.
16. J. J. Buckley. Sugeno type controllers are universal controllers. *Fuzzy Sets and Systems*, 53:299–304, 1993.
17. J. J. Buckley and Y. Hayashi. Fuzzy input-output controllers are universal approximators. *Fuzzy Sets and Systems*, 58:273–278, 1993.
18. M. Burda, M. Štěpnička, and L. Štěpničková. Fuzzy rule-based ensemble for time series prediction: progresses with associations mining. In P. Grzegorzewski, M. Gagolewski, O. Hryniewicz, and M. Á. Gil, editors, Strengthening Links between Data Analysis and Soft Computing, volume 315 of *Advances in Intelligent Systems and Computing*, pages 261–271. Springer International Publishing, Switzerland, 2015.
19. T. Calvo, G. Mayor, and R. Mesiar, editors. Aggregation Operators. New Trends and Applications. Physica-Verlag, Heidelberg, 2002.
20. J. Canny. A computational approach to edge detection. *IEEE Transactions on Pattern Analysis and Machine Intelligence*, PAMI-8(6):679–698, 1986.
21. R. Carnap. Meaning and Necessity: A Study in Semantics and Modal Logic. University of Chicago Press, Chicago, 1947.
22. J. Castro. Fuzzy logic controllers are universal approximators. *IEEE Transactions on Systems, Man, and Cybernetics*, 25:629–635, 1995.
23. P. Cintula, P. Hájek, and C. Noguera, editors. Handbook of Mathematical Fuzzy Logic, Volume 1. College Publications, London, 2011.
24. P. Cintula, P. Hájek, and C. Noguera, editors. Handbook of Mathematical Fuzzy Logic, Volume 2. College Publications, London, 2011.
25. R. B. Cleveland, W. S. Cleveland, J. E. McRae, and I. Terpenning. STL: a seasonal-trend decomposition procedure based on Loess. *Journal of Official Statistics*, 6:3–73, 1990.
26. W. S. Cleveland and S. J. Devlin. Locally-weighted regression: an approach to regression analysis by local fitting. *Journal of the American Statistical Association*, 83:596–610, 1988.
27. D. Coufal. *Radial Implicative Fuzzy Inference Systems*. PhD thesis, University of Pardubice, Pardubice, 2003.
28. D. Coufal. Coherence of radial implicative fuzzy systems. In Proceedings of the 15th IEEE International Conference on Fuzzy Systems, pages 307–314, Vancouver, BC, 2006.
29. D. Coufal. Coherence index of radial conjunctive fuzzy systems. In P. Melin, O. Castillo, L. T. Aguilar, J. Kacprzyk, and W. Pedrycz, editors, Foundations of Fuzzy Logic and Soft Computing, Proceedings of the 12th International Fuzzy Systems Association World Congress, IFSA 2007, Cancun, Mexico, volume 4529 of *Lecture Notes in Computer Science*, pages 502–512. Springer-Verlag, Berlin, 2007.
30. F. Di Martino, V. Loia, I. Perfilieva, and S. Sessa. An image coding/decoding method based on direct and inverse fuzzy transforms. *International Journal of Approximate Reasoning*, 48:110–131, 2008.

31. D. Driankov, H. Hellendoorn, and M. Reinfrank. An Introduction to Fuzzy Control. Springer-Verlag, Berlin, 1996.
32. D. Dubois and H. Prade. Fuzzy Sets and Systems: Theory and Applications. Academic Press, New York, 1980.
33. D. Dubois, H. Prade, and L. Ughetto. Checking the coherence and redundancy of fuzzy knowledge bases. *IEEE Transactions on Fuzzy Systems*, 5(6):398–417, 1997.
34. J. Dunn. A fuzzy relative of the ISODATA process and its use in detecting compact well-separated clusters. *Journal of Cybernetics*, 3:32–57, 1973.
35. A. Dvořák, H. Habiballa, V. Novák, and V. Pavliska. The software package LFLC 2000—its specificity, recent and perspective applications. *Computers in Industry*, 51:269–280, 2003.
36. A. Dvořák and M. Holčapek. **L**-fuzzy quantifiers of type $\langle 1 \rangle$ determined by fuzzy measures. *Fuzzy Sets and Systems*, 160:3425–3452, 2009.
37. A. Dvořák, M. Štěpnička, and L. Štěpničková. On redundancies in systems of fuzzy/linguistic IF-THEN rules under perception-based logical deduction inference. *Fuzzy Sets and Systems*, 277:22–43, 2015.
38. J. Fodor and M. Roubens. Fuzzy Preference Modelling and Multicriteria Decision Support. Kluwer Academic Publishers, Dordrecht, 1994.
39. T.-C. Fu. A review on time series data mining. *Engineering Applications of Artificial Intelligence*, 24:164–181, 2011.
40. M. Gavalec. Solvability and unique solvability of max-min fuzzy equations. *Fuzzy Sets and Systems*, 124:385–393, 2001.
41. S. Gottwald. Fuzzy Sets and Fuzzy Logic. The Foundations of Application—from a Mathematical Point of View. Vieweg, Braunschweig/Wiesbaden and Teknea, Toulouse, 1993.
42. M. Grabisch, J.-L. Marichal, R. Mesiar, and E. Pap. Aggregation Functions, volume 127 of *Encyclopedia of Mathematics and Its Applications*. Cambridge University Press, Cambridge, 2009.
43. D. E. Gustafson and W. C. Kessel. Fuzzy clustering with a fuzzy covariance matrix. In Proceedings of the IEEE Conference on Decision and Control, pages 761–766, San Diego, CA, 1978.
44. P. Hájek. Metamathematics of Fuzzy Logic. Kluwer Academic Publishers, Dordrecht, 1998.
45. P. Hájek and V. Novák. The sorites paradox and fuzzy logic. *International Journal of General Systems*, 32:373–383, 2003.
46. E. Hajičová, B. H. Partee, and P. Sgall. Topic-Focus Articulation, Tripartite Structures, and Semantic Content, volume 71 of *Studies in Linguistics and Philosophy*. Kluwer Academic Publishers, Dordrecht, 1998.
47. J. D. Hamilton. Time Series Analysis. Princeton University Press, Princeton, NJ, 1994.
48. M. Hanss. Applied Fuzzy Arithmetic: An Introduction with Engineering Applications. Springer-Verlag, Berlin, 2005.
49. K. Hirota and W. Pedrycz. Fuzzy relational compression. *IEEE Transactions on Systems, Man, and Cybernetics*, 29:407–415, 1999.
50. M. Holčapek, I. Perfilieva, V. Novák, and V. Kreinovich. Necessary and sufficient conditions for generalized uniform fuzzy partitions. *Fuzzy Sets and Systems*, 277:97–121, 2015.
51. M. Holčapek and M. Štěpnička. MI-algebras: a new framework for arithmetics of (extensional) fuzzy numbers. *Fuzzy Sets and Systems*, 257:102–131, 2014.

52. M. Holčapek and T. Tichý. A smoothing filter based on fuzzy transform. *Fuzzy Sets and Systems*, 180:69–97, 2011.
53. F. Höppner, F. Klawonn, R. Kruse, and T. Runkler. Fuzzy Cluster Analysis. John Wiley & Sons, Ltd, Chichester, 1999.
54. P. Hurtík and I. Perfilieva. Image compression methodology based on fuzzy transform. In Á. Herrero et al., editors, International Joint Conference CISIS'12-ICEUTE'12-SOCO'12 Special Sessions, volume 189 of *Advances in Intelligent Systems and Computing*, pages 525–532. Springer-Verlag, Berlin Heidelberg, 2013.
55. J. Jantzen. Foundations of Fuzzy Control. John Wiley & Sons, Ltd, Chichester, 2007.
56. B. Kedem and K. Fokianos. Regression Models for Time Series Analysis. John Wiley & Sons, Inc., New York, 2002.
57. A. Khastan, I. Perfilieva, and Z. Alijani. A new fuzzy approximation method to Cauchy problems by fuzzy transform. *Fuzzy Sets and Systems*, 288:75–95, 2016.
58. F. Klawonn. Fuzzy points, fuzzy relations and fuzzy functions. In V. Novák and I. Perfilieva, editors, Discovering the World with Fuzzy Logic, pages 431–453. Springer-Verlag, Berlin, 2000.
59. F. Klawonn and R. Kruse. Equality relations as a basis for fuzzy control. *Fuzzy Sets and Systems*, 54:147–156, 1993.
60. F. Klawonn and V. Novák. The relation between inference and interpolation in the framework of fuzzy systems. *Fuzzy Sets and Systems*, 81:331–354, 1996.
61. E. P. Klement, R. Mesiar, and E. Pap. Triangular Norms. Kluwer Academic Publishers, Dordrecht, 2000.
62. V. Kreinovich and I. Perfilieva. Fuzzy transforms of higher order approximate derivatives: a theorem. *Fuzzy Sets and Systems*, 180:55–68, 2011.
63. G. Lakoff. Hedges: a study in meaning criteria and logic of fuzzy concepts. *Journal of Philosophical Logic*, 2:458–508, 1973.
64. G. Leng, T. McGinnity, and G. Prasad. An approach for on-line extraction of fuzzy rules using a self-organising fuzzy neural network. *Fuzzy Sets and Systems*, 150:211–243, 2005.
65. S. P. Lloyd. Least squares quantization in PCM. Unpublished Bell Laboratories technical note, 1957.
66. J. Lu, G. Zhang, D. Ruan, and F. Wu. Multi-objective Group Decision Making. Methods, Software and Applications with Fuzzy Set Techniques. Imperial College Press, London, 2007.
67. E. H. Mamdani and S. Assilian. An experiment in linguistic synthesis with a fuzzy logic controller. *International Journal of Man-Machine Studies*, 7:1–13, 1975.
68. M. Mareš. Computation over Fuzzy Quantities. CRC Press, Boca Raton, FL, 1994.
69. K. Michels, F. Klawonn, R. Kruse, and A. Nürnberger. Fuzzy Control: Fundamentals, Stability and Design of Fuzzy Controllers. Springer-Verlag, Berlin, 2010.
70. A. Mumtaz and A. Masjid. Genetic algorithms and its applications to image fusion. In Proceedings of the IEEE International Conference on Emerging Technologies, pages 6–10, Rawalpindi, 2008.
71. P. Murinová and V. Novák. A formal theory of generalized intermediate syllogisms. *Fuzzy Sets and Systems*, 186:47–80, 2012.
72. P. Murinová and V. Novák. Analysis of generalized square of opposition with intermediate quantifiers. *Fuzzy Sets and Systems*, 242:89–113, 2014.

73. P. Murinová and V. Novák. The structure of generalized intermediate syllogisms. *Fuzzy Sets and Systems*, 247:18–37, 2014.
74. H. T. Nguyen, N. R. Prasad, C. L. Walker, and E. A. Walker. A First Course in Fuzzy and Neural Control. Chapman and Hall, Boca Raton, FL, 2003.
75. V. Novák. Fuzzy Sets and Their Applications. Adam Hilger, Bristol, 1989.
76. V. Novák. The Alternative Mathematical Model of Linguistic Semantics and Pragmatics. Plenum Publishing Corporation, New York, 1992.
77. V. Novák. Towards formalized integrated theory of fuzzy logic. In Z. Bien and K. C. Min, editors, Fuzzy Logic and Its Applications to Engineering, Information Sciences, and Intelligent Systems, pages 353–363. Kluwer Academic Publishers, Dordrecht, 1995.
78. V. Novák. Fuzzy logic deduction with words applied to ancient sea level estimation. In R. Demicco and G. J. Klir, editors, Fuzzy Logic in Geology, pages 301–336. Academic Press, Amsterdam, 2003.
79. V. Novák. On fuzzy type theory. *Fuzzy Sets and Systems*, 149:235–273, 2005.
80. V. Novák. Perception-based logical deduction. In B. Reusch, editor, Computational Intelligence, Theory and Applications, pages 237–250. Springer-Verlag, Berlin, 2005.
81. V. Novák. Which logic is the real fuzzy logic? *Fuzzy Sets and Systems*, 157:635–641, 2006.
82. V. Novák. Mathematical fuzzy logic in modeling of natural language semantics. In P. Wang, D. Ruan, and E. E. Kerre, editors, Fuzzy Logic. A Spectrum of Theoretical & Practical Issues, volume 215 of *Studies in Fuzziness and Soft Computing*, pages 145–182. Springer-Verlag, Berlin, 2007.
83. V. Novák. A comprehensive theory of trichotomous evaluative linguistic expressions. *Fuzzy Sets and Systems*, 159(22):2939–2969, 2008.
84. V. Novák. A formal theory of intermediate quantifiers. *Fuzzy Sets and Systems*, 159(10):1229–1246, 2008.
85. V. Novák. EQ-algebra-based fuzzy type theory and its extensions. *Logic Journal of the IGPL*, 19:512–542, 2011.
86. V. Novák. Reasoning about mathematical fuzzy logic and its future. *Fuzzy Sets and Systems*, 192:25–44, 2012.
87. V. Novák. Linguistic characterization of time series. *Fuzzy Sets and Systems*, 285:52–72, 2016.
88. V. Novák. On modelling with words. *International Journal of General Systems*, 42:21–40, 2013.
89. V. Novák and A. Dvořák. Formalization of commonsense reasoning in fuzzy logic in broader sense. *Journal of Applied and Computational Mathematics*, 10:106–121, 2011.
90. V. Novák, P. Hurtík, H. Habiballa, and M. Štěpnička. Recognition of damaged letters based on mathematical fuzzy logic analysis. *Journal of Applied Logic*, 13:94–104, 2015.
91. V. Novák and J. Kovář. Linguistic IF-THEN rules in large scale application of fuzzy control. In D. Ruan and E. E. Kerre, editors, Fuzzy If-THEN Rules in Computational Intelligence: Theory and Applications, pages 223–241. Kluwer Academic Publishers, Boston, MA, 2000.
92. V. Novák and S. Lehmke. Logical structure of fuzzy IF-THEN rules. *Fuzzy Sets and Systems*, 157:2003–2029, 2006.
93. V. Novák and I. Perfilieva, editors. Discovering the World with Fuzzy Logic, volume 57 of *Studies in Fuzziness and Soft Computing*. Springer-Verlag, Heidelberg, 2000.

94. V. Novák and I. Perfilieva. On the semantics of perception-based fuzzy logic deduction. *International Journal of Intelligent Systems*, 19:1007–1031, 2004.

95. V. Novák and I. Perfilieva. Time series mining by fuzzy natural logic and F-transform. In Proceedings of the 48th Hawaii International Conference on System Sciences, Kauai, HI, 2015.

96. V. Novák, I. Perfilieva, M. Holčapek, and V. Kreinovich. Filtering out high frequencies in time series using F-transform. *Information Sciences*, 274:192–209, 2014.

97. V. Novák, I. Perfilieva, and N. G. Jarushkina. A general methodology for managerial decision making using intelligent techniques. In E. Rakus-Anderson, R. R. Yager, N. Ichalkaranje, and L. C. Jain, editors, Recent Advances in Fuzzy Decision-Making, pages 103–120. Springer-Verlag, Heidelberg, 2009.

98. V. Novák, I. Perfilieva, and V. Krejnovich. F-transform in the analysis of periodic signals. In M. Inuiguchi, Y. Kusunoki, and M. Seki, editors, Proceedings of the 15th Czech-Japan Seminar on Data Analysis and Decision Making under Uncertainty, pages 150–158. Osaka University, Osaka, Japan, 2012.

99. V. Novák, I. Perfilieva, and J. Močkoř. Mathematical Principles of Fuzzy Logic. Kluwer Academic Publishers, Boston, MA, 1999.

100. V. Novák, I. Perfilieva, and V. Pavliska. The use of higher-order F-transform in time series analysis. In Proceedings of the World Congress IFSA 2011 and AFSS 2011, pages 2211–2216, Surabaya, Indonesia, 2011.

101. V. Novák, M. Štěpnička, A. Dvořák, I. Perfilieva, V. Pavliska, and L. Vavříčková. Analysis of seasonal time series using fuzzy approach. *International Journal of General Systems*, 39:305–328, 2010.

102. V. Novák, M. Štěpnička, I. Perfilieva, and V. Pavliska. Analysis of periodical time series using soft computing methods. In D. Ruan, J. Montero, J. Lu, L. Martínez, P. D'hondt, and E. E. Kerre, editors, Computational Intelligence in Decision and Control, pages 55–60. World Scientific, Singapore, NJ, 2008.

103. R. Palm, D. Driankov, and H. Hellendoorn. Model Based Fuzzy Control. Springer-Verlag, Berlin, 1997.

104. J. A. Parker, R. V. Kenyon, and D. E. Troxel. Comparison of interpolation methods for image resampling. *IEEE Transactions on Medical Imaging*, MI-2(1):31–39, 1983.

105. G. Patanè. Fuzzy transform and least-squares approximation: analogies, differences, and generalizations. *Fuzzy Sets and Systems*, 180:41–54, 2011.

106. I. Perfilieva. Fuzzy logic normal forms for control law representation. In H. Verbruggen, H.-J. Zimmermann, and R. Babuška, editors, Fuzzy Algorithms for Control, pages 111–125. Kluwer Academic Publishers, Boston, MA, 1999.

107. I. Perfilieva. Fuzzy relations, functions, and their representation by formulas. *Neural Network World*, 10:877–890, 2000.

108. I. Perfilieva. Fuzzy transform: application to reef growth problem. In R. B. Demicco and G. J. Klir, editors, Fuzzy Logic in Geology, pages 275–300. Academic Press, Amsterdam, 2003.

109. I. Perfilieva. New criteria of solvability of a system of fuzzy relation equations. In Proceedings of the International Conference on Intelligent Technologies InTech03, pages 525–530. Chiang Mai University, 2003.

110. I. Perfilieva. Fuzzy function as an approximate solution to a system of fuzzy relation equations. *Fuzzy Sets and Systems*, 147:363–383, 2004.

111. I. Perfilieva. Fuzzy transforms: theory and applications. *Fuzzy Sets and Systems*, 157:993–1023, 2006.

112. I. Perfilieva. Logical foundations of rule-based systems. *Fuzzy Sets and Systems*, 157:615–621, 2006.

113. I. Perfilieva, M. Daňková, and B. Bede. Towards a higher degree F-transform. *Fuzzy Sets and Systems*, 180:3–19, 2011.

114. I. Perfilieva, M. Daňková, P. Hoďáková, and M. Vajgl. The use of F-transform for image fusion algorithms. In Proceedings of International Conference of Soft Computing and Pattern Recognition (SoCPaR2010), pages 472–477, Cergy Pontoise, France, 2010.

115. I. Perfilieva, M. Daňková, P. Hoďáková, and M. Vajgl. F-transform based image fusion. In O. Ukimura, editor, Image Fusion, pages 3–22. InTech, 2011.

116. I. Perfilieva and B. De Baets. Fuzzy transform of monotonous functions. *Information Sciences*, 180:3304–3315, 2010.

117. I. Perfilieva, P. Hoďáková, and P. Hurtík. Differentiation by the F-transform and application to edge detection. *Fuzzy Sets and Systems*, 288:96–114, 2016.

118. I. Perfilieva, P. Hoďáková, and P. Hurtík. F^1-transform edge detector inspired by Canny's algorithm. In Advances on Computational Intelligence (IPMU2012), pages 230–239, Catania, Italy, 2012.

119. I. Perfilieva and V. Novák. System of fuzzy relation equations as a continuous model of IF-THEN rules. *Information Sciences*, 177:3218–3227, 2007.

120. I. Perfilieva and A. Tonis. Compatibility of systems of fuzzy relation equations. *International Journal of General Systems*, 29:511–528, 2000.

121. I. Perfilieva and R. Valášek. Fuzzy transforms in removing noise. In B. Reusch, editor, Computational Intelligence, Theory and Applications, pages 225–234. Springer-Verlag, Heidelberg, 2005.

122. I. Perfilieva and P. Vlašánek. Influence of various types of basic functions on image reconstruction using F-transform. In Proceedings of the International Conference EUSFLAT-LFA'2013, pages 497–502, Milano, Italy, 2013.

123. I. Perfilieva and P. Vlašánek. Image reconstruction by means of F-transform. *Knowledge-Based Systems*, 70:55–63, 2014.

124. G. Piella. A general framework for multiresolution image fusion: from pixels to regions. *Information Fusion*, 4:259–280, 2003.

125. R. Ranjan, H. Singh, T. Meitzler, and G. R. Gerhart. Iterative image fusion technique using fuzzy and neuro fuzzy logic and applications. In Proceedings of the NAFIPS 2005 Conference, pages 706–710, Detroit, USA, 2005.

126. H. J. Rong, N. Sundararajan, G. B. Huang, and P. Saratchandran. Sequential adaptive fuzzy inference system (SAFIS) for nonlinear system identification and prediction. *Fuzzy Sets and Systems*, 157:1260–1275, 2006.

127. A. Rosenfeld and M. Thurston. Edge and curve detection for visual scene analysis. *IEEE Transactions on Computers*, C-20(5):562–569, 1971.

128. T. J. Saaty. Fundamentals of Decision Making and Priority Theory with the Analytic Hierarchy Process. RWS Publications, Pittsburgh, PA, 2000.

129. E. Sanchez. Resolution of composite fuzzy relation equations. *Information and Control*, 30:38–48, 1976.

130. R. Slowiński, editor. Fuzzy Sets in Decision Analysis, Operations Research and Statistics, *Handbook of Fuzzy Sets Series*. Kluwer Academic Publishers, Dordrecht, 1998.

131. Q. Song and B. S. Chissom. Forecasting enrollments with fuzzy time series—part I. *Fuzzy Sets and Systems*, 54:1–9, 1993.

132. Q. Song and B. S. Chissom. Forecasting enrollments with fuzzy time series—part II. *Fuzzy Sets and Systems*, 62:1–8, 1994.

133. L. Stefanini. F-transform with parametric generalized fuzzy partitions. *Fuzzy Sets and Systems*, 180:98–120, 2011.
134. H. Steinhaus. Sur la division des corp materiels en parties. *Bulletin of the Polish Academy of Sciences*, 4:801–804, 1956.
135. M. Štěpnička, U. Bodenhofer, M. Daňková, and V. Novák. Continuity issues of the implicational interpretation of fuzzy rules. *Fuzzy Sets and Systems*, 161:1959–1972, 2010.
136. M. Štěpnička and A. Dvořák. On perception-based logical deduction and its variants. In Proceedings of the 16th World Congress of the International Fuzzy Systems Association (IFSA) and the 9th Conference of the European Society for Fuzzy Logic and Technology (EUSFLAT), pages 341–350. Atlantis Press, 2015.
137. M. Štěpnička, L. Štěpničková, and M. Burda. Fuzzy rule-based ensemble for time series prediction: the application of linguistic associations mining. In IEEE International Conference on Fuzzy Systems, FUZZ-IEEE 2014, pages 505–512, Beijing, China, 2014.
138. T. Takagi and M. Sugeno. Fuzzy identification of systems and its applications to modeling and control. *IEEE Transactions on Systems, Man, and Cybernetics*, 15:116–132, 1985.
139. A. Telea. An image inpainting technique based on the fast marching method. *Journal of Graphics Tools*, 9:23–34, 2004.
140. F. M. Tseng, G. H. Tzeng, H. C. Yu, and B. J. C. Yuan. Fuzzy ARIMA model for forecasting the foreign exchange market. *Fuzzy Sets and Systems*, 118:9–19, 2001.
141. K. Uhlir and V. Skala. Radial basis function use for the restoration of damaged images. In K. Wojciechowski et al., editors, Computer Vision and Graphics, volume 32 in Computational Imaging and Vision, pages 839–844. Springer-Verlag, Netherlands, 2006.
142. M. Vajgl, I. Perfilieva, and P. Hoďáková. Advanced F-transform-based image fusion. *Advances in Fuzzy Systems*, 2012. Article No. 4.
143. P. Vopěnka. Mathematics in the Alternative Set Theory. Teubner, Leipzig, 1979.
144. P. Wang, editor. Computing with Words. John Wiley & Sons, Inc., New York, 2001.
145. A. M. Yaglom. An Introduction to the Theory of Stationary Random Functions. Revised English Edition. Translated and Edited by R. A. Silverman. Prentice-Hall, Englewood Cliffs, NJ, 1962.
146. L. A. Zadeh. Fuzzy sets. *Information & Control*, 8:338–353, 1965.
147. L. A. Zadeh. A fuzzy–set–theoretic interpretation of linguistic hedges. *Journal of Cybernetics*, 2:4–34, 1972.
148. L. A. Zadeh. A rationale for fuzzy control. *Journal of Dynamical Systems, Measurement, and Control*, 94:3–4, 1972.
149. L. A. Zadeh. Outline of a new approach to the analysis of complex systems and decision processes. *IEEE Transactions on Systems, Man, and Cybernetics*, SMC-3:28–44, 1973.
150. L. A. Zadeh. Quantitative fuzzy semantics. *Information Sciences*, 3:159–176, 1973.
151. L. A. Zadeh. The concept of a linguistic variable and its application to approximate reasoning I, II, III. *Information Sciences*, 8-9:199–257, 301–357, 43–80, 1975.
152. L. A. Zadeh. Fuzzy logic = computing with words. *IEEE Transactions on Fuzzy Systems*, 4:103–111, 1996.
153. L. A. Zadeh. From computing with numbers to computing with words—From manipulation of measurements to manipulation of perceptions. *International Journal of Applied Mathematics and Computer Science*, 12:307–324, 2002.
154. L. A. Zadeh and J. Kacprzyk, editors. Computing with Words in Information / Intelligent Systems 1. Springer-Verlag, Heidelberg, 1999.

INDEX

Insight into Fuzzy Modeling, First Edition. Vilém Novák, Irina Perfilieva, and Antonín Dvořák.

Companion Website: www.wiley.com/go/novak/fuzzy/modeling